AF614801

Methods in Molecular Biology™

Series Editor
John M. Walker
School of Life Sciences
University of Hertfordshire
Hatfield, Hertfordshire, AL10 9AB, UK

For other titles published in this series, go to
www.springer.com/series/7651

Methods in Biobanking

Edited by

Joakim Dillner

Bio Banking and Molecular Resource, Infrastructure of Sweden (BBMRI.se), Karolinska Institutet, Stockholm, Sweden

Editor
Joakim Dillner
Bio Banking and Molecular Resource
Infrastructure of Sweden (BBMRI.se)
Karolinska Institutet
Stockholm, Sweden

ISSN 1064-3745 e-ISSN 1940-6029
ISBN 978-1-58829-995-6 e-ISBN 978-1-59745-423-0
DOI 10.1007/978-1-59745-423-0
Springer New York Dordrecht Heidelberg London

Library of Congress Control Number: 2010938369

Printed on acid-free paper

Springer is part of Springer Science+Business Media (www.springer.com)

Preface

Recent technological advances, primarily in molecular biology and genetics, have greatly improved our ability to investigate how interactions between genes and environment affect our health. Access to reliable information concerning family members, health, and life-style factors that can be linked to biological samples from large numbers of individuals creates an enormous new potential in this area. Although biobanks can be used to study conventional risk markers (such as cholesterol levels and cardiovascular risk), a major emphasis is being placed on the potential for genetic studies. Current studies frequently demonstrate that the importance of genes becomes most evident under circumstances determined by life-style factors. For example, the importance of serum cholesterol for cardiovascular risk can be viewed in a context of genetic variation of lipoprotein genes, receptors, and diet. Modern biobanks are systematically built to allow comprehensive recruitment of cases and matched controls from the same background population and social strata. At the same time, international biobank collaborations allow studies with large number of subjects, where generalizability of findings across populations can be investigated. For such studies, it is of vital importance to establish quality criteria concerning the nature of the sample, conditions of sample storage, and the adequacy of available information. Several collaborative studies and networks are currently actively attempting to develop uniform methods and quality standards – so-called Good Biobanking Practice.

Biobanks that comprise samples stored over a long period of time present the opportunity to investigate accumulated, prospectively occurring disease endpoints – now. New prospective biobanks recruiting participants from a very young age are being designed to contain uniform information and sampling of great future value. Many clinical biobanks consecutively recruit specific clinical cases as they are diagnosed. Current efforts are underway in several countries to produce new well-defined prospective biobanks based on obtaining material from large proportions of the entire population. The visions, organization, and financing of these major efforts differ. Some have received overwhelming popular support, but others are faced by opposition. Biobanking needs to build on public trust, and a high ethical awareness with sound ethical principles governing all use of biobank materials to protect the safety, integrity, and autonomy of sample donors is essential.

We would like this book to contribute to the development of competence in the subject area of biobanking. We discuss how it is possible to use existing collections of biological material to answer significant questions concerning the cause of disease, without violating the personal integrity of participating sample donors. We gain experience from researchers who have succeeded in creating large prospective research biobanks and those who are actively engaged in producing new biobanks. We discuss the ethical issues surrounding biobanks, e.g., the issue of broad consent for the present and future research on biological material. We discuss guidelines for the use of coding systems and the use of biocomputing and registry linkages in research projects. Epidemiological study design is discussed by qualified experts in the field, as is the choice of appropriate technical platforms for different stages of biobank-related research. Finally, several chapters focus on specific clinical topics using biobanks and registries.

Joakim Dillner

Contents

Contributors

KRISTIN ANDERSSON • *Department of Medical Microbiology, University Hospital Malmö, University of Lund, Malmö, Sweden*
MARC ARBYN • *Unit of Cancer Epidemiology, Scientific Institute of Public Health, Brussels, Belgium*
LEONA W. AYERS • *Department of Pathology, College of Medicine and Public Health. The Ohio State University, Columbus, OH, USA*
CHRISTINE BERGERON • *Laboratoire Pasteur-Cerba, Cergy Pontoise, Saint Ouen l'Amone, France*
JOHN-PAUL BOGERS • *Applied Molecular Biology Research Group (AMBIOR), Laboratory for Cell and Tissue Research, University of Antwerp, Antwerp, Belgium*
JOHAN BOTLING • *Rudbeck Laboratory, Department of Genetics and Pathology, University Hospital, Uppsala, Sweden*
ELODIE CABOUX • *International Agency for Research on Cancer, Lyon, France*
JOYCE CARLSON • *Laboratory Medicine Skåne, Department of Clinical Chemistry, University Hospital Lund, University of Lund, Lund, Sweden*
DON CHALMERS • *Faculty of Law, University of Tasmania, Hobart TAS, Australia*
IAN CHANDLER • *Royal Cancer Hospital Cancer Genetics, Sutton, Surrey, London, UK*
I. CHANDLER • *Institute of Cancer Research, Royal Cancer Hospital Cancer Genetics, Sutton, Surrey, UK*
PAOLO DE PAOLI • *Department of Microbiology, Oncological Center, Aviano, Italy*
JOAKIM DILLNER • *Bio Banking and Molecular Resource, Infrastructure of Sweden (BBMRI.se), Karolinska Institutet, Stockholm, Sweden*
ANDERS EKBOM • *Head of Clinical Epidemiology, Department of Medicine, Karolinska Institute, Stockholm, Sweden*
CHRISTER ERICSSON • *Department of Oncology-Pathology, Karolinska Institutet, Stockholm, Sweden*
ASTA FÖRSTI • *Division of Molecular Genetic Epidemiology, German Cancer Research Center (DKFZ), Heidelberg, Germany; Center for Family and Community Medicine, Karolinska Institute, Huddinge, Sweden*
DEBRA L. GARCIA • *Central Operations and Data Coordinating Center, AIDS and Cancer Specimen Resource, University of California, San Francisco, San Francisco, CA, USA*
PIERRE HAINAUT • *International Agency for Research on Cancer, Lyon, France*
GÖRAN HALLMANS • *Department of Public Health and Clinical Medicine, Umeå University, Umeå, Sweden*
MATS G. HANSSON • *Department of Public Health and Caring Sciences, Centre for Research Ethics & Bioethics, Uppsala University, Uppsala, Sweden*

HALLA HAUKSDÓTTIR • *Institute of Laboratory Medicine, Landspítali University Hospital, Reykjavik, Iceland*
KARI HEMMINKI • *Division of Molecular Genetic Epidemiology, German Cancer Research Center (DKFZ), Karolinska Institute, Heidelberg, Germany; K. Hemminki Center for Family and Community Medicine, Karolinska Institute, Huddinge, Sweden*
RICHARD HOULSTON • *Section of Cancer Genetics, Institute of Cancer Research, Sutton, Surrey, UK*
KRISTIAN HVEEM • *HUNT Biobank, Norwegian University of Science and Technology, Trondheim, Norway*
MALIN IVARSSON • *Wallenberg Laboratory, RSKC Malmö, Malmö, Sweden*
KRISTÍN JÓNSDÓTTIR • *Institute of Laboratory Medicine, Landspítali University Hospital, Reykjavik, Iceland*
JAAKKO KAPRIO • *Department of Public Health, Department of Mental Health and Alcohol Research, National Public Health Institute, University of Helsinki, Helsinki, Finland*
MAGNUS VON KNEBEL-DOEBERTITZ • *Institute of Pathology, University of Heidelberg, Heidelberg, Germany*
ESA LÄÄRÄ • *Department of Mathematical Sciences, University of Oulu, Oulu, Finland; Finnish Cancer Registry, Helsinki, Finland*
GÖRAN LANDBERG • *Laboratory Medicine Skåne, Clinical Pathology, University Hospital Malmö, University of Lund, Malmö, Sweden*
JAN-ERIC LITTON • *Department of Medical Epidemiology and Biostatistics, Karolinska Institutet, Stockholm, Sweden*
MICHAEL S. MCGRATH • *Department of Laboratory Medicine, Medicine, and Pathology, University of California, San Francisco, San Francisco, CA, USA*
PATRICK MICKE • *Rudbeck Laboratory, Department of Genetics and Pathology, University Hospital, Uppsala, Sweden*
MONICA NISTÉR • *Department of Oncology-Pathology, Karolinska Institutet, Stockholm, Sweden*
Jan M. Orenstein • *Department of Pathology, School of Medicine and Health Sciences, The George Washington University, Washington, DC, USA*
EERO PUKKALA E. PUKKALA • *Finnish Cancer Registry, Institute for Statistical and Epidemiological Cancer Research, Helsinki, Finland; School of Public Health, University of Tampere, Tampere, Finland*
ELIO RIBOLI • *Division of Epidemiology, Public Health and Primary Care, Imperial College, London, UK*
SABINA RINALDI • *International Agency for Research on Cancer, Lyon, France*
SYLVIA SILVER • *Department of Pathology, School of Medicine and Health Sciences, The George Washington University, Washington, DC, USA*
ROSAMARIA TEDESCHI • *Department of Microbiology, Oncological Center, Aviano, Italy*
ANDRES THORARINSSON • *Vista Engineering, Reykjavik, Iceland*

HRAFN TULINIUS • *The Genetical Committee of the University of Iceland, Reykjavik, Iceland*
JIMMIE B. VAUGHT • *Office of Biorepositories and Biospecimen Research, National Cancer Institute, National Institutes of Health, Bethesda, MD, USA*
BÉATRICE VOZAR • *International Agency for Research on Cancer, Lyon, France*

Chapter 1

Genetic Research and Biobanks

Don Chalmers

Abstract

Human biobanks, and genetic research databases, as referred to by the Organisation for Economic Co-operation and Development (OECD), are essential tools for modern biomedical research. Biobanks may consist in collections created in clinical diagnosis (such as pathology tissue samples in hospitals) or collections created for large-scale longitudinal research (such as the UK Biobank). Human tissue collections are regulated by a patchwork of national laws. However, there is an increasing international uniformity in national privacy laws based on 1980s OECD standards. There are similar uniform standards developing in national research ethics guidelines. As biobanks develop collaborations and linkages, international harmonisation of legislation and human research regulation will be required across jurisdictions. It is essential that international public trust is maintained in biobanking research.

Key words: Biobanks, Regulation, Privacy, Public trust, International governance

1. Introduction

This chapter examines the legal principles and rules for human genetic research with particular emphasis on the development of collections of tissue samples and data held in human genetic research databases. This century has been described by Francis Collins as the Genome Era (1) in science and medicine, acknowledging the volume and intensity of genomic research (2–8) in both the public and private sectors. Human tissue samples are essential tools for genomic research and "translating biomedical research into real improvements in health care" (9). The German National Ethics Council has noted the potential of biobanks for the identification of causes of disease and for breakthroughs in medical and pharmaceutical research and the "particularity of biobanks, which… lies in their twofold character, as collections of both samples and data" (10). Pharmacogenetic research into

Joakim Dillner (ed.), *Methods in Biobanking*, Methods in Molecular Biology, vol. 675,
DOI 10.1007/978-1-59745-423-0_1,

genetic variability in drug response may be substantially advanced by biobanking (11) (see Note 1). Tissue samples, in the form of DNA or RNA samples, cell lines, tissues, cell preparations, or plasma/blood samples, are essential tools for pharmacogenomic research and the analysis that aims to identify potential biomarkers (see Note 2) or drug targets by any of the new generation genomic tests utilising DNA marker, RNA expression level, or protein activity. Unsurprisingly, many pharmaceutical companies operate biobank collections for research purposes and to enrol suitable clinical trial recruits so as to minimise side effects and achieve better results. Biobanks are important resources for medical health research that may benefit current patients but are also aimed at long-term research for future benefits (12).

Ethical and social issues (13) surround biobanks, apart from the technical and scientific issues. Human tissue samples held in human genetic research databases will usually be coded, making the samples potentially re-identifiable (14) (see Note 3). This raises the issue of privacy of the genetic information. Patient identification may be required for the follow-up of results or result validation. There are doubts whether de-identification is realistic as a link back to the patient may be required, particularly in disease identification studies. Such distinctions are critical in the design, conduct and reporting of human genetic research, and pharmacogenomic studies. It has been recognised that complete guarantees of individual privacy are unrealistic in health research. Participant re-contact may also be required by the biobank (15) to collect new information or to seek consent for new approved research uses or a new study. Linkage may also be required to enable re-contact of participants for future research projects, to follow up a participant to pass on clinically significant results or, possibly to recruit for a prospective clinical trial. There has to be an effective balance between individual interests in privacy with the public interest in promoting high quality public health research. Apart from the important legal issue of participant privacy, there is also a mixed range of legal issues (5, 16) (see Note 4) dealing with participant consent, research governance, human tissue, material (tissue) transfer agreements, employee confidentiality, commercialisation, benefit-sharing, and international collaboration.

It is essential that all human research be conducted with integrity and according to the highest ethical standards. This is even more important where large genetic research database collections have been assembled. Public trust (6, 17–20) is an essential pre-condition for the successful operation and future research benefit of human genetic research databases, sometimes referred to as biobanks.

This chapter considers the regulation required to balance individual interests in privacy with the public interest in effective and reliable research. This issue is particularly salient in relation to genetic research databases, where the balance must be made between the proper protections of those recruited as tissue sample providers with the public interest. This chapter does not discuss forensic DNA banks for criminal investigations (21, 22) or "problems that might arise because of other utilisations, for civil or criminal purposes or for employment or insurance" (23). This chapter focuses on the legal responsibilities and obligations of biobank administrators and researchers in dealing with human tissue and data collections and biobank research participants.

1.1. What are Human Genetic Research Databases and Biobanks?

The Organisation for Economic Co-operation and Development (OECD) generally uses the term "Human Genetic Research Databases (HGRDs)" to describe large-scale collections of human tissue for research. The OECD Committee for Scientific and Technological Policy produced *on Human Biobanks and Genetic Research Databases Guideline* in 2009 that provide an excellent outline of the procedures to collect and manage samples, to manage and govern databases and commercialisation aspects (24). Apart from the terms HBGRD and biobanks, the Estonian Genome Project uses "genome database", the Latvian Genome Project uses "genebank", and the French National Ethics Consultative Committee uses "biolibraries" (23). The term "biobank" is used in this sense and is largely synonymous with the term "human genetic research database". All involve the storage of human tissue (see Tutton and Corrigan for a discussion of terminology (25)). For this reason, genetic registers (see Note 5) of personal and family genetic information and histories are usually not included in a discussion of HBGRDs because they generally do not require any collection or storage of human tissue.

A distinction can be drawn between the generic OECD term HBGRD and biobanks (see Note 3). Many existing collections of human tissue were developed primarily for diagnostic and clinical purposes (without consideration of research or with research later considered as a secondary purpose). Moreover, these collections were developed for specific limited research purposes only with specific and limited consent regimes. There are, therefore, some unique considerations in relation to research using existing tissue collections (see Subheading 3 below). In contrast, biobanks have been established generally with the specific aim of conducting research. Biobanks have also been established with careful efforts to ensure that participant's consent has been obtained to cover research generally, including variations to the original purpose for future research. In this sense, if there is a difference between the two terms "genetic research database" and, "biobank" a biobank tends to refer to a collection of human tissue, specifically created for research. However, the term "biobank" is often used interchangeably with "genetic research database" to describe any collection of human tissue, which can and is used for research purpose. Both have the twin goals of facilitation of genomic research balanced with the protection of the welfare of the biobank sample contributors (26–28).

1.2. Genetic Research and Privacy

Genetic tests and research can provide information not only about a person's genes, but also information about the person's parents, siblings, children, and even cousins and other more distant blood relations (see Note 6). For this reason, some forms of genetic information has:

1. predictive potential;
2. implications for family members; and
3. potential to stigmatise.

Genetic research has aroused specific privacy concerns. There are community concerns that personal information disclosed in a genetic research project may be divulged to others, such as insurance companies or employers, to the detriment of not only the research participant, but also the family members and communities that share the participant's genetic profile. Such distinctions are critical in the design, conduct and reporting of human genetic research, and pharmacogenomic studies. The protection of privacy of genetic information was the driver behind the joint Australian Law Reform Commission and Australian Health Ethics Committee Report, *Essentially Yours* (*29*). This Report examined personal genetic information privacy in the context, among others, of anti-discrimination, genetic testing, health service delivery, insurance, employment, law enforcement, and parentage testing.

2. Collection of New Samples for Biomedical Research

2.1. Governance of a Biobank

The establishment of a biobank is a complex task that will involve negotiations with health officials, researchers, governing institution(s), research funding agencies, health consumer/community organisations, and ethics experts. Biobanks can be "staggeringly expensive" (Greely quoted in (30)) (see Note 30) to establish and operate. Some biobanks have been established by national legislation setting up an operating company structure (31) (see Note 8) or using the structure of a foundation (32) (rather than a company). For example, the Estonian database is owned and controlled by the Estonian Genome Project Foundation. In addition, some countries have enacted specific biobank legislation (see Note 9). As biobanks will involve public benefit research, the UK Biobank is managed under the structure of a charitable company (33, 34) with an independent Ethics and Governance Council which is an independent body charged with an oversight of the UK Biobank and to monitor and advise on the UK Biobank's compliance with the Ethics and Governance Framework of the project (35) Similarly, CARTaGENE has an independent Institute for Population, Ethics, and Governance. Some countries have established an oversight body reporting to the relevant government minister. The Scottish Executive has funded Generation Scotland, in large part, and has also established the Generation Scotland Advisory Board with an oversight function.

Apart from considerations of structure, a biobank governing body will introduce guidelines (36, 37) for the ethical operation of the biobank. The issue of participant consent to enrolment in a biobank is and has been the most debated and vexed ethical question (6) The governing body will introduce also standard operating procedures (38).

Biobanks are being established at regional, national, and international levels (39). At the regional level, biobanks have been set up by the Karolinska Institutet (Sweden); CARTaGENE (Quebec); the Western Australia project (40); the National Heart, Lung and Blood Institute (NIH, USA); and the Centre for Integrated Genomic Medical Research (Manchester, UK). At the national level, DeCode (Iceland) was the pioneer programme that has been followed by GenomEUtwin (Finland); Estonian Genome; Danubian Biobank Foundation (involving six countries in Central Europe); KORA-GEN (Germany); LifeGen (Sweden); INMEGEN (Mexico); LifeLines (Netherlands); the UK Biobank and Generation Scotland that will enrol some 500,000 participants; and, the Lifelong Health Initiative (Canada). These regional and national biobanks have been specifically created for large-scale longitudinal genetic research projects. At the international level, the successor to the Human Genome Project, the International Haplotype Mapping Project is a collaboration between the USA, the UK, Japan, Nigeria, China, and Canada to identify and compare genetic similarities and differences in collected human tissue samples to find genes that affect health, disease, and medication responses. Another international collaboration is emerging in the Public Population Project in Genomics (P3G) (see Note 10) that aims to facilitate collaboration between many national biobanks in a not-for-profit initiative to provide a public and accessible knowledge database for the international population genomics community. P3G will enable large-scale epidemiological studies to be undertaken. The regulation of biobanking has, or is being considered in a number of countries and by a range of research or regulatory organisations (23, 41–48). For example, the German National Ethics Council and the French National Consultative Ethics Committee for Health and Life Sciences have produced a joint declaration of the need for a regulatory framework to ensure the development of research balanced with the protection of the individual. The Australian Law Reform Commission (ALRC) published *Essentially Yours: The Protection of Human Genetic Information*, which recommended changes to the regulation of databases and genetic research in general (29) (see Note 11). In the UK, both the UK Biobank and Generation Scotland have developed ethics and governance frameworks (35) to define the scope and limits of the projects, and this has been supplemented with specific human tissue legislation (49).

Once a decision to proceed has been taken, a governing body will be appointed and the governance arrangements instituted. An institution establishing a biobank must establish governance structures appropriate for and consistent with the primary research focus (see Note 12), including a separate independent ethics review board (see Note 13), to scrutinise and assess the ethical acceptability of the project. The governance standards will cover confidentiality and privacy and the management and administration processes of the biobank with transparency and accountability.

Review of governance. Governance and ethical standards in research are not static. Attitudes of today on standards for privacy and consent cannot be assumed to apply to later decade. It is important that the governance arrangements for biobanks are reviewed on a regular basis to ensure compliance with developing governance, ethical, and legal standards (4). These reviews should be conducted with opportunities for community and participant dialogue.

2.2. Public Trust and Transparency

The governance structure for biobanks should enable public scrutiny of processes and promote opportunities for public input (50). The research governance arrangements for biobanks should include public transparency procedures that allow public scrutiny and encourage public trust. For example, the funders (see Note 14) of the UK Biobank have appointed an independent Ethics and Governance Council (EGC) to monitor and advice on the operations of the UK Biobank. Annual reports from both the UK Biobank and the EGC are published and available publicly. The EGC also holds public meetings on its activities and publishes the minutes of all of its deliberations (15). Any specific guidelines or changes in operating procedures should be notified publicly and provide opportunities for public input (see Note 15). Public trust in biobank research is widely accepted as an essential aspect of biobank governance (6, 51) (see Note 16) Public engagement has been a major feature of the development of major public biobanks (52).

2.3. Technical Considerations

There are a number of technical requirements for an effective, secure, and ethical biobank system (38, 53). Some of these can be noted. First, because health data and genetic information are "sensitive" personal information, this information should be protected by encryption codes and only accessible to properly authorised biobank employees and researchers under strict conditions (54) (see Note 17). Computing systems must not only be efficient and reliable, but they must secure confidentiality and privacy of the information derived from the samples. This is a technical as well as an ethical issue. In this respect, a number of privacy enhancement information technology systems are being developed. The computer industry and researchers have invested considerable time and energy in developing specific privacy enhancement technologies (PETs) to protect personal privacy, prevent unauthorised access to this information and, most importantly, to enable authorised access to information, particularly for authenticating and checking information. Secondly, biobank laboratories and collection and testing facilities must comply with prescribed national accreditation standards (55) (see Note 18). Thirdly, the sample collection and storage processes must be quality assured to ensure that the collection, handling, storage, processing, access, and the use of any samples are not tainted by human or process error. Fourthly, beyond the legal requirements for privacy and confidentiality are the technical issues of the number of data points to be collected in relation to each individual sample and then the actual coding of the collected sample. These technical decisions not only provide assurances of the authenticity of the privacy of the collected sample but also, equally importantly, determine the degree of interchangeability of data between biobanks wishing to conduct international research projects (52).

Finally, industry standards for biobanks are developing, through biobank networks (see Note 19) to answer concerns from a Rand Corporation study (56) about inconsistencies in the collection, storage, and access policies of biobank.

2.4. Independent Control of Data and Samples

The control of the biobank samples and data should be under the control of a body or individual independent from the researchers seeking access to the data or samples. Reports (see Note 20) and academic opinion support this general and emerging principle. Biobank governance arrangements should include the appointment of an independent intermediary *between* the researcher and the data or samples. The principle of independent control is specific to the governance of biobanks. The important underlying idea of an independent intermediary is the introduction of a cheque and balance in the governance structure for the date and samples on the biobank. This idea of trusteeship has been described by the Ethics and Governance Framework of the UK Biobank as acting "as the *steward* (emphasis added) of the resource, maintaining and building it for the public good in accordance with its purpose" (57) (see Note 21).

2.5. Information and Consent Procedures for Living Donors

The collection of human tissue samples must be carried out in accordance with legal and accepted ethical standards, particularly the informed consent of the sample donor. The German National Ethics Council Opinion (10) addressed the consent issue and considered that it is essential that explicit information be given to those depositing tissue.

Consistent with established international standards for research generally, consent procedures will emphasise the provision of explicit information to participants, opportunities for further explanation of the information, and time to understand the information.

Consent. The diverse aspects of the consent process for the involvement in a biobank demands that the consent be informed, voluntary and written. Accordingly, the elements of proper consent for the involvement in the biobank should respect participant autonomy (5, 6) and include participant information, understanding, and voluntary consent to the following (37):

- Relevant risks and benefit, if any.
- The types of samples and data to be collected and stored.
- Research may also disclose information about family and relations and whether this will be communicated (see below).
- The nature of the intended research to be undertaken.
- Research projects and purposes (and the data derived) may change to *other* future research.
- Policy on sharing samples and data with other research organisations.
- Policies, guidelines, and procedures for access by researchers to data/samples.
- Permission to collect other data from health-relevant records.
- Procedures for later re-contact.
- Arrangements for privacy security and confidentiality, including restrictions of the release to insurers and employers.
- Anonymisation procedures and restrictions on re-identification.
- Feedback of research results and how they will be reported.
- The right to withdraw.
- Arrangements for the data/samples in the event of incapacity or death.
- Policy on benefit-sharing.
- IP prospects.
- Potential commercial involvement.
- Absence of any personal financial gain for any participant (see Note 22).

Consent is a *process* that must ensure that proper informed and voluntary consent is obtained. The rights of sample donors must be clearly set out in the consent form to be signed before donating the sample. These rights include the voluntary nature of the consent, the right to obtain one's own information, and the right to withdraw from the database. Proper consent may extend to re-contact by the biobank to collect new information or tissue/data for research in the future (35). Consent in the case of biobanking goes beyond the legal form of the original consent and raises wider issues of the public interest and public good. Any discussion of privacy and autonomy raises the issues of human rights and the principle of human dignity that, it has been argued, underpins human rights provisions in national constitutions and international conventions (58, 59).

The consent process must also recognise and respect cultural, social, and religious differences. National research codes generally include special guidelines for indigenous communities. So the Canadian Institutes of health research guidelines (60) that provide explicit consent is always required and that the transfer of data and samples also requires consent of the other original parties (see Note 23). In such cases, consent may be required from the community and/or its leaders. Care in this type of research is essential to avoid some of the controversies that accompanied the earlier Human Genome Diversity Programme (HGDP) (61, 62) that aimed to construct the history of development, migrations, and expansion of human population. The HGDP encountered considerable opposition and suspicion from indigenous peoples (63).

Consent to future research. Biobanks are established with the express aim of conducting long-term research, where human tissue collected and the data derived will be stored and used for future research. In contradistinction, existing collections of samples and data may be limited for use in particular research projects, depending on the original participant consent (generally limited for research in specific projects). With existing collections of samples and data, further follow-up consent is generally required, if there is to be any new extension or substantial variation from the original research project ("re-consent" or "follow-up consent"). In cases of existing collections, an Ethics Review Board (ERB) would review the original participant consent to be satisfied, after proper consideration, that the participant had consented to the project or whether follow-up consent is required for the new variation or extension in the research

Biobanks, at the time of initial enrolment, aim to provide full information to the participants and to obtain broad consent for the research purposes of the biobank. This "broad" (65) consent requires full information to and the voluntariness of the participant to enrol. The biobank obtains participant consent for its approved and planned purposes, but not for an unrestricted ("blanket") consent for participation and the use of tissue in any research project in the future (see Note 24).The consent is to the actual published research purposes of the biobank. So the Ethics and Governance Framework of the UK Biobank states "The consent...will apply...unless the participant withdraws. Further consent will be sought for any proposed activities that do not fall within the existing consent" (emphasis added). (15, 35) This UK biobank consent is not a "blanket" consent. Often misleading called "blanket" consent, is not common in health research and is the subject of continuing debate and some controversy (5). There have been suggestions that the uniqueness of long-term commitment to a biobank requires not only a focus on the voluntariness of the original consent but some rethinking of the traditional ideas of consent to specific research projects . There should be a focus also on the governance arrangements of the biobank. (66). Governance is a new term, that refers not only to initial participant consent but also the ongoing and long-term formal regulation and processes involved in the biobank, including national legislation, management policies, legal and risk assurance, insurance, institutional approvals, formal recording and reporting, monitoring etc. These are all critical to the welfare and protection of the participants (see Note 32).

In all cases, participant consents must be reviewed on an on-going and routine basis that the biobank protocols ensure that the collection, use, storage, and the release of information are consistent with the actual consent given and the approved purposes and governance of the biobank.

Health-related information. Biobank research will involve health and genetic research that has the potential to reveal medically relevant information about the health or future health of participants and possibly, participant's offspring or relations. It is essential that the research project include a clear policy on whether such information will be disclosed to the participants and the procedures to be followed for disclosure (see Note 25). Consent processes should clearly communicated in writing to the participant at the recruitment stage whether health relevant information will or will not be, disclosed to the participant, participant's off-spring or relations (67) (see Note 26).

2.5.1. Competent Adults

Recruitment into a biobank should ensure the *voluntariness* of consent and participation in conformity with general ethical principles and specific information above. Recruitment into a biobank should ensure non-discrimination (see Note 27), the voluntariness of consent, and participation in conformity with accepted research ethics principles (see Note 28). Many biobanks, such as the UK Biobank, have decided to concentrate on the recruitment of competent adults in the higher age groups.

2.5.2. Incompetent Adults

There may be advantages for the inclusion in research of incompetent adults, suffering from cognitive impairment, intellectual disability or mental illness because they suffer from specific and hereditary genetic diseases that may be better understood through long-term research on their disease or disorder. However, many biobanks are not recruiting incompetent adult participants. The inclusion of incompetent adults in research (including others highly dependent on medical care or dependent or unequal relationships) is governed by legislation or research codes in all countries (see Note 29). Broadly, these guidelines establish that:

- Special considerations and responsibilities attach to incompetent adults in research.
- The research project and ethical approval should pay due regard to the best interests of the incompetent adult.
- Consent procedures and ethical review must address these special considerations and responsibilities for each specific research project.
- Ethical review should recognise that some incompetent adults may have some level of understanding of the research project, but not to provide consent.
- There should be no harm to the incompetent adult's safety and emotional psychological security.
- The research project should not involve any more than low risk (which is usually the case with biobanks) to the incompetent adult.
- The research project should involve a research question that could not be carried out on other competent research participants.
- The guardian or other required legal representative's consent must be obtained.

2.5.3. Children

The practice of recruitment of children is variable between biobanks. The issue is no settled practice norm. Some studies are specifically aimed at children (68) and some biobank studies have decided not to recruit children as participants, but others recruit. For example, the trans-genomic research in the African Diaspora (TgRIAD) has been implemented by the Howard University National Human Genome Centre to study diseases common among African Americans and other populations of Africa and the Caribbean (see Note 30). This study recruits whole households, including children. In the case of the Latvian legislation, the inclusion of children is permitted.

There may be considerable advantages for the inclusion of children in research. The inclusion of children is likely to assist in research into genetic diseases affecting the young and in understanding the development of late onset genetic diseases and other health problems from childhood to maturity. Similarly, the inclusion of children in research is governed by research codes in most countries (see Note 31). Broadly, the guidelines in these codes establish:

- That special consideration and special responsibilities be attached to child research.
- That there is a requirement that consent procedures and ethical review must be developed for the specific research project.
- That children have developing levels of maturity from being unable to understand the research project, to understand some other relevant information, to understand information but not being old enough to provide proper informed consent.
- The research project should not involve any more than low risk to the child (by and large in biobank inclusion there should be no more than low risk).
- There should be no harm to the child and the child's safety and emotional psychological security and wellbeing should be included in the signed consent and conduct of the research.
- Parental or guardian consent should be obtained.
- Overall, the project and ethical approval should pay due regard to the best interests of the child (even though there may be no direct benefit).

2.6. The Role of Ethical Review Boards in Selection of Appropriate Information and Consent Procedures

Biobank participants will receive the range of information set out in Subheading 2.5 before they are asked to consent to participate in the project. Once established, the biobank oversight body and governance arrangements are critical (4). In addition, ERB (see Note 32) will review and assess applications for the access to its resource. The oversight body will ensure that the application complies with the purposes and ethical frameworks of the biobank and national legislation, guidelines, and policies. Many biobanks have developed their own guidelines, supplementing national guidelines (35, 38). In addition, the oversight body or ERB will approve and monitor all research access applications. The role of the ERB is the traditional protection of the interests of the participants. When the project is independently reviewed for approval, the ERB will ensure that the project complies with the participants' consent. Apart from ensuring that the consent process addressed the consent matters set out at 2.5 above, the ERB should also ensure that the proposed project.

- involves a valid research question.
- addresses confidentiality and privacy.

- involves whether collection and storage of new samples or data.
- explains any changes to original access or release conditions.
- involves research in other institutions, including overseas.

2.7. Requirements for Privacy

Biobanks have legal duties to ensure the privacy and confidentiality of samples and data. The governing institution must assume responsibility for maintaining legal and ethical standards of confidentiality and privacy in the overall governance of its biobank Privacy legislation (see Note 33) is fairly standard in most countries because of the original OECD privacy principles developed in the early 1980s. Most countries have privacy legislation; some also have specific biobank legislation or other specific access to health records legislation.

Constitutional rights to privacy. Rights to privacy are constitutionally guaranteed in some countries. These constitutional and legislative privacy rights are not absolute and are usually subject to exceptions and conditions determined by law. Constitutional rights to privacy, for historical reasons, usually apply to privacy of communications and have little relevance to modern biobanks (see Note 34). Many countries do not include constitutional rights to privacy but have judicial recognition of such rights (see Note 35).

Non-discrimination and freedom of information. Anti-discrimination laws may also apply to some of the research and governance arrangements of biobanks. Biobanks should implement appropriate measures to avoid discrimination of stigmatisation of participants, their families and social groups (37). Similarly, freedom of information legislation allows access to government-held information, but is not generally relevant to biobanks (see Note 36).

Data Protection. The protections introduced in the computer age to protect personal data are important for biobanks and their sample donors. European nations must implement legislation to comply with the European Union *(EU) Data Protection Directive* (95/46/EC) (see Note 37). The two major North American nations have complex data protection regulation arising from their federal arrangements (see Note 38). Some Asian countries also have introduced data protection by legislation (see Note 39).

Privacy legislation. Privacy of personal information is an accepted legal and ethical principle. Originally, privacy law was aimed towards government record keepers and credit providers. By the 1990s, greater concerns were being expressed about privacy in telecommunications and electronic record linkage, including health information in general, and genetic information in particular. Privacy law now has a major influence in the regulation of medical research generally and biobanks, in particular.

Privacy legislation applies across a range of principles from the collection through to the storage and use of data as follows:

- *Principle 1* – Personal information should be collected for a lawful purpose and collected in a lawful and fair manner.
- *Principle 2* – Where personal information is collected for a record or solicited, the collector must ensure that the individual concerned is aware of the purpose of the collection (at the time or as soon after as practicable), if the collection is authorised by law and the persons or agencies that could have the information disclosed or passed on to them.
- *Principle 3* – The collection or solicitation of personal information should generally be relevant to the purpose for which it is collected.
- *Principle 4* – Records of personal information should be stored with "such security safeguards as... reasonable in the circumstances" to prevent loss or unauthorised access, use or disclosure.
- *Principle 5* – A record-keeper of personal information should take reasonable steps to enable persons to ascertain the existence of record about them and details about the nature and purposes of the record.
- *Principle 6* – Person should have access to records about them, except if restricted by law.
- *Principle 7* – A record-keeper to allow reasonable alteration of records containing personal information by the person and, if not, may attach a statement of correction, deletion or addition by the person.
- *Principle 8* – A record-keeper to check that personal information accurate and up-to-date before use.
- *Principle 9* – A record-keeper cannot use personal information except for relevant purposes.
- *Principle 10* – Limits are placed on a record-keeper not to use personal information unless the person consents; authorised by law; there is reasonable belief of a threat to life or health; for law enforcement; or use is directly related to the purpose for which the information was collected.
- *Principle 11* – Limits are placed on a record-keeper not to disclose personal information unless individual aware information likely to be passed on; individual consents; disclosure authorised by law; there is reasonable belief of a threat to life or health; for law enforcement; or disclosure is to an agency that will not use it for a purpose other than that for which the information was given.

These principles are general in most jurisdictions. Privacy is required and personal information must not be disclosed unless

- Person consents, expressly or by implication;
- Disclosure necessary to lessen/prevent serious/imminent threat to person (life, health, safety) or serious threat to public health/safety;
- Required or authorised by law; or
- Law enforcement.

The major additions to this list have been the development of privacy principles dealing with

- trans-border data flows; and
- sensitive information exceptions.

Sensitive information. This last principle is important as "sensitive information" covers health information in general and biobank data in particular. Tissue samples, subject to genetic analysis provide *information* on sample donor are "sensitive information" and attract the privacy protection and enforcement procedures of the privacy legislation.

Enforcement. Most privacy legislation is described as "light-touch" avoiding a strict enforcement regime in favour of the introduction of specific industry codes developed by the industries themselves and approved by an appointed Privacy Commissioner/Ombudsman. Generally, complaints do not go to court but are dealt with administratively by the Privacy Commissioner/Ombudsman (see Note 40), according to the following steps:

- Person may complain (no costs) to Privacy Commissioner
- Privacy Commissioner investigates/conciliates
- Privacy Commissioner may impose fine or award compensation.

Access to information. Privacy legislation generally includes a right of access to and correction of personal information (see principles 6 and 7 above). In addition to the general privacy legislation, some countries (and states within federal systems) have supplemented the privacy with specific statutory rights to patients, and particularly in relation to access medical records (see Note 41). The Estonian legislation extends full access rights to sample donors (see Note 42). There can also be court-authorised access to personal information, where access is refused for improper reasons.

Ethical and legal duties of confidentiality. Finally, biobank staff are usually bound by the codes of ethics, incorporated as terms of their contracts of employment. Similarly, researchers are usually bound by ethical and legal duties of confidentiality in

MTAs (*see* **Note 43**) or in research access agreements. These duties require staff and researchers to maintain confidentiality of information acquired in the course of biobank work or research. Breaches of duties of confidentiality can lead to dismissal from employment. Where biobanks are established by legislation, the act usually includes a statutory offence for unauthorised disclosure of information (*see* **Note 44**).

2.8. Research Guidelines

The "hard law" privacy legislation is supplemented by "soft law" research guidelines and policies that establish ethical duties for privacy of information and data in research. The *Declaration of Helsinki* (1964 and subsequent revisions) is the international foundation for the common framework for the regulation of human experimentation and established the key pillars for ethical review in medical research (voluntary consent of the research participant; independent review of the project; assessment of the risk; involvement of competent researchers of integrity and research merit). These guidelines are contained in national codes of ethical conduct in research in most countries (see Note 45). The trend in most countries is towards greater the regulation of human research and away from earlier self-regulation (69). Importantly, the approval processes of ERBs must ensure "... provisions to protect the privacy of subjects and to maintain the confidentiality of data" (see Note 46) be in place.

Overarching these national codes, most biobanks have special ethics and governance oversight frameworks in place that have been introduced in legislation (see Note 47) or in guidelines and policies. The OECD proposes it as the best practice to establish such an oversight body (see Note 48), as was done by the UK Biobank Ethics and Governance Framework. Similarly, the Department of Health and Human Services, the National Institutes of Health, and the National Cancer Institute (38, 53) have developed jointly a comprehensive template set of guidelines, policies, and procedures for biorepositories in the USA that support such oversight (see Note 49).

2.9. Using Biological Material from Deceased Donors

Death of a biobank participant raises the issue of withdrawal from the biobank. Critically, the right to withdraw may become technically difficult after the data is anonymised. The UK Biobank has decided to exclude and not to enrol participants who express the view that they would want to withdraw in the event of death or incapacity (see Note 50).

The consent process and any instruction of the participant determine the use of biobank data/samples after the death of the participant. The information is provided and consent forms should state explicitly what may be done with the samples after death. These forms should be retained and available to ensure compliance with the actual consent. Generally, next-of-kin have no property in the tissue of a deceased and no rights of removal from a biobank, unless conferred and stipulated in the consent form. However, there may be some privacy interests that may be pursued (70) (see Note 51).

2.10. Using Biobank Data for Research

The biobank governing body will establish clear policies, guidelines, and procedures consistent with the governance aims of the biobank, for access by researchers to data/samples. First, access must be consistent with participants' consent and will require ERB approval and undertakings that privacy and confidentiality will be guaranteed. Secondly, biobanks will have privacy enhancement technology systems for anonymisation of data, including systems for re-anonymisation of tissue samples after later re-identification of a participant (provided consent permits such re-contact for future research projects). Access by researchers will be recorded and may be granted under a licence setting out the duties and obligations of the researcher (see Note 52).

Biobank governance arrangements will prescribe the proper and allowable research purposes for the data/samples (see also Subheading 2.7 above). Nevertheless, concerns exist about the possible misuse of biobank data focus on possible improper access to the data by enforcement authorities or, possibly private health care providers, interested in direct marketing. In addition, insurance companies, employers, litigants in paternity disputes, or immigration departments could be interested in biobank records. As noted at Subheading 2.7 above, privacy legislation provides that the information collected for one purpose should not be used for other purposes. While biobanks may not prevent access by law enforcement agencies, their governance arrangements should specify that access would be for approved research purposes and not for other purposes.

2.11. Transfer of Samples and Data Within and Between Countries

Transnational Recognition of Research Ethics Approvals.

It is also becoming common and technologically feasible for data collections to be linked through formal exchange and co-operation agreements to facilitate research and to enable large-scale research and comparative work on the collaborating datasets. In these cases, the collaborating partner institutions should develop formal exchange agreements. These exchange agreements between collaborating institutions should also include reciprocal access and release agreements. Importantly, licences or materials transfer agreements (MTAs) should be in place and each MTA recorded (see Note 53). All access to and the release of information from data collections should be strictly recorded, and so providing a guaranteed, continuous "chain of responsibility" (71) for all access and the release dealings in relation to the storage, handling and the use of body material and personal data. Access to and the release of information must be able to be tracked and audited (53).

OECD *Guidelines on the Protection of Privacy and Transborder Flows of Personal Data.* 1980 was influential in the revisions of national privacy legislation to ensure conformity to standards for transborder flows of data.

The MTA should set out conditions on the processes for transfer of the data, data security, the use and release of the data, approved research uses, intellectual property rights and duties, liability arrangements, termination and, finally, requirements for the data on completion of the project (38). As a general ethical principle, a researcher should not transfer tissue or data to another research group unless an ERB has approved the research and the genetic material and data is provided in a form, which ensures that participants cannot be identified. Some national codes for ethical research recognise a system of centralised ethical review for multi-centre research. Under these arrangements, guidelines usually allow the acceptance of a central ethical assessment or adoption of the decision of another research review committee. This avoids duplication and enables common monitoring and reporting responsibilities to be undertaken.

With the growth of biobanks and cross-border collaborations, there is a need for greater international harmonisation of regulation (2, 72, 73). There is already considerable harmonisation between the codes of research guidelines of most nations (74, 75), The CIOMS *International Ethical Guidelines for Biomedical Research Involving Human Subjects*, 2002 establishes standards, but the operation of national codes often uncovers significant divergences in practice. International cooperation has already been well established with the international HapMap Consortium and the development of the P3G Consortium. In addition, statements of general principles are developing in the Human Genome Organisation (HUGO) *Statements on Human Genomic Databases on DNA Sampling: Control and Access, on the Principled Conduct of Genetic Research,* UNESCO's *International Declaration on Human Genetic Data* (2003), and *Declaration on the Human Genome and Human Rights* (see Note 54).

2.12. Collaboration Between Academic and Commercial Partners

Biobanks have a primary public research focus. This does not preclude private companies that may apply, subject to conditions, to use biobank data and resources. The pharmaceutical industry is interested in biobanking with hopes that pharmacogenomic research may herald a new generation of medicines tailored to individual needs. If not individualised medicine, this research may enable better patient stratification thus achieving better patient outcomes from the drug administration. Commercial collaborations may arouse, in the words of the German National Ethics Council, "anxiety and distrust" (76). Similarly, the Australian Law Reform

Commission public consultation process uncovered public scepticism about the continuing "heavy degree of commercialisation of [medical and genetic] research" and that people did not want their "altruism to lead to billion dollar profits for multinational pharmaceutical companies" (77). Recognising that commercialisation challenges public trust in science (17, 18), a policy of transparency and public engagement by biobanks in relation to their commercial activities is advisable. The Generation Scotland project is carrying out an on-going programme of public engagement, focussing especially on issues and concerns about commercialisation.

Commercialisation. Some biobanks have been established as research platforms to support both public and private research. Some of this research, therefore, may have commercial outcomes (see Note 55). There is evidence of community concerns with commercialisation of research that must be tackled by demonstrating the public benefits that may flow from this research (78). A distinction can be drawn between the intellectual property rights in the databases and intellectual property arising from research using these databases. In the former case, the European Union Directive on the *Legal Protection of Databases* (96/9/EC) provides that the ownership of the intellectual property in the database vests in the "maker" of the database, giving 50 years protection for work and costs in compiling, verifying, and presenting data. So, the governing foundations of some biobanks (e.g. Iceland, Estonia and UK) establish that the intellectual property accruing from the creation and development of the database accrues to the biobank. In the case of intellectual property arising from research using these databases, the arrangements for access and use of the biobank data will set out the intellectual property arrangements. Generally, the biobank will expect some share of the IP rights with the researcher/research organisation. Generally, IP rights are clearly stated by the biobank to remain with the researcher/research organisation (or, in some cases shared with their assignees) and not with the participant.

Conflicts of interest. Potential conflicts of interest must be audited and managed in collaborations and partnerships between commercial organisations and biobanks. The general principle of disclosure of interest is recognised in national codes for the responsible conduct of research (see Note 56). There are also well-established policies of science and medical research journals requiring declarations of financial associations with commercial organisations before, and as a condition of, publication.

Ownership of samples. The question of ownership of body parts and tissue remains unsettled in both common and civil law jurisdictions (19). The better view is that a biobank is trustee/steward of the samples for the purposes set out in the consent. In any case, the data created from the research will be owned by the researcher or subject to some special agreement between the biobank and the researcher. The sample donor does not have any claims in the eventual product of the research. Some biobanks have tried to clarify these positions. The UK Biobank states that participants "will have no property rights in the samples" (see Note 57), and this will be explained in the consent process. Similarly, the Estonian Genome Projects states that ownership to samples vests in the Project. This does not preclude the capacity of sample donors to have agreed rights of access to information or to withdraw from the project, or, in some cases, have the sample destroyed. Importantly, consent documents will clarify that the sample donor does not have and will not obtain any intellectual property rights in the database, in research results or in any product arising from the research use of the biobank.

The commercialisation of biobank results is quite separate from the issue of fees for service. Many biobanks have a tiered pricing system for different researcher categories.

2.13. Public Dissemination of Research Results

As a general ethical standard, participants should be provided with information about the results of the research (see Note 58). As a general accepted ethical principle, the results of research should normally be published and disseminated to contribute to the advancement of public knowledge (see Note 59). Biobanks should commit to this principle and encourage research to be published in the scientific literature or in other ways that allow the assessment and scrutiny of the results. The International Haplotype Mapping Project (see Note 60) and GenBank (see Note 61) accept this publication policy. On the other hand, where a biobank is operated as a private resource, for example, by a pharmaceutical company, there may be policies or restrictions on publication and dissemination of results (18).

2.14. Requirements Regarding Coding and Anonymisation

Privacy and confidentiality of data are critical for biobanks. Biobanks should have explicit policies about coding and data linking to sample donors to safeguard privacy and confidential handling of and access to the data. Standard operating procedures for biobanks will include explicit conditions for maintaining privacy by coding and de-identifying data (38). The use of unique identifiers and security access codes for authorised users are essential. Computing programmes will also include password and other restricted access systems to limit or block data access only to authorised users.

National codes of research ethics distinguish generally between *identified*, *de-identified*, and *re-identifiable* information, but the use of these terms are not consistent and may pose difficulties for developing an international framework (2, 14, 79). In the latter, the tissue and data are coded, but the code can be reversed and the participant's identity revealed. The UNESCO *International Declaration on Human Genetic Data* (2003) adopts similar distinctions between "(ix) *Data linked to an identifiable person*: Data that contain information, such as name, birth date and address, by which the person from whom the data were derived can be identified; (x) *Data unlinked to an identifiable person*: Data that are not linked to an identifiable person, through the replacement of, or separation from, all identifying information about that person by use of a code; (xi) *Data irretrievably unlinked* to an identifiable person: Data that cannot be linked to an identifiable person, through destruction of the link to any identifying information about the person who provided the sample".

The NBAC referred to *unidentified samples* that can sometimes be termed "anonymous" human biological specimens; *unlinked samples* that can sometimes be termed "anonymised" because they lack identifiers or codes that can link a sample to an identified person; *coded samples* that can sometimes be termed "linked" or "identifiable" that link identified specimens to a code and then to personally identifying information; and, *identified samples* that include a personal identifier (such as a name or patient number) to link the biological information directly to the individual from whom the material was obtained (26, 80).

2.15. Withdrawal of Consent and Its Effect on Research

Biobank standards, policies, and procedures generally allow participants to withdraw from biobank studies and projects. This is consistent with accepted international ethical research standard requiring participants be free, at any time to withdraw consent and to withdraw from further involvement in the project. In the case of a biobank research, it will not be possible to withdraw data from previously completed studies. Therefore, the ethical (and possibly contractual) right to withdraw must be contextualised to biobanks and may involve the withdrawal of consent, samples, and data at different levels, depending on the consent and choice of the participant. These levels of withdrawal are:

- *No further contact* – with the participant directly but allowing retention and use of previously provided data/samples with permission to obtain health-relevant records.
- *No further access – allowing* retention and use by the biobank of the data/sample but no participant contact and no permission to obtain health-relevant records.
- *No further use* – no further contact with, or information from, the participant, including the destruction of samples and health-related information (but not data already used) (35).

2.16. The Completion of a Project and Its Effect on Samples and Data

As a general principle, biobanks should have policies and guidelines dealing with the possibility of transfer, closure of assets, and these should be communicated to the participants at the time of recruitment. Similarly, any variation in the arrangement for the maintenance or storage or stewardship of the data for samples should be communicated during the currency of the biobank.

3. Use of Previously Collected Samples

Collections of human tissue (81) (see Note 62) have been a common place in hospitals and specialist clinics from the nineteenth century when preservation techniques were introduced (82). In 1998, the former National Bioethics Advisory Committee (NBAC) estimated that there were more than 282 million specimens stored in the USA and further estimated that the accumulation rate from blood tests, surgery, and other medical procedures was probably in the region of 20 million specimens per year (26, 83). This NBAC report outlined the types of existing collections of human tissues as follows

- Pathology samples – clinical/diagnostic purposes;
- Researchers'/pharmaceutical company collections for unique/ longitudinal research studies;
- Newborn screening tests (Guthrie cards);
- Forensic DNA banks;
- Umbilical cord blood banks;
- Organ, sperm, embryo, and now stem cell banks; and
- Blood banks.

To this list should be added specialised human tissue collections, particularly of cancer tissue, used for specialist research (84). Each of these samples can be further divided into slides, paraffin blocks, frozen or formalin-fixed or extracted DNA. DNA test results from these divided samples forms another further dataset.

3.1. Using Samples and Data Without Consent or Without Renewed Consent

These collections of tissue and data, held in long-term storage, are often not covered by patient consent. However, it is common for these tissue collections that were originally collected for clinical or diagnostic purposes, to be used for other undefined research. This is frequently the case with hospital pathology samples that were usually collected for routine diagnostic and clinical purposes but may now used for research. Historically, hospitals and other institutions holding tissue did not presume refusal, or implied refusal, of consent by patients but presumed, in the absence of consent that it was "consistent with good stewardship to allow reasonable and respectful use [in research] of such legacy tissue collections for the greater public good" (85) (see Note 63). The debates about biobanking have focused discussion on how existing tissue collections may be best managed ethically recognising that in such cases, the issue of participant consent may be problematic. Generally, most countries allow stored tissue to be used in research provided the project is scientifically assessed, approved by an ERB and the samples de-identified (see Note 64).

The distinction between these *existing* collections of human tissue and *future* collections developed specific research purposes is significant in law. Under existing privacy legislation discussed at Subheading 2.7, the privacy rules and principles restrict data and information from being used *except* for the purpose for which it was collected. In effect, this rule of privacy precludes the use of data and information for any *secondary* purpose.

Waiver of consent. The use of human tissue samples in existing collections for research purposes will usually be accompanied by the express consent of the participant. However, ethical approval for the research may be granted by an ERB, in the absence of express consent. In such case, the ERB may waive express consent (86). Where researchers propose to use existing collections for other *secondary* research purposes, national codes of research ethics generally allow researchers to apply to an ERB for the approval of a project. In these cases, the ERB may *waive* the requirement for individual consent. Waiver of consent is not uncommon in epidemiological research and human tissue research. In such cases, ERBs may waive consent after carefully considering a number of factors. Generally, the most important factor is whether the public benefit interest (8) in the value of the research *outweighs* the private interest in personal privacy. The types of factors that will be considered are:

- the nature of existing consents relating to the collection;
- the justification presented by the researcher for the waiver;
- the extent to which it is impossible or difficult or intrusive to obtain specific consent;
- the proposed arrangements to protect privacy;
- the extent to which the proposed research poses a risk to the privacy and wellbeing of the individual;
- whether the research proposal is an extension of, or closely related to, a previously approved research project;
- the relationship of the project to an existing project;
- the possibility of commercial exploitation of the sample;
- statutory provisions; and
- most importantly, whether the public interest in the value of the research *outweighs* the requirements of personal privacy.

Where a research project is approved and allows the project to proceed without individual consent, the ethics committee may impose conditions on the methods for the data collection, use, and protection. Most obviously, the ERB may require that the data be only accessed in a de-identified form. The access to the data may be restricted to certain researchers only. Certainly, the research data must only be used for the research purposes specified in the ethics approval and cannot be used for further research projects without a new ethics approval.

3.2. Role of Ethics Review Boards in Selection of Appropriate Information and Consent Procedures

For existing collections, ERBs also have the traditional role, discussed at Subheading 2.6 above, of protection of the welfare of the sample contributors. The ERB has the usual role to scrutinise and assess the ethical acceptability of submitted research projects using the existing stored data and tissue and decide whether the project involves proper participant consent and ethical conduct before deciding whether a researcher is permitted to carry out the research.

4. Solidarity, Dignity, and Benefit-Sharing

Biobanking research will involve large-scale population cohorts. The scale of this type of research will challenge traditional notions of individualistic research and many social ideas (87). New ideas within the new trilogy of "solidarity" and "benefit-sharing" are emerging. So, the UNESCO *International Declaration on Human Genetic Data* (2003) aims "(a) ... to ensure the respect of *human dignity* and protection of human rights and fundamental freedoms in the collection, processing, use and storage of human genetic data, human proteomic data and of the biological samples ...in keeping with the requirements of equality, justice and *solidarity*..."(emphasis added).

The term "solidarity" invites discussion about the social, family, political, legal, and other factors that promote and maintain integration and trust in society. However, social solidarity in some countries can be used "in a somewhat stronger and more egalitarian sense, [to] require that so much help is provided that the gap between the under-privileged and the others is reduced or eliminated" (71). Biobanking is also about social trust, as discussed in Subheading 2.2 above. This should require biobanks to consider ways in which public trust and engagement can be maintained to promote social solidarity. Similarly, there are deeper issues of the ethical principles to apply to biobanking research. In particular, there are genuine questions about a rigid adherence to individual rights and autonomy in the pursuit of the long-term public health goals of these research tools. Some conventional conceptions of consent may be difficult to accommodate. In some biobanking research (88) (see Note 65), the traditional individualist principle of autonomy may be at odds with Asian (89), Melanesian, and Pacific approaches to decisions made harmoniously within the family and group.

"Benefit Sharing" has found expression in guidelines prepared by UNESCO (90) and HUGO (91). The principle of benefit-sharing promotes the equitable distribution of benefits from research. UNESCO's *International Declaration on Human Genetic Data* is one of the most emphatic assertions of the principle and states that "benefits...from the use of human genetic data... should be shared with the society as a whole and the international community". However, the principle is amorphous, particularly in relation to the operation of intellectual property protections and licencing (40, 72, 92, 93). Nevertheless, the principle encourages researchers and research organisations to consider ways in which the benefits of the biobank research may be equitably distributed. It has been argued (94) that the rhetoric of this principle should be replaced with the implementation of appropriate and practical mechanisms for benefit-sharing. Benefit-sharing also arises in relation to the public or private benefits (7) to be derived from biobanking research and whether those benefits will accrue for the public good (8). The French National Ethics Committee has commented that "resources used by private genomic laboratories, ...are not to be compared with those of public sector activity... private laboratories tend to keep their biological resources and their data banks to themselves. [and] the powerful bio-computerised genomic analysis tools are mainly developed in the private sector (using for the most part, for that matter, data and algorithms produced by the public sector). Such a situation could lead to a form of capture of this research domain by the private sector, and, because public and private strategies differ, the risk of impoverishment of scientific or conceptual quality" (23). This comment emphasises that there are general advantages from the public and private research that can lead to specific development of new health care products (40, 78).

5. Conclusion

Biobanks have the potential to enable a dramatic increase in the quantity of genomic research, as well as significantly improving the quality of the research outcomes. Public trust (17–20) will be an imperative for biobanks. Public trust is a fundamental cornerstone in genetic science and biobanking. Equally importantly, good research data should inform discussion on the development of biobanking (95). Generally, the limited empirical research that has been undertaken indicates a cautious level of public confidence in favour of the development of databases for medical research. Empirical research (96) supporting this view of public support has been undertaken in Canada, Iceland (12, 97), Ireland (98), Australia (99, 100), and Sweden (101). Two projects in

Britain have been especially concerned about public engagement. The funders of the UK Biobank and the project's Ethics and Governance Council have commissioned public opinion surveys, while the set-up of the Generation Scotland project includes a specific branch dedicated to public engagement. Biobanks must commit to their duties of good governance, probity, transparency, and security (see Note 66). There are a host of other unique questions raised by biobanks, including autonomy and consent, public engagement, data-sharing, benefit-sharing, and international harmonisation. There should be a renewed debate on ideas about the public good (5–8) with particular focus on compulsory participation, even a duty (102) to participate, in research for public health purposes and benefits (7). Appropriate and effective regulation is a pre-requisite to the development of the research potentialities of genetic research biobanks and, to a similar degree, the development of the genomics industry. International harmonisation and consistency of biobank regulation on access to database information, transfer between countries, and privacy regimes and policies are essential to realise the promise of biobank research (103).

6. Notes

1. The Generation Scotland project, which is run by a consortium of the medical schools in Scotland with Scottish Executive funding, has this as an explicit objective: http://www.generationscotland.org.
2. A bio-marker is a physiological response or a laboratory test that occurs in association with a pathological process that has possible diagnostic and/or prognostic utility.
3. The term is not precise as has been noted by Knoppers and Saginur (14).
4. These papers provide a helpful list of ethical tensions and issues in biobanking, including consent, ownership and IP, governance, public engagement, data-sharing, research access, security, privacy, benefit-sharing, commercialisation, discrimination, public good, cultural sensitivity, and international harmonisation.
5. Special health registers may include the Perinatal Registers, Cancer Registers, and Mental Health Registers. Some registers may be governed under specific legislation, which defines the type of data to be collected, the method of collection, and restrictions on its use and availability.
6. Or possibly unrelatedness in the case of, say, parentage testing.

7. There are also critics of biobanks who question their methodological soundness and research value.
8. The Icelandic Supreme Court November 27, 2003, judgment No. 151/2003 suggested that the 1998 Health Sector Database Act might be unconstitutional. In 2000, the *Act on Biobanks* No. 110/2000 was introduced for the "collection, keeping, handling and utilisation of biological samples from human beings".
9. For example, Sweden: *Biobanks in Medical Care Act* 2002 information may only be used for research purposes.
10. The P3G motto is "transparency and collaboration".
11. Dr Francis Collins (Head, US National Human Genome Research Institute and Chair, Human Genome Project and International Haplotype Mapping Project) described the Report as "a truly phenomenal job, placing Australia ahead of what the rest of the world is doing" – News release during the XIX International Congress of Genetics Melbourne 5–9 July 2003.
12. In this respect, there is a fundamental divergence between the commercial company structure and the research governance structure. Under a company structure, the accepted legal standard demands that the company owes its principal duties to the shareholders.
13. The term ERB will be used generically in this chapter to refer to research ethics committees that approve human research proposals. These are national variously called, as examples, Institutional review Boards, human Research Ethics Committees, Local Ethics Review Committees. All have a broadly similar composition, including community members, lawyer, religious/ethicists member; researcher with no affiliation with the research project to be considered; and, independent chair.
14. The Medical Research Council, the Wellcome Trust and the Health Department.
15. In Australia, there is a statutory requirement, under the *National Health and Medical Research Council Act*, 1992, for two stages of public consultation before the publication of ethical guidelines for medical research. Similarly, in GMO licencing compulsory public consultation at the application and assessment stages are required, *Gene Technology Act*, 2000 S 52).
16. The National Institutes of Health, National Institute of General Medical Sciences (NIGMS), Human Genetic Cell Repository in the Coriell Institute, has produced a *Policy for the Responsible Collection, Storage and Research Use of Samples from Named Populations*, 2004. Note the Nolan Principles of

Public Life covering responsibility, merit, independent scrutiny, equal opportunities, poverty, openness and transparency, and proportionality. Office of Science and Technology, see http://www.ost.gov.uk/policy/advice/copsac/annex.htm.

17. This is not to under-estimate the complexity of information technology reliability and sometimes exaggerated claims about the new information technology era, see (54).
18. Increasingly, national accreditation standards align with international standards developed by bodies, such as the International Organisation for Standardisation (ISO). "Global integration (through the facilitation of world trade by the WTO) is also forcing greater use of international Standards," with a concomitant reduction in the need for national Standards, Ministry of Economic Development *Review of New Zealand's Standards and Conformance Infrastructure* Wellington NZ September 2005 at 36.
19. In Australia and New Zealand, the voluntary, not-for-profit Australasian Biospecimen Network is developing standardisation advice http://www.abrn.net/.
20. The Australian Law Reform Commission in Report 96, 2003 recommended that best practice in genetic research involving genetic databases require the appointment of an independent intermediary between the researcher and the data and samples (a gene trustee) to protect the privacy of samples and information.
21. See also the "custodian" proposal by the Ireland Law Reform Commission *The Establishment of a DNA Database* Report 78-2005 at Chapter 4. This principle will involve changes in practice and organisation for researchers and for some groups such as hospital-based pathologists.
22. Similarly, the HUGO Ethics Committee *Statement on Human Genomic Databases* in December 2002 declares that human genomic databases are a public resource (1[b]) and all should have access to the benefits of such databases (1[c]) declared that individuals should have choice with regard to donation storage and use of the sample and information derived from it. The participants were also to be informed of a degree of identifiability and the possibility of information from the database might be shared with other researchers in other countries or commercial entities.
23. See Article 12.2 of (60). See also Article 12.3 Secondary use of data or biological samples requires specific consent from the individual donor and, where appropriate, the community. However, if the research data or biological samples cannot be traced back to the individual donor, then consent for secondary use need not be obtained from the individual.

24. Specific requirements of consent are specified on general research ethics guidelines (e.g. (64)) or specific biobank guidelines (e.g. (15, 35)).
25. An important consideration is whether a qualified genetic counsellor will disclose the information or whether such a counsellor will be available to explain the significance of the results.
26. Johnston and Kaye (67) argue that, in the case of the UK and other EU countries, there may in fact, be not only an ethical duty to disclose, but also a legal duty by Article 2 of the *European Convention on Human Rights.*
27. The Council of Europe's "Convention on Human Rights and Biomedicine" provides in Article 11 that "any form of discrimination against a person on grounds of his or her genetic heritage is prohibited".
28. The UK Biobank, Ethics and Governance Framework (35) provides that the selection process reflects inclusion of a wide variety of participants from minority groups and reflecting socially diverse cultural and functionally incapacitated groups.
29. See, for example, Chapter 4 Subheading 5: People with a Competent Impairment, an Intellectual Disability, or a Mental Illness, *National Statement on Ethical Conduct involving Human Research* (64).
30. OECD Creation and Governance of Human Genetic Research Databases, 2007 Subheading 2.1.7.
31. See, for example, Chapter 4 Subheading 2: Children and Young People, *National Statement on Ethical Conduct in Human Research* (64).
32. See OECD *Guidelines on Human Biobanks and Genetic Research Databases* 2009 (37, 48, 52), Principles 3A-, 3D, and Best practices 3.1–3.4.
33. For example, in Australian the *Privacy Act* 1988 is similar to the New Zealand *Privacy Act*, 1993. See also Victoria: *Information Privacy Act* 2000; *Health Records Act* 2000; NSW: *Privacy and Personal Information Protection Act* 1998; *Health Records Information Privacy Act* 2002 ACT: *Health Records (Privacy and Access) Act* 1997.
34. Belgium: Constitution recognises the right of privacy (Article 22); Estonia: Constitution 1992 recognises the right of privacy and data protection (Article 42); Finland: Constitution of Finland The right to privacy (Section 10); Iceland: the 1944 Constitution was amended in 1995 for personal privacy (Article 72); Spain: Constitution recognises the right to personal privacy; the UK: The *Human Rights Act* includes a right of privacy.

35. *Grundgesetz*, the German Constitution does not include a right to privacy. Similarly, Ireland and Canada, Singapore and India (Constitution 1950) have no express rights to privacy in their Constitutions. However, in France; Constitutional Court ruled in 1995 that the right of privacy was implicit in the Constitution by decision 94-352DC du Conseil constitutionnel, 18 January 1995. So too in Japan in 1963, the Supreme Court recognised a right to privacy. There is no explicit right to privacy in the US Constitution.
36. The original legislation was in the USA *Freedom of Information Act* (FOIA) 1966 that allows access to federal government records. See Thailand: *Official Information Act* (OIA) 1997 rights to government information.
37. Belgium: *Act concerning the Protection of Privacy with regard to the Treatment of Personal Data Files*, 8 December 1992 updated 11 December 1998; Estonia: *Personal Data Protection Act*, 1996; Finland: *Personal Data Act* 1999; France: *Data Protection Act* 1978 *amended by Data Protection Act* 2004 for the EU Directive; Germany: 1997 Federal Data Protection Act (*Bundesdatenschutzgesetz* or BDSG) amended in 2002 to be in line with the EU Data Protection Directive; Iceland: 2000, *Act on the Protection of Individuals with regard to the Processing of Personal Data* for compliance with the EU Directive; Ireland: *Data Protection Act*, 1998; Spain: *Data Protection Act* (LOPD), 1999; Sweden: *Personal Data Act* (PDA) or *personuppgiftslagen* (PUL) 1998; Switzerland: *Federal Data Protection Act* 1992; the UK: *Data Protection Act* 1998.
38. The *Privacy Act* 1985, Canada regulates the federal public sector. The *Personal Information Protection and Electronic Documents Act* 2000 (PIPEDA) applies to private sector commercial activities throughout the country, three provinces (Alberta, British Columbia, and Quebec) that have enacted "substantially similar" provincial legislation. Four provinces have legislation for the protection of health information. Ontario (*Personal Health Information Protection Act* 2004), Manitoba (*Personal Health Information Act*), Saskatchewan (*Health Information Protection Act*), and Alberta (*Health Information Act*). The USA: *Privacy Act* 1974 protects records of the US government agencies.
39. For example in Taiwan: Computer-Processed Personal Data Protection Law 1995.
40. Finland: Data Protection Ombudsman (DPO); France: *Commission nationale de l'informatique et des libertes* (CNIL) enforces the Data Protection Act; Spain: Data Protection Agency (*Agencia Espanola de Proteccion de Datos*, or AEPD) enforces the LOPD; Sweden: monitored by the Data

Inspection Board (DIB), *Datainspektionen.*; Canada: both the Privacy Act and PIPEDA are overseen by the independent federal Privacy Commissioner of Canada; New Zealand: Office of the Privacy Commissioner; the UK: The Office of the Information Commissioner enforces the *Data Protection Act*; the USA: there is no independent privacy oversight agency in the USA.

41. For example, the USA: Protections for medical records are found in the *Health Insurance Portability and Accountability Act* (HIPAA) of 1996. In April 2003, Standards for Privacy of Individually Identifiable Health Information (the HIPAA Privacy Rule) were introduced; Finland: *Act on the Status and Rights of Patients* 1993 and *Medical Research Act* 1999; Sweden: Health and medical sector regulated by *Health Care Register Act* 1998 and *Patients' Records Act* 1985. DNA use in law enforcement, Chapter 28 of the Code of Judicial Procedure and the rules in the *Police Data Act* of 1998.
42. For example, France genetic data, under the *Internal Safety Law* Loi n2003-239, 18 march 2003 extended for the DNA National Computerised File of Genetic Data (*Fichier national automatisé des empreintes génétiques* or FNAEG).
43. For example, "The Recipient will in no way attempt to identify or contact the person(s) associated with the biospecimen(s) that make up the MATERIAL under this Agreement. Furthermore, Recipient will not attempt to obtain or otherwise acquire any private identifiable information associated with the biospecimen(s) that make up the MATERIAL under this Agreement" Clause 8 Appendix A2-1 (38).
44. For example, Estonia, *Human Genes Research Act* 2001.
45. For example, *National Statement on Ethical Conduct in Human Research* 2007 prepared by the Australian Health Ethics Committee under the relevant provisions of the *National Health and Medical Research Council Act*, 1992 (Cth).
46. This is the US Common Rule formulation Department of Health and Human Services Policy for the Protection of Human research Subjects 45 CFR 46.111(a)(7). See also Bioethics Advisory Committee (BAC) of Singapore Report on genetic testing and genetic research 2005 on privacy and the confidentiality at http://www.bioethics-singapore.org/resources/reports4.html. Japan published, *Guidelines for the Protection of Personal Information in Businesses that Use Human Genetic Information* in December 2004.
47. See for example, in Singapore, the *Human Tissue Research* (2002), *Genetic Testing and Genetic Research* (2005) and

Personal Information in Biomedical Research (2007), The Bioethics Advisory Committee, Singapore (http://www.bioethics-singapore.org/resources/reports.html).

48. See OECD *Guidelines on Human Biobanks and Genetic Research Databases* 2009, Principles 3B-and Best practices 3.2 and annotations paras 21–25.
49. The most persuasive justification for these oversight bodies is assurance of public trust and confidence, rather than novelty of ethical, research or research governance questions (acknowledging comments from Professor Laurie).
50. The UK Biobank has so decided "because this would reduce the value of the resource for research" (35 at Section 1, B, 7). The OECD *Guidelines on Human Biobanks and Genetic Research Databases* 2009 (37, 52), are silent on this issue.
51. *See Ragnhildur Guomunsdotirv State of Iceland* 920030 Supreme Court of Iceland No151/2003. The Estonian act allows relatives access. For comment, see Gertz, R "An Analysis of the Icelandic Supreme Court Judgment of the Health Sector Database Act", (2004) 1(2) SCRIPT-ed 241-258, available: http://www.law.ed.ac.uk/ahrc/script-ed/issue2/iceland.asp.
52. The UK Biobank has so decided "because this would reduce the value of the resource for research", (35) at 14–16.
53. See Appendix 2 "Material Transfer Agreement for Human Biospecimens" (38). And International Cancer Genome Consortium at www.icgc.org/
54. Promulgated by the General Conference of UNESCO at its 29th Session on 11 November 1997.
55. See (52) Chapter 6 *Commercialisation Considerations.*
56. See, for example, Australian Code for the Responsible Conduct of Research 2007.
57. (35) Section II -A "Stewardship of Data and Samples" at 12
58. See for example, in Australia, *National Statement on Ethical Conduct in Human Research* 2007, Section 1.5 "Research outcomes should be made accessible to research participants". However, with large-scale biobanks, such as the proposed 500,000 volunteers on the UK Biobank, such participant consent may become difficult and impractical. Some biobanks, and the UK Biobank is an example, of chosen, that they will not provide "participants with information, genetic or otherwise, derived from the examination of the database or samples by research undertaken after enrolment". See the UK Biobank *Ethics and Governance Framework* at 8. However, the initial laboratory analysis results will be provided to participants at the physical assessment preliminary stage.

59. Australia, *National Statement on Ethical Conduct in Human Research*, Section 1.3(d) "disseminating and communicating, whether favourable or unfavourable, in ways which permit scrutiny and contribute to public knowledge".
60. The successor to the Human Genome Project, see http://www.hapmap.org/ Access attracts a "clickwrap" licence to protect the data from bogus patent claims.
61. The Human Genome Project's public domain sequence data site at http://www.ncbi.nlm.nih.gov/Genbank/.
62. For example, the tens of millions of cervical cell samples collected each year are invaluable archival samples for research that can be linked to cancer registries, (81).
63. See the helpful discussion on this point in BAC in Singapore, Report on *Human Tissue Research* (2002) (85) at paras 9.1–9.6. This Report interestingly describes existing collections as "legacy tissue".
64. The Report on *Human Tissue Research* (paras 9.1–9.6) (85) felt that it was unjustified to equate the absence of consent with the refusal of consent, and therefore allowed research if the stipulated safeguards of IRB approval and anonymisation were in place.
65. The author proposes some procedural and substantive rules for the basis of an international multi-cultural bioethics (the rule of peaceful dialogue; rule against xenophobia; rule of respect for cultural pluralism; rule of the common good; rule of cultural apprehension; rule of respect for persons in context; and the rule of existential A Prioris).
66. Other suggestions for the regulation of biobanks have included possible national registration. For example, the Australian Law Reform Commission report *Essentially Yours* (29) recommended that the registration of these databases on the public register (Recs 18.1, 18.3). This would enable the NHMRC not only to track the genetic research undertaken in Australia, but also ensure greater transparency and accountability for the biobanks. Registration would provide an effective and inexpensive audit trail in annual reports to the NHMRC.

Acknowledgements

This article has been prepared with the support of Australian Research Council Discovery Grant DP 0559760. Acknowledgement also to Professors T Caulfield, AV Campbell, GL Laurie, M Arbyn, and Associate Professor T Kaan Sheun-Hung for their invaluable contributions, insights, and comments.

References

1. Collins, F. (2003) Keynote address, XIX International *Congress of Genetics*– Melbourne July 7 reported in *Australian Biotechnology News*. 8.
2. Knoppers, B.M., Ma´n H, A.R. and Karine, B. (2007) Genomic Databases and International Collaboration. *King's Law Journal*. **18**, 291–311.
3. Knoppers, B. and Chadwick, R. (2005) Human Genetic Research: Emerging Trends in Ethics. *Nature Reviews Genetics*. **6**, 75–9.
4. Kaye, J. and Stranger, M. *Principles and Practice in Biobank Governance* Surrey, Ashgate Publishing, 2009 and Gibbons, S. and Kaye, J. (2007) Governing Genetic Databases: Collection, Storage and Use. *King's Law Journal*. **18**, 201–8. See also Gottweis H and Petersen A *Biobanks–Governance in Comparative Perspective* Oxford Routledge 2008.
5. Caulfield, T. (2007) Biobanks and Blanket Consent: The Proper Place of the Public Good and Public Perception Rationales. *King's Law Journal*. **18**, 209–26.
6. Campbell, A. (2007) The Ethical Challenges of Genetic Databases: Safeguarding Altruism and Trust. *King's Law Journal*. **18**, 227–45.
7. Brownsword, R. (2007) Genetic Databases: One for All and All for One? *King's Law Journal*. **18**, 247–73.
8. Beyleveld, D. (2007) Data Protection and Genetics: Medical Research and the Public Good. *King's Law Journal*. **18**, 275–89.
9. Aldridge, S. (2005) Biobanking Emerging as a Key Growth Area. *Genetic Engineering News*. **25**, 1.
10. Opinion on Biobanks for Research. Berlin: Nationaler Ethikrat; 2004 March 17.
11. Shastry, B. (2006) Pharmacogenetics and the Concept of Individualized Medicine. *The Pharmacogenomics Journal*. **6**, 16–21.
12. Kaiser, J. (2002) Biobanks: Population Databases Boom, from Iceland to the U.S. *Science*. **298**, 1158–61.
13. Cambon-Thomsen, A. (2004) The social and ethical issues of post-genomic human biobanks. *Nature Reviews Genetics*. **5**, 866–73.
14. Knoppers, B. and Saginur, M. (2005) The Babel of Genetic Data Terminology. *Nature Biotechnology*. **23**, 925–7.
15. Wellcome Trust, Medical Research Council and Department of Health UK, UK Biobank Ethics and Governance Framework Version 3.0 October 2007.
16. Cambon-Thomsen, A., Ducournau, P., Garraud, P.A. and Pontille, D. (2003) Biobanks for Genomics and Genomics for Biobanks. *Comparative and Functional Genomics*. **4**, 628–34.
17. Stranger, M., Chalmers, D. and Nicol, D. (2005) Capital, Trust & Consultation: Databanks and Regulation in Australia. *Critical Public Health*. **15**, 349–58. And Kaye J and Stranger M *Principles and Practice in Biobank Governance* Surrey, Ashgate Publishing, 2009.
18. Chalmers, D. and Dianne N. (2004) Commercialisation of Biotechnology: Public Trust and Research. *International Journal of Biotechnology*. **6**, 116–33 and Gottweis, H. and Petersen, A. *Biobanks–Governance in Comparative Perspective Oxford* Routledge 2008.
19. Bovenberg, J.A. (2004) Inalienably Yours? The New Case for an Inalienable Property Right in Human Biological Material: Empowerment of Sample Donors or a Recipe for a Tragic Anti-Commons? *SCRIPT-ED*. **1**, 591–616.
20. Bovenberg, J. (2005) Towards an International System of Ethics and Governance of Biobanks: A "Special Status" for Genetic Data? *Critical Public Health*. **15**, 369–83 and Kaye J and Stranger M *Principles and Practice in Biobank Governance Surrey*, Ashgate Publishing, 2009.
21. Chalmers, D., ed. (2005) Genetic Testing and the Criminal Law. London: UCL Press.
22. Criminal Investigations (Blood Samples) Act (NZ); 1995.
23. Ethical issues raised by collections of biological materials and associated information data: "biobanks" and "biolibraries", Comité consultatif national d'éthique pour les sciences de la vie et de la santé, France; 2003. Report No.: Opinion 77.
24. OECD Committee for Scientific and Technological Policy: Working Party on Biotechnology. Tokyo Workshop Report: Human Genetic Research Databases: Issues of Privacy and Security; 2005. Report No.: DSTI/STP/BIO.
25. Tutton, R. and Corrigan, O., ed. (2004) Genetic Databases: Socio-Ethical Issues in the Collection and Use of DNA. London: Routledge. And see UK Biobank Ethics and Governance Framework Version 3.0 October 2007.
26. National Bioethics Advisory Commission, Research Involving Human Biological Materials: Ethical Issues and Policy Guidance Volume I: NBAC; 1999 August.

27. National Bioethics Advisory Commission. Ethical and Policy Issues in Research involving Human Participants Volume II: Commissioned Papers. Bethesda, Maryland; 2001.
28. National Bioethics Advisory Commission. Ethical and Policy Issues in Research involving Human Participants Volume I: Report and Recommendations of the National Bioethics Advisory Commission. Bethesda, Maryland; 2001.
29. Australian Law Reform Commission. Essentially Yours: The Protection of Human Genetic Information in Australia; 2003. Report No.: 96.
30. Longtin, R. (2004) Canadian Province Seeks Control of Its Genes. *Journal National Cancer Institute.* **96**, 1567–69.
31. Health Sector Database Act (Iceland); 1998.
32. Human Genes Research Act 2001 (Estonia); 2001.
33. Winickoff, D.E. and Winickoff, R.N. (2003) The Charitable Trust as a Model for Genomic Biobanks. *The New England Journal of Medicine.* **349**, 1180.
34. Boggio, A. (2005) Charitable Trusts and Human Research Genetic Databases: The Way Forward? *Genomics, Society and Policy.* **1**, 41–9.
35. Wellcome Trust, Medical Research Council and Department of Health UK. UK Biobank, Ethics and Governance Framework, Version 3.0; October 2007.
36. Trouet, C. (2004) New European guidelines for the use of stored human biological materials in biomedical research. *Journal of Medical Ethics.* **30**, 99–103.
37. OECD Working Party on Biotechnology. Draft Guidelines for Human Genetic Research Databases. Paris; 2007. Report No.: DSTI/STP/Bio (2007) 17/REVI (see OECD, *Guidelines on Human Biobanks and Genetic Research Databases* 2009).
38. National Cancer Institute, National Institutes of Health and U.S. Department of Health and Human Services. First-Generation Guidelines for NCI-Supported Biorepositories; 2006 April.
39. Kaye, J., Helgason, H., Nomper, A., Sild, T. and Wendel, I. (2004) Population Genetic Databases: A Comparative Analysis of the Law in Iceland, Sweden, Estonia and the UK. *TRAMES.* **8**, 15–33.
40. Nicol, D. (2006) Public Trust, Intellectual Property and Human Genetic Databases: The Need to Address Benefit Sharing. *Journal of International Biotechnology Law.* **3**, 89–103.
41. Steering Committee on Bioethics. Draft Recommendations on Research on Biological Materials of Human Origin. Strasbourg; 2005 November.
42. Cambon-Thomsen, A., et al, (2003) Ethical and Legal Aspects of Biological Sample Banks: Synthesis, Practical Questions and Proposals [Aspects ethiqués et réglementaires des collections d'échantillons biologiques: Synthèse, questions pratiques et propositions]. *Revue d'Epidemiologie et de Sante Publique.* **51**, 99.
43. Working Group on DNA and Epidemiology (TUKIJA). DNA Samples in Epidemiological Research: National Advisory Board on Health Care Ethics (ETENE); 2002 August.
44. Swedish Medical Research Council (MFR). Research ethics guidelines for using biobanks, especially projects involving genome research; 1999 June.
45. European Group on Ethics in Science and New Technologies. Opinion of the European Group on Ethics in Science and New Technologies to the European Commission, Ethical Aspects of Human Tissue Banking: European Commission; 1998 July.
46. ESRC Research Ethics Framework. Discussion Paper 2: The International Dimension to Research Ethics: The Significance of International and Other Non-UK Frameworks for UK Social Science; 2004 April.
47. Department of Health & Human Services, Public Health Service, National Institutes of Health and National Cancer Institute. 133rd National Cancer Advisory Board, Summary of Meeting; 2005 February.
48. Bioethics Advisory Committee of the Israel Academy of Sciences and Humanities. Population-Based Large-Scale Collections of DNA Samples and Databases of Genetic Information; 2002 December.
49. Human Tissue Act (UK); 2004.
50. Kaye, J. and Stranger, M. *Principles and Practice in Biobank Governance* Surrey, Ashgate Publishing, 2009 and Tutton, R. (2007) Constructing Participation in Genetic Databases: Citizenship, Governance, and Ambivalence. *Science Technology and Human Values.* **32**, 172–95.
51. Hansson, M.G. (2005) Building on Relationships of Trust in Biobank Research. *Journal of Medical Ethics.* **31**, 415–8.
52. OECD, *Guidelines on Human Biobanks and Genetic Research Databases* 2009.
53. International Society for Biological and Environmental Repositories (ISBER) (2005) Best Practices for Repositories I: Collection, Storage, and Retrieval of Human Biological

Materials for Research. *Cell Preservation Technology.* **3**, 5–48.
54. Blumenthal, D. and Glaser, J. (2007) Information Technology Comes to Medicine. *New England Journal of Medicine.* **356**, 2527–34.
55. Ministry of Economic Development. Review of New Zealand's Standards and Conformance Infrastructure. Wellington, New Zealand; 2005 September.
56. Eiseman, E., Bloom, G., Brower, J., Clancy, N. and Olmsted, S.S. Case Studies of Existing Human Tissue Repositories: "Best Practices" for a Biospecimen Resource for the Genomic and Protemic Era: Prepared for the National Cancer Institute, National Dialogue on Cancer; 2003.
57. Wellcome Trust, Medical Research Council and Department of Health UK. UK Biobank, Ethics and Governance Framework, Version 3.0; October 2007.
58. Brownsword, R. (2003) Bioethics Today, Bioethics Tomorrow: Stem Cell Research and the Dignitarian Alliance. *Notre Dame Journal of Law Ethics and Public Policy.* **17**, 15.
59. Beyleveld, D. and Brownsword, R. (2001) Human Dignity in Human Ethics and Bio-law. Oxford: Oxford University Press.
60. CIHR. CIHR Guidelines for Health Research involving Aboriginal People; 2007 May.
61. Fleming, J. (1996) Ethics and the Human Genome Diversity Project. *Law and the Human Genome Review.* **4**, 141.
62. Calderon, R. (1996) The Human Genome Diversity Project: Ethical Aspects. *Law and the Human Genome Review.* **4**, 107.
63. (1996) Declaration of Indigenous Peoples of the Western Hemisphere Regarding the Human Genome Diversity Project. *Law and the Human Genome Review.* **4**, 209.
64. NHMRC. National Statement on Ethical Conduct in Human Research 2007; 2007.
65. Hansson, M. (2006) Should Donors be Allowed to Give Broad Consent to Future Biobank Research? *The Lancet Oncology.* **7**, 266–9.
66. Kaye, J. (2004) Abandoning Informed consent: The Case of Genetic Research in Population Collections. In: Tutton R, Corrigan O, eds. Genetic Data Bases: Socio- Ethical Issues in the Collection and Use of DNA. London: Routledge. And Kaye, J. and Stranger, M. *Principles and Practice in Biobank Governance* Surrey, Ashgate Publishing, 2009.
67. Johnston, C. and Kaye, J. (2004) Does the UK Biobank have a Legal Obligation to Feedback Individual Findings to Participants? *Medical Law Review.* **2**, 239–67.
68. The British Avon Longitudinal Study of Parents and Children (ALSPAC).
69. Chalmers, D. (2004) Research Involving Humans: A Time for Change? *The Journal of Law, Medicine & Ethics.* **32**, 583–95.
70. An analysis of the Icelandic Supreme Court judgement on the Health Sector Database Act. Script-ed, 2004. (Accessed 7 March 2006, at http://www.law.ed.ac.uk/ahrb/script-ed/issue2/iceland.pdf.)
71. (2005) The European Group on Ethics in Science and New Technologies the European Commission. *EGE Newsletter "Ethically Speaking".* **5**, 27.
72. Knoppers, B.M. (2005) Biobanking: International Norms. *Journal of Law, Medicine and Ethics.* **33**, 7–14.
73. Kaye, J. (2006) Do We Need a Uniform Regulatory System for Biobanks Across Europe? *European Journal of Human Genetics.* **14**, 245–8.
74. Quebec Network of Applied Genetic Medicine. Ethical Conduct of Human Genetic Research Involving Populations; 2003.
75. Chalmers, D. (2006) Ethical Principles for Research Governance of Biobanks. *International Journal of Biotechnology Law.* **3**, 221–30.
76. German National Ethics Council. Biobanks for Research; 2004.
77. Weisbrot, D. Public Conspiracy, Genetic Counselling and the Required Legal Infrastructure; 2005 August.
78. Haddow, G., Laurie, G., Cunningham-Burley, S. and Hunter, K.G. (2007) Tackling Community Concerns About Commercialization and Genetic Research: A Modest Interdisciplinary Proposal. *Social Sciences & Medicine.* **64**, 272–82.
79. Elger, B. and Caplan, A. (2006) Consent and Anonymization in Research Involving Biobanks: Differing Terms and Norms Present Serious Barriers to an International Framework. *EMBO reports.* **7**, 661–6.
80. National Bioethics Advisory Commission. Research Involving Human Biological Materials: Ethical Issues and Policy Guidance: Volume II Commissioned Papers; 2000 January.
81. Medical Research Council policy and guidance on human tissue.
82. Scott, R. (1981) The Body as Property. London: Alan Lane.
83. Knoppers, B. (2002) DNA Banking: A Retrospective-Prospective. In: Burley, J. and

Harris J, eds. A Companion to Genethics. Oxford: Blackwell Publishing: 379–86.

84. Knoppers, B.M., ed. (1997) Human DNA: Law and Policy – International and Comparative Perspectives. The Hague: Kluwer Law International.
85. Bioethics Advisory Committee, S. Human Tissue Research; 2002.
86. Zeps, N., Iacopetta, B.J., Schofield, L., George, J.M. and Goldblatt, J. (2007) Waiver of Individual Patient Consent in Research: When do Potential Benefits to the Community Outweigh Private Rights? *Medical Journal of Australia*. **186**, 88–90.
87. Glasner, P., Atkinson, P. and Greenslade, H. (2006) New Genetics, New Social Formations. London: Routledge.
88. Thomasma, D. (2001) Proposing a New Agenda on Bioethics and International Human Rights. *Cambridge Quarterly of Health Care Ethics*. **10**, 299–310.
89. Sleeboom-Faulkner, M. (Ed.) (2009) *Human Genetic Biobanks in Asia: Politics of Trust and Scientific Advancement*. Oxford: Routledge, and see Jing-Bao, N. (2007) The Specious Idea of an Asian Bioethics. Chapter 19 In: Ashcroft, R. et al, eds. Principles in Health Care Ethics: John Wiley. London at 144–149.
90. UNESCO. Universal Declaration on Bioethics and Human Rights; 2005.
91. HUGO, E.C. Statement on Benefit Sharing: The Council of the Human Genome Organisation; 2000.
92. Simm, K. (2005) Benefit-sharing: an inquiry regarding the meaning and limits of the concept in human genetic research. *Genomics, Society and Policy*. **1**, 29–40.
93. Chadwick, R. and Berg, K. (2001) Solidarity and Equity: New Ethical Frameworks for Genetic Databases. *Nature Reviews Genetics*. **2**, 318–21.
94. Knoppers, B.M. and Sheremeta, L. (2003) Beyond the Rhetoric: Population Genetics and Benefit-Sharing. *Health Law Journal*. **11**, 89.
95. Hirtzlin, I., Dubreuil, C., Préaubert, N., Duchier, J., Jansen, B., Simon, J., Lobato de Faria, P., Perez-Lezaun, A., Visser, B., Williams, G.D., Cambon-Thomsen, A. and EUROGENBANK Consortium. (2003) An Empirical Survey on Biobanking of Human Genetic Material and Data in Six EU Countries P/C. *European Journal of Human Genetics*. **11**, 475–88.
96. Caulfield, T. and Outerbridge, T. (2002) DNA Databanks, Public Opinion and the Law. *Clinical and Investigative Medicine*. **25**, 252–6.
97. Caulfield, T. (2002) Perceptions of Risk and Human Genetic Databases: Consent and Confidentiality Policies. In: Armason, G., et al, eds. Blood and Data: Ethical, Legal and Social Aspects of Human Genetic Databases: University of Iceland Press and Centre for Ethics: Reykjavik: 283–9.
98. Cousins, G., McGee, H., Ring, L., Conroy, R., Kay, E., Croke, D. and Tomkin, D. Public Perceptions of Biomedical Research: A Survey of the General Population in Ireland: Health Services Research Centre, Royal College of Surgeons in Ireland; 2005.
99. Williams, C. (2005) Australian Attitudes to DNA Sample Banks and Genetic Screening. *Current Medical Research and Opinions*. **21**, 1773–5.
100. Fleming, J. (2007) Issues with Tissues: Perspectives of Tissue Bank Donors and the Public Towards Biobanks and Related Genetic Research. *Biobanks: Centre for Law and Genetics Symposium*.
101. Kettis-Lindblad, A., Ring, L., Viberth, E. and Hansson, M.G. (2007) Perceptions of Potential Donors in the Swedish Public Towards Information and Consent Procedures in Relation to Use of Human Tissue Samples in Biobanks: A Population-Based Study. *Scandinavian Journal of Public Health*. **35**, 148–56.
102. Harris, J. (2000) Research on Human Subjects. In: Freeman, M. and Lewis, A., eds. Law and Medicine, Current Legal Issues. Oxford: Oxford University Press: 379–97.
103. Reymond, M., Steinert, R., Escourrou, J. and Fartainer, G. (2002) Ethical, Legal and Economical Issues Raised by the Use of Human Tissue in Postgenomic Research. *Digestive Diseases*. **20**, 257–65.

Chapter 2

The Need to Downregulate: A Minimal Ethical Framework for Biobank Research

Mats G. Hansson

Abstract

There are currently multiple international bodies suggesting legal and ethical frameworks for regulating international biobank research. One will for obvious reasons find inconsistencies in terminology and differences in procedures suggested for biobank research among all those guidelines, emanating from many different moral and legal traditions. A central question is whether this constitutes a threat to making progress in international biobank research, as some have argued. In this book, Chapter 1 suggests that there are sufficient and well-established instruments and ethical principles available to guide research in this area. Basically I argue that there is no need for a top-down superstructure of detailed rules and guidelines to be imposed on biobank researchers. With the existing ethical review boards (ERBs) playing a central role guided by well-established ethical guidelines (e.g., the Helsinki Declaration) and solutions to specific ethical problems suggested in the literature, self-regulation by researchers providing arguments for balancing of interests in association with different research initiatives and protocols will be sufficient. Traditional information and consent procedures suffice and data protection implies a sovereign right of the individual citizen to grant the use of biobank material and personal data that is needed for biobank research. Clearly, there may still be inconsistencies in terminology when researchers of different nationalities meet in common enterprises, but both they and the ERBs are well equipped to sort out what is actually meant and propose different instruments for, for example, coding following recently established nomenclatures. The existing ERBs should play the key role, guided by the sound argumentation of the researchers in their applications to the board.

Key words: Ethics, Informed consent, Autonomy, Privacy, Public trust

1. Introduction

As has recently been described by Knoppers et al., there are currently multiple international bodies suggesting legal and ethical frameworks for regulating international biobank research (1). UNESCO issued its universal declaration on human genome and human rights in 1997. The European Council agreed on a convention on biomedicine and human rights in 1996, a document that has been

Joakim Dillner (ed.), *Methods in Biobanking*, Methods in Molecular Biology, vol. 675,
DOI 10.1007/978-1-59745-423-0_2,

a beacon to many legislators. A follow-up came in 2006 regarding research on biological material. WHO issued in 2003 a report on genetic databases. OECD and its working party on biotechnology provided a draft of guidelines for human genetic research databases in July 2007. Don Chalmers has in his chapter in this book (Chapter 1) provided a comprehensive account with full references to these and other official documents. Different academic bodies have taken several initiatives, notably the HUGO Ethics Committee in its statement on human genomic databases from 2002. National biobank consortia provide their own guidelines (e.g., the UK Biobank) and the recent initiative called P^3G has the ambition to suggest a comprehensive global framework of guidelines for genetic research using human biological material.

One will for obvious reasons find inconsistencies in terminology and differences in procedures suggested for biobank research among all those guidelines, emanating from many different moral and legal traditions. A central question is whether this constitutes a threat to making progress in international biobank research. Knoppers et al. conclude that in the absence of "common … norms, laws and approaches within a properly harmonized international framework, international collaboration will remain an empty platitude" ((1) p. 311). I seriously doubt that this is the case. As witnessed in this book, there are already many ongoing successful international collaborations using biobank material. I will in this chapter suggest that there are sufficient and well-established instruments and ethical principles available to guide research in this area. Basically I will argue that there is no need for a top-down superstructure of detailed rules and guidelines to be imposed on biobank researchers. With the existing ethical review boards (ERBs) playing a central role guided by well-established ethical guidelines (e.g., the Helsinki declaration) self-regulation by researchers providing arguments for balancing of interests in association with different research initiatives and protocols will be sufficient. Taking into consideration the low risks for sample donors associated with biobank research, something most participants in the discussion seem to agree on (see, for example, Chapter 1), the current efforts to create long and complex lists of "principles" and "best practices" looks like trying to kill a mosquito with a baseball bat. Before suggesting the components of a more appropriate, minimal framework, I will go through some of the central questions in the current discussion.

2. The Claim That Biobank Research Implies "New" Challenges

It is often claimed that genetic research using human biological material together with personal data and different medical records gives rise to a number of "new" ethical issues to be handled by the research community. Gibbons and Kaye state that "genetic databases raise a host of challenging issues, many of which test our traditional legal concepts, governance provisions and bioethical principles" ((2) p. 204). However, as recently argued by Ruth Chadwick and Mark Cutter:

> the concept of collection of information into databases is not a new phenomenon; similarly, the collection and use of genetic information is not a new practice. The use of "family history" in determining life insurance, assurance and relative premiums is well documented, as is its use in diagnosis during genetic counseling sessions. Equally the storage of human genetic material and information in the form of medical records is not unusual or new. Arguably, since Gregor Mendel's original experiments with the hereditary characteristics of pea plants, through to James Watson and Francis Crick's identification of the double helix of DNA, the biological sciences have been on a trajectory that seems naturally to culminate in the creation of human genetic research databases or biobanks ((3) p. 225).

This view is shared by Thomas Murray who early on questioned the view that genetic information is something exceptional in comparison with other kinds of medical information (4). Along the trajectory of genetic research, ERBs and data protection authorities seem to have managed quite well to keep up with new research initiatives to balance the different interests at stake. Chadwick and Cutter suggest that it is the negotiation between the individual and public interests that cause population-based genetic databases to be something special. I will come back to this claim in a discussion of the concept of autonomy.

3. The Role of Patient and Public Surveys

Wendler has recently made an overview of 30 studies published in English that reported the views of individuals on consent for research with human biological samples (5). He concludes that:

> Data from more than 33,000 people around the world support offering individuals a simple choice of whether or not their samples can be used for research purposes, with the stipulation that an ethics committee will decide the studies for which there samples are used. This approach offers a method that could be adopted across institutions and around the world ((5) p. 547).

Wendler admits that framing effects can affect survey results and that some questions may not have been fully understood by the respondents. However, the data seem to be consistent across many different studies using different questions and different methodologies in different cultures. We have in similar studies acquired the same results (6). Caulfield is skeptical to this use of surveys (7). He claims, rightfully so, that at best they represent the majority view and there are examples of individuals wanting other information and consent procedures. However, in the majority–minority negotiation, it should be observed that whether there is an instrument available to protect the minority view (e.g., those individuals wanting specific information and consent for each new research project) one may feel more comfortable in acting on behalf of the majority view (e.g., broad consent with surrogate decision by an ERB). As a matter of fact, there is such an instrument available that can serve this purpose and that is the right of an individual to withdraw his or her consent. This is part of the information and consent procedure to be decided by the ERB. As Caulfield argues, majorities may change so there is a continuous demand on all involved parties to secure public understanding and public trust.

4. The Role of Commercial Interests

Caulfield argues, furthermore, "there is evidence that some members of the public are uneasy about the involvement of private interests" (op. cit. p. 220). There seems to be support for such a conclusion from several studies. However, the picture is complex and one question here concerns what conclusions that may be drawn from public surveys. A question that was discussed above in connection to the selection of appropriate information and consent procedures. Caulfield mentions an Australian study as a source of evidence: "Thus, an Australian study exploring public attitudes to biobanking found that '75% indicated concerns over commercialization' of the research process and access to information by health insurance companies" (Ibid. 221). Williams performed the study with 358 patients attending a cardiology department who were given a questionnaire while registering for a gene bank, thus a highly select group and not representative of the general public (8). Williams concludes that "75% indicated concerns over commercialization and access to information by health insurance companies" (p. 1774), so Caulfield's quote is partly right even if it was not the public view as he claimed. However, a closer look at the questionnaire that is presented in the article shows that Williams' conclusion does not follow from her data. Question 10 was phrased: "Do you think insurance companies should be allowed access to your genetic information?" 7% answered *yes*, 74% answered *no*, and 9% were unsure. There is no question regarding commercial interests involved in the research process presented in the questionnaire.

It is well known that people are concerned about insurance companies getting access to genetic information through medical databases. Whether and under what conditions they should have access is a complex question that I will not go into here. As indicated by Chadwick and Cutter earlier, the question is not new since insurance companies have access to other kinds of medical data. However, I think one should distinguish between access to information by private insurance companies and by pharmaceutical companies. If properly informed I believe that most people will understand the need of partnership between academic and commercial interests. Scientists at the universities have simply no possibility to assume responsibility for the whole chain of research and development from a basic scientific finding to a new medical product. In practice, however, the question about the access of pharmaceutical companies to biobanks may not be so difficult to resolve since they often have their own biobanks, collected under very strict conditions.

5. The Importance of Public Trust

To realize the potential of biobanks, efficient collaboration between many actors is essential and the practice as a whole rests upon the confidence of patients and healthy persons donating blood and tissue samples. Trust must be established both within the medical and the research community and with the general public. Decreased patient confidence in biobanking practice may have damaging consequences. If individuals start revoking their consents the banks will not be complete, the possibility to draw scientifically valid conclusions will decrease and the potential for follow-up examinations and medical treatment will not be fulfilled. In Sweden, there is an efficient legal instrument available for those individuals losing confidence in the system through the Biobank Act which gives each sample donor or sample source such a right to withdraw their consent and have the sample destroyed or stripped of identification possibilities, strongly decreasing the potential by precluding the important possibility to match the information of the sample with information in different medical and personal registries ((9), 3 kap, Subheading 6).

Conflicts between the researchers and between the universities and hospitals are not instrumental for increasing the trust essential for the success of biobank research (10). The main victims of the distrust are the actual and future patients waiting for improved methods in diagnosis and treatment. The success of core facilities for biobank research and collaborative projects depend on appropriate acknowledgment of the different contributions to these facilities and research results. Collaboration should be based on a transparent organization of the research and on legally binding agreements. Such agreements should also include policies and rules regarding the sharing of samples, data, and research results.

Patient confidence in biobank research is maintained by keeping strict rules for privacy protection and respecting patient–physician relationship. However, it should be observed that ERBs and regulatory bodies setting up rules for biobank research are themselves subjects to public trust. Patients and healthy donors have interests at the beginning of the research line, for example, being assured about the protection of their integrity and providing tissue material and access to personal data for good scientific reasons, but they have also general research interests connected to the potential of providing new treatment and new medical products (11, 12). A too strict interpretation of the legal principles governing this kind of research, for example, regarding the possibility to use previously collected samples without a renewed consent, may be detrimental to their research interests. They may have good reasons for wanting to waive the right to be informed. As shown by Wendler and others, it has in fact been shown in public surveys that a majority want broad information and consent procedures and want to waive their right to provide an explicit and specific informed consent for each research project, handing over the decision to an ERB (5, 6, 13).

6. The Concept of Autonomy

In our research team, we have often argued for different practical solutions regarding biobank research on the basis of a respect for autonomy. McQuillan et al. have suggested that "specific consent must be obtained if an individual's autonomy is to be respected in all aspects of the research, both current and future" ((14) p. 40). This represents indeed a very limited view of autonomy and, as O'Neill has pointed out "there are many distinct conceptions of individual autonomy, and their ethical importance varies" ((15) p. 4). However, I do not entirely agree with Knoppers and Chadwick that we need to "move away from autonomy as the ultimate arbiter," even if we should pay attention to other fundamental notions related to biobank research, such as solidarity, reciprocity and citizenry ((16) p. 75).

I have at length recently discussed the notions of autonomy and privacy elsewhere and shall just briefly mention some important points here (17). It seems that the view taken by McQuillan et al. about the research subject's autonomy is shaped by a political concept that basically derives from the ancient world. In ancient Greece, autonomy was a political concept that emphasized independence. An individual is autonomous when he takes charge of his own affairs and is protected from external interference, even if its price is isolation from other people and from the world around. It was first with Kant that autonomy was defined as a moral concept (18). Respect for people's autonomy entails, according to Kant, a respect for their capacity to participate in the formulation of the moral principles that every human being would wish to endorse. In this sense, human beings are self-legislators, but it is a question of laws and rules with, in principle, a universal sphere of application.

Making autonomous decisions in accordance with the Kantian tradition thus involves taking account of the well-being of others through a judgment of how one's own decisions affect other people's ability to act in a morally responsible way and to attain their own goals. Kant has, in his concept of autonomy, incorporated an element of intersubjectivity. The individual is a member of a moral community of beings and is expected to take into account how one's own interests may affect other individuals. Autonomy is inherently social, with the implication that the working out of legal protection for self-determination and integrity in association with biobank research must simultaneously do justice to both the research subject's independence and this individuals' dependence on others for fulfilling mutual interests. Furthermore privacy interests should not, as it is commonly understood be set in direct opposition to public interests (for one example of this confusion see (3) pp. 225f). The individual wishes simultaneously to enjoy a private sphere protected from insight but also to participate and to be a member of society. This view implies the importance of protecting private information, for example, through different coding measures in association with biobank research while at the same time ensures that the individual can take part in a common enterprise such as the production of medical knowledge and treatment opportunities that is provided through large population-based biobank research platforms (12).

O'Neill has suggested that respect for autonomy implies control over how one's samples are used (15). As she acknowledges, this includes a possibility to affirm requests for broad and future consents without the opportunity to be approached in the future. However, in my view it does not necessarily imply that there in addition must be an opportunity for individual control after the initial sampling has taken place so that those who wish should have a possibility of being recontacted for new research projects, something O'Neill suggests. Taking the Kantian view on moral autonomy in consideration where the individual is called upon to take also other individual's interests into consideration (e.g., future members of society), it may be sufficient if there is a democratic instrument available that ensures the individual citizen insight into how the biobank is organized and that principles for balancing of interests at the ERBs take all relevant interests into account. It may for instance be openly declared that in some cases public health interests have been judged to be of overriding importance compared with individual interests.

An example of when this level of democratic control is appllied is medical registries, for example, cancer registries, which are instituted by the parliament and under the care and supervision of public authorities and do not allow any possibility for individuals to withdraw their data.

7. The Selection of Appropriate Information and Consent Procedures

Timothy Caulfield argues that "biobanks have created some of the most difficult legal and ethical dilemmas within modern biomedicine" and that "maintaining traditional consent norms may harm the social utility and scientific value of large-scale biobanking initiatives" ((7) p. 210). However, as I argued already in 1998, ERBs have in their tool box several information and consent procedures that are all legitimate and that are appropriate for different purposes (19). The key task for the ERB is to select an appropriate procedure that represents a reasonable balancing of the risks and benefits associated with a specific research protocol.

For competent adults the rules of informed consent are rather straightforward. Incompetent research subjects constitute a greater problem. Informed consent cannot be a general solution. I have recently argued that one should also apply a "safety principle," which take into consideration patient safety with regard to diagnosis, treatment, care, and prevention, implying that research may be conducted on these individuals even if no consent is available (and cannot be) (20).

Rules of informed consent are based on a respect for the moral authority and autonomy of individual research subjects. In the practice of medical research, this implies that research subjects should never be exposed to a risk in association with a research project without their consent. It does not follow that research subjects should never be exposed to any risks. There are few, if any, research protocols that do not carry a potential risk to the research subject. The researcher has to control as far as possible for short- and long-term risks. After informing the research subject about the purpose of the research, its expected benefits, the risks associated with it, and how these risks will be managed, informed consent is obtained from the subject – a way of handing over the decision to the research subject – Are you willing to assume the remaining risk (indeed in Phase 1 and 2 clinical trials the unknown risk)? Information is also given about stopping rules and procedures for control of the risk and about the opportunity to withdraw from the study without this having any effect on evidence-based treatment provided, and care is taken to make sure that the research subjects are not object for exploitative incentives of any sort.

In practice, there are many pieces of legitimate information and consent procedures available (19). The appropriate procedure is selected on the basis of balancing the scientific value against the risk entailed by the project. It is not reasonable that the rule of obtaining an informed consent shall be the same in situations of ordinary treatment, in clinical trials and in protocols of epidemiological biobank research where no personal identification is possible or both the biological material and the personal data are coded and strictly protected. I have earlier argued that: "The quality of consent needs to be balanced against the different values that are at stake in different contexts. The kind of information, the way it is given, the degree of voluntariness and the format of authorization must be adjusted accordingly" (Ibid. p. 182). According to the model I have suggested, "appropriate information and consent procedures vary depending on context between extensively informed consent with written and oral information to informed refusal with only a limited amount of information given. At the other end it should just be a matter of making relevant information available" (Ibid). In biobank research, one has to distinguish between two fundamentally different kinds of research protocols, those using only previously collected samples and those associated with the collection of new samples for future research.

Against this view Caulfield argues that "most large-scale biobanks should be thought of not as discrete research projects, but as 'research platforms' that will be used by a number of researchers, for various research initiatives, over many decades, which are not fully known when the genetic information is obtained from participants. As a result, it is impossible to obtain truly informed consent from biobank participants" (Op. cit. p. 213). Biobank research implies broad consent to future research and this cannot be a "truly" informed consent. Caulfield's view is shared by Vilhjálmur Arnason who argues against the use of broad or generally formulated consent forms. Arnason argues that:

> If we are to preserve a meaningful notion of informed consent for participation in research, it should only be used about specified research where the participants are informed about the aims and methods of a particular research proposal. … There is no such thing as "general informed consent." The more general the consent is, the less informed it becomes. It is misleading to use the notion of informed consent for participation in research that is unforeseen and has not been specified in a research protocol ((21) p. 41).

The success of biobank research implies that large repositories of human tissue material are collected together with well-described and managed clinical and personal data. As described in the previous chapter, there are now several large national biobanks working in this way. The specific nature of the research is unknown and only general descriptions about the goals of these biobanks are possible, for example, for biomedical research or research on large groups of common diseases. A specific consent to a narrowly described research protocol is not possible and there is a need to ask for a broad consent covering future research. Caulfield and Arnason argue that the traditional meaning of informed consent cannot accommodate these broad and future consents. Consent should be based on specific information otherwise it is not a valid consent. However, as we have pointed out earlier this only raises the question: "What is appropriate information? If the information covers all aspects relevant for a person's choice, then that person's consent is appropriately informed. If the essential risk and benefit levels are general to a number of studies, then general information on these studies may be sufficient for the donor of the sample to make an informed decision" (22). As has been described there are many pieces of legitimate information and consent procedures that balance the scientific value of the biobank, the nature of research and the risks that are believed to be at stake.

We have recently argued that "accepting broad and future consent implies a greater concern for autonomy than if such consents are prohibited. Respect for autonomy does not imply total self-governance when a decision also affects others such as family members. However, infringement on autonomy should only be done with good cause. Under the condition that information is coded and safely handled and that secrecy is maintained, both donors and families are protected from harm, no limitation of autonomy is necessary" ((22) p. 267). Asking for a broad consent to future research, for example, biomedical research, implies a respect for each individual to decide for him- or herself if the general information is sufficient. A mechanism that allows individuals to change their minds and withdraw their consent will provide an extra protection. There are different mechanisms for this, for example, withdrawal allowing further use (with or without de-identification) and withdrawal prohibiting further use. Accepting broad and future consent is consistent with a policy where the ERBs will examine and give permission to each new research project using these large biobanks. "In order for en ERB to evaluate the risk/benefit relationship for a donor, it must review the coding measures, information security and other potential risks for the donor that may arise from, for example, changes in legal status, principal investigators or organization of the original biobank" (Ibid. p. 269). Broad consent, not broad permissions, is the favorable policy. This policy of broad consent seems now to emerge internationally as the generally preferred solution according to a recent review of the literature (23).

It is not at all implausible that donors to biobanks may understand the medical importance of creating such research platforms, including the cost of returning for renewed consent. Biobank research has been going on for some time and many patients and research subjects seem to be willing to take part also for broadly described purposes. Furthermore, as argued by Campbell, to safeguard altruism and trust in biobank research one should refrain from "suggesting that individual donors have ongoing rights to exercise control over uses of their donated materials and the resource itself " ((24) p. 242). Campbell emphasizes that maintaining trust is essential and this includes also a requirement on those issuing rules and guidelines not to impose too many restrictions that will constitute a hindrance in fulfilling important donor interests related to the production of new medical knowledge and treatment opportunities (11). For some examples of how biobanks in association with good clinical data are vital assets for understanding the underlying mechanisms of human diseases and for providing medical care and for treatment of current and future patients see Sigstad et al. (25) Kaijser (26), Lindberg (27), and Sundstrom et al. (28).

The use of previously collected samples seems to constitute a special problem in international collaboration. Recontacting donors who earlier have contributed to pathology biobanks or to a research biobank to obtain a renewed informed consent for a new research project may not be practically feasible. However, the major ethical reason for abstaining from asking again is the cost in scientific value it implies, and consequently decreased potential for providing new biomedical knowledge and medical treatment. Asking again may be seen as an act of respect for autonomy but if the donor learns to know that this is detrimental to his/her general research interests they may very well instead feel a disrespect. The European Council has acknowledged the need of balancing in a commentary to article 22 in the European Convention on Biomedicine and Human Rights (29) where they state that: "information and consent arrangements may vary according to the circumstances, thus allowing for flexibility since the express consent of an individual to the use of parts of his body is not systematically needed" ((29), Commentary 137 to Article 22).

When potential risks of a breach of privacy and unauthorized use of samples and personal data is kept low by applying strict coding procedures, the use of previously collected samples should be permitted without the need for a renewed consent. An opt-out scheme with information in national media or advertising in local newspapers with an associated right to withdraw from the study may be used when feasible. We have recently provided a template for handling consent issues related to the use of different sample collections where the original information and consent arrangements vary (30). An expressed *no* to any future research in the original consent form should always be respected as a respect for autonomy and in line with the importance of preserving trust in biomedical research. "Specific considerations apply to the case of a donor who once agreed to participate in a research study, when the donor is no longer alive and therefore no longer available for either informed consent, opt out, dissent, or reports of results. This may frequently be the case, for example, in cancer research. Systematic exclusion of deceased participants would introduce a significant selection bias abolishing the chances for objective scientific studies. Inclusion of the donor's sample cannot impose harm on the donor, and therefore the sample may be included, with the single exception that the donor's survivors have specifically requested that the donor's samples not be used for research – in which case the sample should be excluded, while maintaining a record for future statistics that this has occurred" (30). To let relatives have a veto when the deceased earlier has affirmed his or her willingness to donate tissue for research would constitute a breach of respect for autonomy. However, when the attitude of the deceased is not known, using the tissue against the expressed wish of the relatives would jeopardize the trust in research.

8. Benefits and Harms

Due to long lead times in biomedical research aiming at providing better treatment and new medical products there are seldom, if at all, any direct benefits for the actual donors in biobank research. However, all patients depend for their medical treatment on previous research results and, accordingly, on the fact that earlier generations of patients and healthy volunteers have participated as research subjects and donated tissue samples both to the pathology biobanks and to the biomedical research projects (20).

8.1. Breach of Privacy

The major risk of harm in biobank research is associated with the processing of sensitive personal data. Such processing may be seen as a breach of privacy and if unauthorized parties access information this may put the donor at risk. Insurance companies, employers, and other third parties may have a great interest in information acquired through human tissue sampling. Maintaining strict coding and secrecy procedures controls potential risks of damage of this kind. These coding procedures must, as was the case regarding information and consent procedures, be sensitive to the interests and risks that are at stake. In its latest Report on *Personal Information in Biomedical Research* (2007) (http://www.bioethics-singapore.org/resources/reports.html – in Subheading 4), the Singapore Bioethics Advisory Committee argued that protection measures should be proportional to the sensitivity of the information, so that not every kind of information need be protected with the same vigor, for example, a database of children with myopia (very common among children in Singapore) would obviously need much less protection than a database on HIV/AIDS patients. As argued by Terry Kaan Sheung-Hung in his comments to my first draft of this chapter, this avoidance of a mechanical broad-brush approach requires data custodians to apply more intelligent rules and measure for the protection of information.

8.2. Misuse by Third Parties

According to Swedish legislation, there has in addition been a shift of attention from putting cumbersome restrictions on research to prevent unauthorized use to making such use in itself unlawful. The new law on genetic integrity (31) which came into effect 1 July 2006 laid down that nobody may stipulate as a condition for entering into an agreement, that another party should undergo a genetic examination or submit genetic information about themselves. There should also be a general prohibition to the effect that without support in law, genetic information may not be sought after or used by anyone other than the person that the information is about. This applies even if the person concerned has given his or her consent to such an investigation or use, but not if they themselves have requested it. The proposed prohibition is not to be applicable to genetic information that is sought for medical purposes, for scientific or genealogical research or to obtain evidence in legal proceedings. For criminal investigations and for insurance purposes, there is regulation in place or suggested. Illegitimate requests of or uses of information may still be a problem, but this risk is minimized since such actions will according to the new law constitute criminal offences. A scale of penalties that includes fines or a term of imprisonment not exceeding 6 months will enforce the proposed prohibitions (Law 2006:351).

8.3. Harm to Groups

There may also be a risk of harm to a group of individuals associated with a specific biobank-related research protocol, for example, when a linkage is suggested between an ethnic group and the prevalence of a specific disease, for example, a sexually transmitted disease or a psychiatric condition. The individuals pointed out may experience a harm done to them just by the information being revealed of them as members of this group. This problem is, however, complex (see (10) for discussion). When genetic factors are revealed for multifactorial conditions such as alcoholism, sexual identity, and cognitive capacity and psychiatric disorders such as schizophrenia, dyslexia, ADHD, and autism, individuals belonging to these groups may feel stigmatized. However, such consequences of increased knowledge must be dealt with on a societal level and political decisions have to be made to protect exposed groups, for example, to provide equal opportunities for a good life, not by limiting the search for knowledge. "Through biobank research a linkage may (also) be established between sensitive medical information and groups of individuals that without much difficulty can be identified after the results of the research have been published, for example, a geographically distinct group of individuals, persons with a certain job position, education, income, etc. However, this is not an entirely new phenomenon. In order to minimize the risk of damage done, the researcher and the research ethics committee may decide that the information should be disguised or coded in a way that makes it impossible or very difficult to identify the group being studied" (Ibid. p. 417).

8.4. Dignitary Harms

Regarding research that uses previously collected human tissue samples an ERB has to select an appropriate information and consent procedure. Under certain conditions, for example, strict coding measures are applied and it may not be practically feasible to ask for a renewed consent, the board may decide that the research may be carried out without an informed consent or decide that an opt-out scheme shall be used. If individuals who should not want research to be carried out on their samples, or are negative to a specific kind of research, learn to know that research is carried out without their consent they may feel disrespect. I call this kind of harm "dignitary" harm. They may feel that their dignity as political citizens with moral authority has been violated. However, this kind of harm would arise in many other situations as well when a decision is taken on behalf of a public interest but at the price of not honoring the interests of each individual. An analogous example to biobank research is the establishment of national medical registries, such as a cancer registry or a death cause registry. These decisions are taken by the parliament or by a public authority to protect vital public health interests. Because of their public interest importance they do not need an approval by each individual and they do not admit any right on the part of individuals to have their information removed.

At the end, dignitary harms, as well as other kinds of harm, must be balanced against the scientific value of each research project and the potential benefits of doing research. It is quite conceivable that some individuals have strong personal reasons for not wishing to participate in a certain type of medical research. "These interests should be respected as far as possible, but legislators and the authorities concerned must also apply a balancing principle which weighs one interest against others and where ultimately it is those that are worst off in society who should be favored in the outcome. In this case, the interests current and future patients have in access to new medical treatment must also be taken into account. This interest can be one of which a person who is ill or someone with a relative, who died from cancer, can be acutely aware" (20). If, therefore, it is the case that allowing people to exercise their right to consent when only dignitary harms are at stake, or to withdraw their consent, has particularly negative effect on those who are already worst off in society, there is reason to abstain from this possibility. "The interest of the sick in being cured should be given higher priority than a healthy donor's opportunity to have his attitude to a certain type of medical research respected. Protection of the sample donor's privacy is still respected in the sense that the information is, and remains, strictly confidential" (Ibid.).

9. Using Personal Data

In data protection legislations and in regulations of biobank research, the patient/donor has the sovereign right to decide whether and how personal data and tissue material may be used, for example, a yes to use of personal data must be respected by an ERB and by the data protection authorities. These authorities may in some instances grant permission to do research using sensitive data without consent from the donor. However, the individual has normally a right to grant such use. This implies that it is essential that the information to patients and research subjects include all possible uses of personal data associated with a research project or the collection of human tissue samples, as well as the measures taken to protect the privacy of the individual donors. It should for example include information about genetic analyses and international collaboration that implies the transfer of biological material and data across borders. If the research may involve commercial partners and interests, for example, future patenting, this should also be included in the information.

Since it is the combination of human tissue material and clinical and personal data that carries the promise of providing understanding of underlying mechanisms of diseases and their treatment, data should as a general rule not be anonymized. Anonymization "precludes accumulative assessments for which multiple inclusions of the same participant must be avoided, and prevents retroactive validation and demonstration of reproducibility. That would preclude the possibility to make important links in the future. As a general strategy, anonymization can therefore not be recommended," coding is preferred (32). To evade confusion about the different coding alternatives and what "anonymization" means I suggest that the recommendations by EMEA are used (33). They recommend that regarding *anonymous* samples there are no links to the individual donor (although there may be general descriptions like "man, age 50–55, Cholesterol level >240 mg/dl"). *Identified* samples are linked to the individual in a way that makes them immediately identifiable. A *simple code* is a direct link to the individual, usually through a random set of numbers or letters, or a bar code. A *double code* implies that to link the sample and the data to the individual a second code is needed. *Anonymized* are samples that earlier have been identified or coded but the identification, or the code and the code key have been destroyed so there is no longer any link to the individual. The International Conference on Harmonization of Technical Requirements (ICH) has in November 2007 adopted this nomenclature for the Registration of Pharmaceuticals for Human Use. In the European Union, the Committee for Human Medical Products has endorsed the guidelines, which came into operation in May 2008. This nomenclature is then an important part of an already existing international Charter regarding coding in biobank research.

10. Feedback Concerning Results of Research Studies

As a general rule, information about the progress of research from a biobank is made available through publication in scientific publications. General information may also be made through national media. Specific information to individual donors is generally not advisable since it implies assuming a responsibility for the clinical significance for an individual based on information about the odds ratio expressing risk only for a study population. Research groups may not be equipped for assuming such a responsibility. Communicating genetic information implies skills in genetic counseling and the information may be of direct concern to genetic relatives who also must be informed. "Misinterpretation can cause potential psychological, social, and economic harm – especially before validation of the clinical significance of the findings. This is particularly true if no relevant treatment or prevention modality to combat the investigated risk is yet available" (30). If clinically significant findings are expected to emanate from the research this implies that a close collaboration has to bet set up from the start together with clinical departments and wards that can provide counseling and advice about treatment. As pointed out to me by Campbell in his comments to my first draft of this chapter, there was a debate in the UK biobank about avoiding the idea that participation would render a "health check," as this would be a false promise. It should be clearly understood and stated that the only benefit for large population-based biobanks is the health of future generations, including information about the long lead times before scientifically significant results become clinically significant.

There may also be incidental findings associated with a biobank project or a research protocol, for example, a mutation in a breast cancer gene where treatment is available. These incidental findings should be handled in a manner that also implies collaboration with clinical departments that can give information and provide treatment to affected individuals. A detailed guide for researchers has recently been provided (34). A model has also been suggested for the communication of genetic information that has not been asked for by the individual (17). It takes account of the character of the information and the possibility to provide treatment and could be used when organizing the feedback of incidental findings in association with biobank research and entails that an individual is informed first when certain conditions are satisfied. Such conditions might include one or more of the following: (1) that the information is reliable according to medical science or tested experience (2) that the information is linked to a reasonably certain risk of illness, (3) that the illness is of a reasonably serious kind or is at least nontrivial (4) that the genetic component has high penetrance, (5) that there is an effective prevention or treatment, (6) that personal support and regular checkups are offered.

11. Is There a Need of Separate Ethics Boards?

As described by Don Chalmers in his chapter of this book (Chapter 1), several of the large biobank initiatives have separate ethics review boards as part of their governance structure. This organization is believed to promote public trust and also be necessary for controlling that data are securely handled. I tend to disagree with this development. Since the first Helsinki Declaration, which among other things requested that an independent body of scientists and laypeople should review all human subjects research, a strong tradition has been established with groups of scientists and lay people well experienced in handling different kinds of research protocols and making the ethical balancing. The procedures for electing them and securing relevant scientific expertise are well established and the boards have a clear mandate. In Sweden law regulates them and the government elects the members. There is also an "ethics board of appeal" which can discuss and suggest how new issues should be handled. Under the condition that both the initiation of a new biobank and each new research project emanating from this biobank are examined by the ordinary ethics review boards there is no need of extra independent bodies. Their mandate is unclear with members often elected by parties directly involved in the biobank effort. For the scientists they create a new bureaucratic level and they cost money that could be used for research. In our research group, we argued recently for broad consent (not "blanket" as Caulfield asserts (7)) but emphasized that this did not imply broad approvals to many research projects (22). There is a need for the ethics review board to check the nature of the new research project, that the legal status of the biobank is the same and that the data protection measures initially agreed upon are still applicable.

As pointed out to me by Terry Kaan Sheung-Hung in his comments to my first draft of this chapter it is essential that the review boards guard themselves from the instinctive response to apply ethical principles evolved from the setting of therapeutic care in the relationship of doctor–patient to the quite different relationship between researcher and research subject. Also doctors participating as researchers in randomized clinical trials sometimes have problems to uphold the distinction between therapeutic ethics and research ethics. As Peter Armitage has pointed out, investigators in the same trial may sometimes move away from the region of uncertainty implied in a randomized design at different rates depending on their prior judgements, the weights attached to different criteria and psychological characteristics (35). The tensions between the two relationships are obvious in the Helsinki Declaration but they cannot be solved by simply putting the doctor–patient relationship absolutely above that of the researcher–subject relationship.

12. Conclusion – A Minimal Ethical Framework

When taking into consideration the actual interests at stake and the possibility of balancing these interests in an ethically appropriate way it seems clear that the attempt by different international bodies to create global frameworks with long lists of principles and best practices for biobank research represent an overkill of some magnitude. Traditional information and consent procedures suffice and data protection implies a sovereign right of the individual citizen to grant the use of biobank material and personal data that is needed for biobank research. Clearly, there may still be inconsistencies in terminology when researchers of different nationalities meet in common enterprises, but both they and the ERBs are well equipped to sort out what is actually meant and propose different instruments for, for example, coding. The existing ERBs should play the key role, guided by the sound argumentation by the researchers in their application to the board.

There are of course important and difficult questions remaining to be solved, for example, on sharing of results and how to design intellectual property rights, how to handle data protection in a way that acknowledges the sensitivity of the information acquired (not giving in to the legal definition that all health information is sensitive in the same sense), the way research on minors and incompetent persons may be conducted, and how to handle informed consent in longitudinal studies including minors (9, 36). However, these matters are complex and need to be the focus of sound research, not be a matter for considered opinions by different groups. In conclusion, I suggest that researchers and ERBs should have the following points to consider in mind when designing a project, informing the sample donors, applying for approval by an ERB, conducting the research, and reporting research results. As indicated with several references in the text, the framework is based on previous research published in international peer-reviewed scientific journals.

13. Points to Consider

1. The initial collection of human tissue samples and personal data should be based on an informed consent by the sample donor.
2. The ERB has to balance the interests at stake and select an appropriate information and consent procedure for each research project that is using a biobank.
3. ERBs may under certain conditions grant research without consent on previously collected samples and may permit researchers to ask for broad consent to future research.
4. Personal data and genetic information should be protected by coding and accessible only by authorized persons.
5. An individual donor may grant permission to the researchers to handle personal information, for example, to perform genetic analyses, engage in an international collaboration that implies the transfer of biological material and data across borders and collaborate with commercial partners. This kind of information should therefore be included in the information to the sample donor.

Acknowledgement

I am grateful to Alastair Campbell, Don Chalmers, and Terry Kaan Sheung-Hung for valuable comments to an earlier version of this chapter.

References

1. Knoppers, B.M., Abdul-Rahman, M.H., Bédard, K. (2007) Genomic databases and international collaboration. *Kings Law Journal.* **18**, 291–311.
2. Gibbons, S.M.C., Kaye, J. (2007) Governing genetic databases: collection, storage and use. *Kings Law Journal.* **18**, 201–208.
3. Chadwick, R., Cutter, M. (2007) The impact of biobanks on ethical frameworks. In *The ethics and governance of human genetic databases* (Häyry, M., Chadwick, R., Árnason, V., Árnason, G., eds.), Cambridge University Press, New York, pp. 219–226.
4. Murray, T.H. (1997) Genetic exceptionalism and "Future Diaries": is genetic information different from other medical information? In *Genetic secrets: protecting privacy and confidentiality in the genetic era* (Rothstein, M.A., ed.), Yale University Press, New Haven, pp. 60–73.

5. Wendler, D. (2006) One-time general consent for research on biological samples. *BMJ.* **332**, 544–547.
6. Kettis-Lindblad, Å., Ring, L., Viberth, E., Hansson, M.G. (2007) Perceptions of potential donors in the Swedish public towards information and consent procedures in relation to use of human tissue samples in biobanks: population based study. *Scandinavian Journal of Public Health.* **35**(2), 148–156.
7. Caulfield, T. (2007) Biobanks and blanket consent: the proper place of the public good and public perception rationales. *Kings Law Journal.* **18**, 209–226.
8. Williams, C. (2005) Australian attitudes to DNA sample banks and genetic screening. *Current Medical Research and Opinions.* **21**, 1773–1775.
9. Helgesson, G., Ludvigsson, J., Gustafsson Stolt, U. (2005) How to handle informed consent in longitudinal studies when participants have a limited understanding of the study. *Journal of Medical Ethics.* **31**, 670–673.
10. Rose, H. (2003) An ethical dilemma. The rise and fall of human genomics – the model biotech company? *Nature.* **425**, 123–124.
11. Hansson, M.G. (2005) Building on relationships of trust in biobank research. *Journal of Medical Ethics.* **31**, 415–418.
12. Hansson, M.G. (2006) Combining efficiency and concerns about integrity when using human biobanks. *Studies in History and Philosophy of the Biological and Biomedical Sciences.* **37**, 520–532.
13. Hoeyer, K., Olofsson, B-O., Mjörndal, T., Lynöe, N. (2004) Informed consent and biobanks: a population-based study of attitudes towards tissue donation for genetic research. *Scandinavian Journal of Public Health.* **32**, 224–229.
14. McQuillan, G., Porter, K.S., Agelli, M., Kington, R. (2003) Consent for genetic research in a general population: the NHANES experience. *Genetics in Medicine.* **5**, 35–42.
15. O'Neill, O. (2003) Some limits of informed consent. *Journal of Medical Ethics.* **29**, 4–7.
16. Knoppers, B.M., Chadwick, R. (2005) Human genetic research: emerging trends in ethics. *Nature Reviews Genetics.* **6**, 75–79.
17. Hansson, M.G. (2008) *The private sphere. An emotional territory and its agent.* In Philosophical Studies in Contemporary Culture, Monograph, Springer, p. 182.
18. Hansson, M.G. (1991) *Human dignity and animal well-being. A Kantian contribution to biomedical ethics.* Acta Universitatis Upsaliensis. Uppsala Studies in Social Ethics 12, Uppsala.
19. Hansson, M.G. (1998) Balancing the quality of consent. *Journal of Medical Ethics.* **24**(3), 182–187.
20. Hansson, M.G. (2007) For the safety and benefit of current and future patients. *Pathobiology.* **74**, 198–205.
21. Árnason, V. (2004) Coding and consent: moral challenges of the database project in Iceland. *Bioethics.* **18**, 27–49.
22. Hansson, M.G., Dillner, J., Bartram, C.R., Carlsson, J., Helgesson, G. (2006) Should donors be allowed to give broad consent to future biobank research? *The Lancet Oncology.* 7, 266–269.
23. Hansson, M.G. (2009) Ethics and biobanks. *British Journal of Cancer.* **100**, 8–12.
24. Campbell, A.V. (2007) The ethical challenges of genetic databases: safeguarding altruism and trust. *Kings Law Journal.* **18**, 227–246.
25. Sigstad, E., Lie, A.K., Luostarinen, T., Dillner, J., Jellum, E., Lehtinen, M., Thoresen, S., Abeler, V. (2002) A prospective study of the relationship between prediagnostic human papillomavirus seropositivity and HPV DANN in subsequent cervical carcinomas. *British Journal of Cancer.* **87**(2), 175–180.
26. Kaijser, M. (2003) Examples from Swedish biobank research. In: *Biobanks as resources of health* (Hansson, M.G., Levin, M., eds.), Uppsala University, Uppsala, pp. 33–50.
27. Lindberg, B.S. (2003) Clinical data – a necessary requirement for realizing the potential of biobanks. In: *Biobanks as resources of health* (Hansson, M.G., Levin, M., eds.), Uppsala University, Uppsala, pp. 21–32.
28. Sundstrom, P., Juto, P., Wadell, G., Hallmans, G., Svenningsson, A., Nystrom, L., Dillner, J., Forsgren, L. (2004) An altered immune response to Epstein-Barr virus in multiple sclerosis: a prospective study. *Neurology.* **62**(12), 2277–2282.
29. Council of Europe (1997) Convention for the protection of human rights and dignity of the human being with regard to the application of biology and medicine: Convention on Human Rights and Biomedicine. Oviedo, ETS No. 164.
30. Helgesson, G., Dillner, J., Carlson, J., Bartram, C.R., Hansson, M.G. (2007) Ethical framework for previously collected biobank samples. *Nature Biotechnology.* **25**, 973–976.
31. Lag om genetisk integritet m.m. (Act on Genetic Integrity), 2006, p. 351.
32. Eriksson, S., Helgesson, G. (2005) Potential harms, anonymization, and the right to withdraw consent to biobank research. *European Journal of Human Genetics.* **13**, 1071–1076.

33. EMEA (2002) Position paper on terminology in pharmacogenetics. Committee for Proprietary Medicinal Products (CPMP), EMEA/CPMP/3070/01.
34. Wolf, S.M., Lawrenz, F.P., Nelson, C.A., Kahn, J.P., Cho, M.K., Clayton, E.W., Fletcher, J.G., Georgieff, M.K., Hammerschmidt, D., Hudson, K., Illes, J., Kapur, V., Keane, M.A., Koenig, B.A., Leroy, B.S., McFarland, E.G., Paradise, J., Parker, L.S., Terry, S.F., Van Ness, B., Wilfond, B.S. (2008) Managing incidental findings in human subjects research: analysis and recommendations. *The Journal of Law Medicine & Ethics* **36**(2), 219–248.
35. Armitage, P. (1998) Attitudes in clinical trials. *Statistics in Medicine.* **17**, 2675–2683.
36. Helgesson, G. (2005) Children, longitudinal studies, and informed consent. *Medicine, Health Care and Philosophy.* **8**, 307–313.

Chapter 3

Nordic Biological Specimen Bank Cohorts as Basis for Studies of Cancer Causes and Control: Quality Control Tools for Study Cohorts with More than Two Million Sample Donors and 130,000 Prospective Cancers

Eero Pukkala

Abstract

The Nordic countries have a long tradition of large-scale biobanking and comprehensive, population-based health data registries linkable on unique personal identifiers, enabling follow-up studies spanning many decades. Joint Nordic biobank-based studies provide unique opportunities for longitudinal molecular epidemiological research. The Nordic Biological Specimen Banks working group on Cancer Causes and Control (NBSBCCC) has worked out very precise quality assurance principles for handling of the *samples*, based on the tradition in biobank culture. The aim of this paper is to demonstrate how high standards of quality assurance can also be developed for the *data* related to the subjects and samples in the biobanks. Some of the practices adopted from the strong Nordic cohort study experience evidently improve quality of nested case-control studies nested in biobank cohorts. The data quality requirements for the standardised incidence ratio calculation offer a good way to check and improve accuracy of person identifiers and completeness of follow-up for vital status, which are crucial in case-control studies for picking up right controls for the cases. The nested case-control design applying incidence-density sampling is recommended as an optimal design for most biobank-based studies. It is demonstrated how some types of biobanks have a period immediately after sampling, when the cancer risk is not comparable with the cancer risk in the base population, and how many of the biobanks never represent the normal average population of the region. The estimates on the population-representativeness of the biobanks assist in interpretation of generalisability of results of the studies based on these samples, and the systematic tabulations of numbers of cancer cases will serve in study power estimations. The well over 130,000 prospective cancer cases registered among subjects in the NBSBCCC biobank cohorts have already offered unique possibilities for tens of strong studies, but for rare exposure-outcome combinations predictions on future numbers of cases improve the chance to select the right moment when the study will have accurate statistical power.

Key words: Biobanks, Cancer incidence, Cohort study, Record linkage, Control selection, Selection bias, Inverse causality

Joakim Dillner (ed.), *Methods in Biobanking*, Methods in Molecular Biology, vol. 675,
DOI 10.1007/978-1-59745-423-0_3,

1. Introduction

In the Nordic countries, there exists a series of established biological specimen banks with many decades of follow-up that enable performing prospective epidemiological studies with adequate statistical power even for diseases and exposures that are not common. Between 1995 and 2006, more than 30 joint articles ((1–33); Table 1) were published by the Nordic

Table 1
Use of serum samples in studies published by the Nordic Biological Specimen Banks for Cancer Causes and Control (NBSBCCC) network. Update 21 Nov 2006

NBSBCCC study number	Study (references)	Sera used in the study																	
		Finland				Iceland		Norway (Janus)		Sweden									
											Sweden Health and Disease Study					Malmö Microbiology Biobank			
		Maternity cohort (females)	Helsinki Heart Study (males)	ATBC (males)	Mobile clinic	Maternity cohort (females)	Heart preventive clinic	Health examinations	Blood donors	Northern Sweden Maternity Cohort (females)	VIP	MONICA	Mammography (females)	Diet and cancer	Preventive medicine	Maternity cohort (females)	Blood-borne viral screening	Other	Infectious disease control
1	Dillner et al. [1]				X														
2	Lehtinen et al. [2]		..	..	X														
3	Lehtinen et al. [3]				X														
4	Dillner et al. [4]		..	..															X
5	Bjørge et al. [5]	X						X	X										
6	Dillner et al. [6]	X	..	..				X			X	X							
7	Bjørge et al. [7]							X	X										
8	Dillner et al. [8]	..			X	..				..			..			..			
9	af Geijersstam et al. [9]	X	..	..															
10	Luostarinen et al. [10]	X	..	..				X			X	X							
11	Lehtinen et al. [11]	X	..	..							X	X							
12	Kibur et al. [12]	X	..	..															

(continued)

Table 1 (continued)

NBSBCCC study number	Study (references)	Sera used in the study																	
		Finland				Iceland		Norway (Janus)		Sweden									
											Sweden Health and Disease Study					Malmö Microbiology Biobank			
		Maternity cohort (females)	Helsinki Heart Study (males)	ATBC (males)	Mobile clinic	Maternity cohort (females)	Heart preventive clinic	Health examinations	Blood donors	Northern Sweden Maternity Cohort (females)	VIP	MONICA	Mammography (females)	Diet and cancer	Preventive medicine	Maternity cohort (females)	Blood-borne viral screening	Other	Infectious disease control
13	Koskela et al. [13]	X	..	..				X			X	X							
14	Sigstad et al. [14]	X	..	..				X			X								
15	Mork et al. [15]	X	X					X			X								
16	Anttila et al. [16]	X	..	..				X			X	X							
17	Stattin et al. [17]	..	X			..		X	X	..	X	X	..			..			
18	Bjørge et al. [18]	X						X	X										
19	Stattin et al. [19]	..	X			..		X	X	..	X	X	..			..			
20	Lehtinen et al. [20]	X	..	..				X			X	X							
21	Lehtinen et al. [21]	X	..	..				X			X	X							
22	Youngman et al. [22]	X	..	..				X			X	X							
23	Paavonen et al. [23]	X	..	..				X			X	X							
24	Lehtinen et al. [24]	X	..	..	..	X	..	..	..		..	..	..	..	..		..	..	..
25	Tuohimaa et al. [25]	..	X			..		X	X	..	X	X	..			..			
26	Luostarinen et al. [26]	X	..	..				X			X	X							
27	Lehtinen et al. [27]	X	..	..	..	X	..	..	..		..	..	..	..	..		..	..	..
28	Anttila et al. [28]	..	X			..		X	X	..	X	X	..			..			
29	Stolt et al. [29]	X	..	..	..		..	..	..		..	..	..	..	..		..	..	..
30	Hakama et al. [30]	X	..	..				X			X	X							
31	Tedeschi et al. [31]										X								
32	Tedeschi et al. [32]	X	X	X				X	X		X								
33	Korodi et al. [33]	..	X			..		X	X	..	X	X	..			..			

Participating serum banks marked with **X** and those which do not include sera from relevant persons with two dots (**..**). *Columns shaded with gray* indicate that sera from these serum banks have not been used in any NBSBCCC study so far. The table includes only studies that have got the internal NBSBCCC study number, i.e., officially accepted as network studies

Biological Specimen Banks working group on Cancer Causes and Control (NBSBCCC). The majority of studies so far were aimed at elucidating infections such as Human Papillomavirus (HPV) as causes of cancer. In addition to the joint Nordic studies, the biobanks operate independently with several hundred publications based on one or several of the biobanks described in this paper. Major subject areas for study have been hormones, nutrition, smoking, organochlorine compounds and genetic polymorphisms as causes of cancer in addition to a number of studies evaluating tumour markers. Still, the first systematic evaluation of characteristics and quality of the biobank cohorts or features of cancer risk pattern among the donors was done just recently and published in 2007 (34). This book chapter borrows much of the text of that publication, modified to give practical insight of thinking and methodology normally used in quality assurance (QA) of other types of epidemiological study cohorts than biobank cohorts.

This paper includes systematic descriptions of the participating biobanks: background, organisation, size, years of sample collection and administrative aspects. Numbers of cancer cases found among persons in the serum banks after serum drawing are given, advertising the unique possibilities of the national cancer registration systems in the Nordic countries. Population representativeness of the serum bank cohorts is estimated by comparing cancer incidence in the biobank cohorts with the respective national rates. Finally, issues to be taken into account in designing case-control studies nested in the Nordic biobanks are discussed.

In their classical assessment of the quantitative importance of avoidable causes of cancer, Doll and Peto estimated that a majority of human cancer was attributable to avoidable causes (35). They concluded that most of these avoidable causes remained unidentified. For risk factor identification and causality inference as well as for studies searching for mechanisms behind increases or decreases in cancer incidence, they recommended the use of prospective studies nested in cohorts of stored biological specimens.

This paper introduces the Nordic biobank network NBSBCCC to new potentially interested partners and serves as a general reference for specific studies based on these biobanks. NBSBCCC is a network of excellence that contains 17 independent biobank cohorts, five cancer registries and numerous expert user groups. The purpose of the network is to provide a concerted resource for etiologic studies of cancer, with a focus on longitudinal studies addressing unexplained causes and trends over time.

People who have donated samples to a biobank can also be considered as classical study cohort that is in most aspects technically comparable with, e.g., cohorts of occupationally exposed persons. Therefore, methods used in quality assurance and evaluation of accuracy of other types of cohorts can be adapted to biobank cohorts as well. Because that kind of approach has not been tradition in biobank culture – most principles of quality assurance have their roots in laboratory sciences – this paper describes some of the practices typical to cohort studies that evidently also improve quality of nested case-control studies in biobank cohorts.

2. Participating Biobanks

The first crucial characteristic of any cohort study is to understand the *history* of the cohort collection:

- definition of the cohort; which type of persons were included (in the following I use the term "being exposed" to mean people who fulfil the inclusion criteria, although in biobank context the "exposure" simply means donation of sample),
- region of coverage; this can be a geographical area, or one institution (such as a factory in occupational studies, or a hospital in clinical cohorts),
- years of coverage; the cohort can be cross-sectional (including everyone under exposure at a point of time), or dynamic one (including everyone being exposed during a given period, no matter if the exposure started before that period or ended after that period),
- exclusions; there may be exclusion by purpose (defined in study protocol) or by accident (e.g., exclusion of deceased persons from the cohort because of lack of storage base; this would totally ruin the possibilities to use the cohort in any study on disease risk),
- other selection mechanisms; in biobank context one of the most important ones is selective participation that may decrease the population-representativeness of the cohort meant to be random sample of a population,
- variables collected, with full descriptions of principles in coding and input into the database; in the context of biobank samples such data may have existed but they have not always (e.g. if the samples are taken for clinical purposes) been collected systematically into databases in such format that they can be used afterwards in a scientific studies together with the sample,
- accuracy check-ups of the data in the context of storing them at the baseline; most importantly were the identification information of the person (in the Nordic countries: person ID codes) confirmed,
- collection of follow-up information; how were the data on possible deaths, migration out of the follow-up region and outcome events achieved; was the linkage procedure (key) fully complete, where there temporal or spatial holes in follow-up.

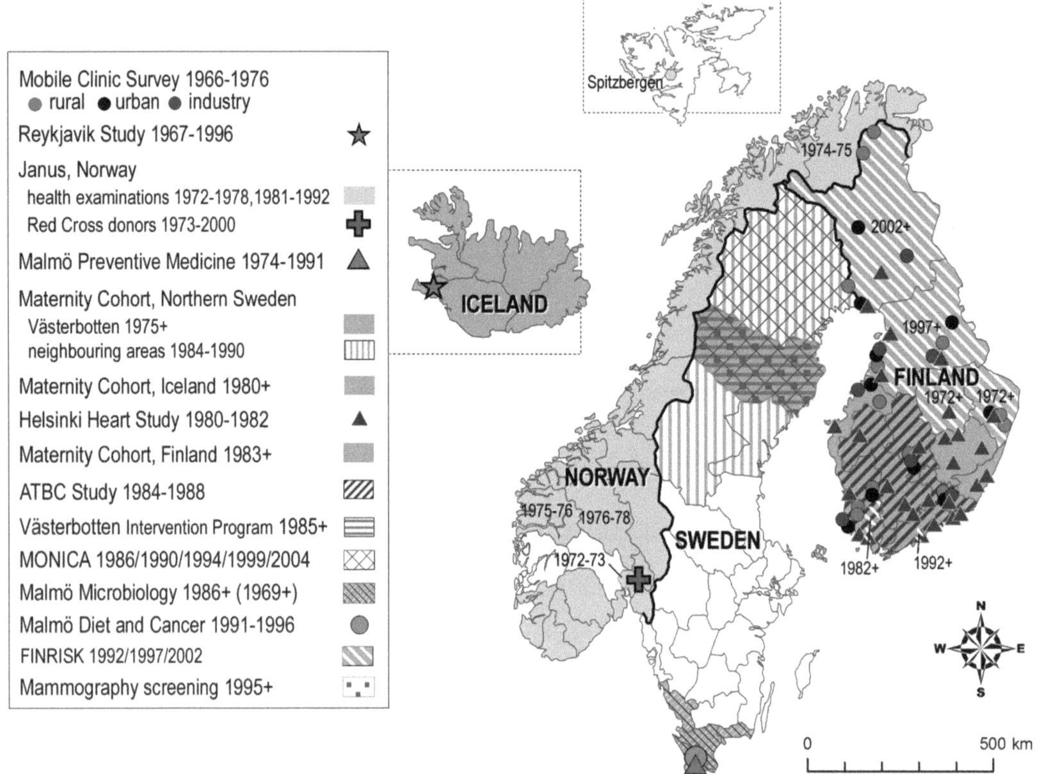

Fig. 1. Map of Nordic countries indicating the coverage areas of the serum banks.

In the following, some key characteristics of the background of the Nordic biobanks belonging to the NBSBCCC programme are described in a systematical way. The network so far consists of 12 biobanks in Finland, Iceland, Norway and Sweden, three of which are split into two to three independent subcohorts (Fig. 1, Table 2). Participating biobanks are independent entities that make their own decisions, but are committed to facilitate joint studies by working towards similar policies for quality assurance, logistics and study designs as well as for permission and terms of collaboration. NBSBCCC is funded by the Nordic Council of Ministries and as a European Union sixth framework programme Network of Excellence.

Research projects using the biobanks need appropriate permissions from the national Data Protection Authorities, National or Local Ethical Committees and from the boards of the biobanks. Informed consent is obtained from all persons donating samples, making it clear to the donors that the material will be used for future research purposes. Details of the permission procedure can be obtained via contact email addresses given in Table 2.

All samples have been stored at –20°C to –25°C except those of the Alpha-Tocopherol, Beta-Carotene Cancer (ATBC)

Table 2
Characteristics of the serum banks included in the Nordic Biological Specimen Banks for Cancer Causes and Control (NBSBCCC) network. Status as of June 2005

Name, country (contact address)	Type	Years of first serum donation and subsequent samples of same individuals	Number of persons (+annual increase)	Number of sampling occasions (+annual increase)	Closing year in this study (complete cancer incidence & vital status)	Number of person-years
Finnish Maternity Cohort (helja-marja.surcel@ktl.fi)	S	1983+	722,000 women (August 2005) (+30,000/year)	1.47 million (+60,000/year)	2005	9.71 million
Helsinki Heart Study, Finland (leena.tenkanen@ktl.fi)	S	1980–1982	18,900 men	117,000	2005	419,000
Alpha-Tocopherol-Beta-Carotene (ATBC) Study, Finland (jarmo.virtamo@ktl.fi)	S	Baseline sera 1984–1988	29,200 male smokers	55,000 {follow-up sera from all 1986–1993, annual sera from 800 men}	2005	423,000
Finnish Mobile Clinic Health Examination Survey (paul.knekt@ktl.fi)	R	1966–1976	50,400	60,000	2005	1.42 million
FINRISK, Finland (pekka.jousilahti@ktl.fi)	R	1992/1997/2002 (samples from years 1972/1977/1982/1987 incomplete)	22,900	22,900	2005	185,000
Icelandic Maternity Cohort (arthur@landspitali.is; Arthur Löve)	S	1980+	53,000 women (+1,700/year)	96,000 (+2,500/year)	2005	768,000

(continued)

Table 2 (continued)

Name, country (contact address)	Type	Years of first serum donation and subsequent samples of same individuals	Number of persons (+annual increase)	Number of sampling occasions (+annual increase)	Closing year in this study (complete cancer incidence & vital status)	Number of person-years
Reykjavik Study, Icelandic Heart Association (v.gudnason@hjartavernd.is; (Vilmundur Guđnason))	R	1967–1996	19,300	60,000	2005	457,000
Janus, Norway (randi.elin.gislefoss@kreftregisteret.no)		1972–2005	332,000	493,000	2006	
Health examinations	R	1972–1978, 1980–1992 (+ few from Finmark and Troms counties in 2002)	304,000	377,000	2001	6.96 million
Blood donors	S	1973–1991, 1998–2000	31,900	117,000 (last samples 2005)	2001	1.27 million
Sweden						
Northern Sweden Health and Disease Study (goran.hallmans@nutrires.umu.se)	R	1985+	86,000	114,000	2003	
VIP	R	1985+	70,000 (+2000/year)	83,000	2003	560,000
MONICA	R	(1986)/1990/1994/1999/2004	9,000	14,000	2003	51,000
Mammography	R	1995+	27,500 women (+1,500/year)	48,000	2003	158,000

Northern Sweden Maternity Cohort (goran.wadell@climi.umu.se)	S	1975+	86,000 women (+2,000/year)	118,000 samples	2003	1.24 million
Preventive Medicine in Malmö, Sweden (goran.berglund@medforsk.mas.lu.se)	R	1974–1991	33,000	8,000	1999	560,000
Malmö Diet and Cancer, Sweden (goran.berglund@medforsk.mas.lu.se)	R	1991–1996	29,100	–	1999	159,000
Malmö Microbiology, Sweden (joakim.dillner@med.lu.se)		1986+ (incomplete 1969+)	454,000 (+40,000/year)	1.24 million (+120,000/year)	1999	1.84 million
Malmö Maternity Cohort	S	1985, 1989+ (incomplete 1969+)	70,000 women	115,000	1999	
Bloodborne virus screening	C	1986+			1999	
Other virus testing	C	1990+			1999	
Swedish Institute for Infectious Disease Control (joakim.dillner@med.lu.se)		1957+ (complete 1977+)	358,000 in computerised files	>900,000 (629,000 computerised)	2003 (test)	–
Population sample	R	1968, 1977–1978, 1990–1991, 1997	12,000	12,000	2003 (test)	–
Diagnostic microbiological testing	C	1990+	346,000	617,000 computerised	2003 (test)	–

Type: *R* random sample of population or other systematic invitation based on population register, *S* specific group with clearly defined enrolment criteria, *C* Clinical samples

Prevention Study, the Northern Sweden Health and Disease Study, the Malmö Diet and Cancer study and FINRISK Study (since 1997), which are stored at -70°C. Malmö Diet and Cancer biobank also has aliquots stored at -135°C.

Every resident of the Nordic countries has a unique personal identification code (PID) that is used in all main registers in these countries. The PID allows automatic and precise linkage of registers, without the need to use names. For meaningful research use, the PIDs have to be available for each person in the biobanks. Biobank cohorts are typically linked with the population-based cancer registries shortly before a new case-control set will be extracted for a specific study.

2.1. Finnish Maternity Cohort

Sera collected during the first trimester of pregnancy (two-thirds at 8–12 weeks) for screening of congenital infections and rubella immunity have been stored since late 1983 by the National Institute for Health and Welfare (THL). The biobank covers more than 98% of all pregnant women in Finland. So far, basic data for the sera up to 21 August 2005 have been transferred to Finnish Cancer Registry to be used in case-control studies.

Up to about 2005, The Finnish Cancer Registry took care of quality control of the data and also developed programs for precise random case-control selection within the FMC cohort. Record linkages for both incident cancers with cancer registry data and causes of death through Statistics Finland were administered from the Finnish Cancer Registry. This biobank has been used in more NBSBCCC studies than any other (Table 1). In the latest years, necessary quality assurance and record linkage routines have been developed at THL and no external consultancy is needed any more.

2.2. Helsinki Heart Study

The sera were collected during 1980–1982 for a trial to test the hypothesis that lowering serum LDL-cholesterol and triglyceride levels and elevating serum HDL-cholesterol levels with gemfibrozil (a fibric acid derivative) reduces the incidence of coronary heart disease (CHD) in middle-aged dyslipidaemic men (34). The volunteers for the trial were selected from men aged 40 through 55 years, employed by two government agencies and five industrial companies and living in different parts of Finland. Approximately 19,000 men participated in the first screening, and to be selected for the trial the participants (N=4,081) had to have their non-HDL cholesterol ≥5.2 mmol/l and no evidence of CHD or other major diseases.

Serum samples were collected from the participants at the first screening and from the participants in the trial at each follow-up visit during the trial. As the participants were followed up four times per year during the 5-year trial and twice a year during a subsequent extension of the trial, there are 28 serial samples from about 3,500 of those 4,081 who initially attended the trial. Also from the last trial follow-up visit, blood samples were stored. This biobank has participated in numerous NBSBCCC studies since 1999 (Table 1).

2.3. Alpha-Tocopherol-Beta-Carotene Cancer Prevention Study

The Alpha-Tocopherol-Beta-Carotene (ATBC) study was a randomised, double-blind, placebo-controlled, primary prevention trial conducted in Finland by the National Institute for Health and Welfare in collaboration with the U.S. National Cancer Institute. The main aim of the study was to evaluate whether daily supplementation with alpha-tocopherol or beta-carotene would reduce the incidence of lung cancer and other cancers (35).

In 1985–1988, a questionnaire on current smoking and willingness to participate in the trial was sent to the total male population of 50–69 years living in south-western Finland (n=290,000). Of them, 43,000 men smoked at least five cigarettes per day and were willing to participate. Men with prior cancer (except non-melanoma skin cancer and carcinoma-in-situ), severe angina pectoris, chronic renal insufficiency, alcoholism, or liver cirrhosis as well as those taking anticoagulants, beta-carotene, or vitamin A/E supplements in excess of defined doses were excluded. After exclusions and written informed consent, 29,133 eligible men were randomly assigned to receive either alpha-tocopherol 50 mg per day, or beta-carotene 20 mg per day, or both alpha-tocopherol and beta-carotene, or placebo.

At baseline, serum samples were collected. New serum samples were collected from all participants at the 3-year follow-up visit, and from about 800 randomly selected men a serum sample was collected annually throughout the trial. A whole blood sample was collected from the participants at the end of the trial between August 1992 and April 1993. This biobank was used for the first time in an NBSBCCC study just lately (33).

2.4. Finnish Mobile Clinic Health Examination Survey

The Mobile Clinic Health Examination Survey was carried out by the Social Insurance Institution during 1966–1972 in 34 rural, industrial or semiurban subpopulations (Fig. 1). Total populations aged 15 years or older or random samples of them were invited to participate in the study. On average 83% (57,400 men and women) participated in the health examination. Blood samples have been stored from 40,200 individuals in the baseline examination and from all 19,500 individuals in the re-examination survey of 12 subpopulations four to seven years later (1973–1976). This biobank participated particularly in early NBSBCCC studies (Table 1).

2.5. FINRISK

The National FINRISK Study has been conducted in Finland every 5 years since 1972. At the beginning, the Study was done only in eastern Finland as part of the North Karelia Project. The study area was expanded gradually. The serum samples are systematically available since 1992. In 1992, the Study was carried out in four areas: North Karelia and Kuopio Provinces in Eastern Finland,

Turku-Loimaa region in Southwest Finland, and cities of Helsinki and Vantaa in Southern Finland (Fig. 1). Oulu province in Northern Finland was included in 1997 and Lapland province in 2002. In each study year, a random sample of 2,000 individuals aged 25–64 years (stratified by sex and 10-year age group) has been taken in each study area according to the WHO MONICA protocol. Since 1997, a sub-sample of 1,500 men and women aged 65–74 years was included. Total cumulative sample size since 1992 is 33,000 and of them 22,900 (69%) have participated in the Study. DNA samples are available for most participants.

Study cohorts have been followed up through computerised register linkage of the National Causes of Death Register, the Hospital Discharge Register and the Finnish Cancer Register. The samples of the FINRISK Study have not been used in any NBSBCCC studies so far, but the general principle of the Study is that the collected samples can be utilised in large-scale collaborative studies that according to the FINRISK Steering Group are scientifically important.

2.6. Icelandic Maternity Cohort

Sera generally collected at 12–14 weeks of pregnancy for rubella screening from all of Iceland have been stored since 1980 in the centralised Department of Medical Virology, Landspitali University Hospital. About 6% of the cohort members cannot be used in studies because they have moved out of the country, but the date of emigration is not registered. This biobank has participated in two NBSBCCC studies (Table 1).

2.7. Icelandic Heart Association, the Reykjavik Study

The Reykjavik Study by the Heart Preventive Clinic and Research Institute of the Icelandic Heart Association is a prospective cardiovascular cohort study carried out in the Reykjavik capital area in 1967–1996. Selected birth cohorts of 14,923 men and 15,872 women in the Reykjavik area born in 1907–1935 were divided into six equally sized subgroups according to the date of birth and recruited systematically for collection of sera. The first subgroup was recruited in 1967–1969 and has attended altogether six times. The second one (first invited in 1970–1972) has attended twice. The later birth cohorts have been invited once (1974–1996) or never. Altogether 19,300 persons actually provided samples (annual participation rates between 71% and 76%), but about 200 of them cannot be used in analyses because of lacking dates of emigration. This biobank has not yet been used in any of the published NBSBCCC studies.

2.8. Janus Project (Norway)

A project to collect and store blood samples from healthy persons for later scientific use was initiated in the 1960s and named Janus after the Roman god with two faces, one looking backward, and the other one looking forward (symbolising the retrospective and prospective directions of epidemiological research). The first collection, related to a survey of risk factors for cardiovascular disease in ages of 35–49 years, covered four counties (Oslo 1972–1973, Finmark 1974–1975, Sogn og Fjordane 1975–1976 and Oppland 1976–1978; see Fig. 1). More subjects were added during 1985–1992 in the context of cardiovascular health examination of 40–42 years old Norwegians from all of the country except two counties (Hordaland and Buskerud; Fig. 1).

Red Cross blood donors in capital Oslo and surrounding areas were enrolled in 1973–1991 and 1999–2000. Every second year, these Janus donors donated 20 ml of extra blood to the biobank. Collection of later samples from these individuals ended in spring 2005.

The Janus bank consists of serum samples from 331,801 persons, 10% of them Red Cross donors. The average is two to three samples per donor, but some donors have given samples more than ten times. The Janus biobank is also collecting follow-up samples from cohort members who develop cancer. Before any treatment, a sample is collected when the donor is hospitalised at the Radium Hospital in Oslo (a nationally centralised cancer treatment hospital).

The Janus Project is funded by the Norwegian Cancer Registry which is also responsible for the data handling. This also allows frequent updates for incident cancer cases; several thousands of new prospective cancer cases have been registered after the closing date used in this study, and the addition in 2004 exceeded 3,000. Samples of the Janus health examination cohort have been used in 20 NBSBCCC publications, and the blood donors' sera in eight studies (Table 1).

2.9. The Northern Sweden Health and Disease Study Cohort

The Northern Sweden Health and Disease Study (NSHDS) Cohort contains three subcohorts: the Västerbotten Intervention Program (VIP), the MONICA (Monitoring Trends and Determinants in Cardiovascular Disease) and the Mammography Screening in Västerbotten. The cohorts represent a population-based sample of the county of Västerbotten in Northern Sweden (254,000 inhabitants). The Monica study also contains a population-based sample from the adjacent county of Norrbotten.

The VIP is a long-term project intended for health promotion. Since 1985, all individuals of 40, 50 and 60 years of age are invited for screening. They are also asked to donate a blood sample for later research purposes. In June 2004, the cohort included 74,000 individuals, of whom 70,000 had donated blood. A second sample is taken after 10 years; this has produced 13,000 re-sampling occasions.

Samples taken in the context of the population-based mammography screening have been stored since 1995. Screening is done every second year among all women in the age group 50–69 years in the county. There have been 48,000 sampling occasions from 27,500 women. About 50% of the women in the mammography cohort have also attended VIP.

The Northern Sweden MONICA project contains material from population-based screenings for risk factors of cardiovascular diseases that were carried out in 1986, 1990, 1994, 1999 and 2004. There are 14,000 sampling occasions of 9,000 individuals, 50% of whom are also included in VIP. Samples from 1986 have not been used in NBSBCCC studies and they are not included in this standardised incidence ratio (SIR) analysis, either.

The VIP cohort has been used most frequently out of the numerous Swedish biobanks in NBSBCCC studies, and also MONICA cohort in 15 studies (Table 1).

2.10. Northern Sweden Maternity Cohort

Northern Sweden Maternity cohort consists of sera collected since 1975 from pregnant women screened for rubella immunity during week 14 of pregnancy in the Västerbotten county and especially in the 1980s also for some of the adjacent counties in Northern Sweden. So far, almost 120,000 samples from 86,000 women have been stored at the virus laboratory of Umeå University. This biobank has not yet been used in any of the published NBSBCCC studies.

2.11. Preventive Medicine in Malmö, Sweden

The prospective, population-based Preventive Medicine study, with main focus on cardiovascular disease, diabetes and cancer, includes sera from a population-based sample of 33,400 persons 40–60 years of age, resident in the city of Malmö. The samples were donated at baseline examination in 1974–1991. The biobank is owned by Lund University.

2.12. Malmö Diet and Cancer Study, Sweden

The prospective population-based Malmö Diet and Cancer study started with a baseline examination in 1991–1996. Main focus is on cancer and cardiovascular diseases. All men born between 1923 and 1945 and all women born between 1923 and 1950 living at the time in the city of Malmö were invited to participate. The participation rate was 40% (28,100 participants). Mean age at enrolment was 58.2 years. The biobank is owned by Lund University.

2.13. Malmö Microbiology Biobank, Sweden

The Malmö Microbiology Biobank is owned by the County Council of Skåne and contains samples submitted for clinical microbiological analyses to the University Hospital in Malmö that today serves the entire county of Skåne in southernmost Sweden. Samples have been saved for clinical diagnostic and documentation purposes, the majority of them taken for diagnosis of blood-borne viral infections, such as hepatitis viruses. The oldest samples are from 1969 and were submitted from the city of Malmö. The annual number of samples increased in 1986 when HIV testing started and the catchment area extended to cover most of the Skåne county (Fig. 1). Since 1990, also the samples submitted for virus serology (typically because of clinical suspicion of virus infection or desire to investigate viral immunity) have been stored.

In recent years, a large number of samples have been submitted from the microbiology laboratories of adjacent counties in southern Sweden (Blekinge and Halland), raising the annual number of samples added to the biobank to about 60,000.

The Malmö Microbiology Biobank also includes samples of the population-based serological screening for virus infections and rubella immunity during pregnancy scheduled to be taken during week 14 of pregnancy (Malmö Maternity Cohort). The maternity cohort contains all samples from 1986 and from 1989 onwards, altogether more than 100,000 samples from 74,000 mothers.

Malmö Microbiology Biobank was computerised in 1997. NBSBCCC studies with MMB participation have as yet not been published.

2.14. Swedish Institute for Infectious Disease Control Biobank

The Swedish Institute for Infectious Disease Control (SIIDC) has performed a series of population-based, nationwide investigations on the immunity against infections in the Swedish population.

A small fraction of the biobank consists of randomly selected persons sampled in 1968 (3,000 subjects), 1977–1978 (1,845), 1990–1991 (4,800) and 1997 (2,400) and analysed to estimate age-specific population immunity rates of, e.g., polio, parotitis, measles, rubella, diphtheria and tetanus.

Most of the about 900,000 biological samples in the SIIDC biobank are diagnostic ones, submitted for microbiological analyses from all over Sweden. The oldest stored samples are from 1957, and complete series exist since 1977.

The information on the samples has been transferred from paper documentation to computerised files for about 629,000 samples. The biobank has recently been linked with the Swedish Cancer Registry, and the quality control of the result of the linkage is on-going. Samples of the Swedish Institute for Infectious Disease Control have been utilised in one NBSBCCC study (Table 1).

3. Some Quality Control Tools for Study Cohorts

The same methods that are used to check accuracy of any new study cohort, e.g., in occupational cancer epidemiology, can and should also be used for biobank cohorts. The following types of evaluations – presented below as a cookbook type list – were made for the NBCBCCC biobanks systematically in the context of a specific quality assurance study (34). For some of the biobank that kind was never done before and numerous gaps in the quality were revealed (and corrected).

3.1. Irregularities in the Entry?

The first tabulation to control completeness of any study cohort is to count numbers of cohort members by gender, year of entry

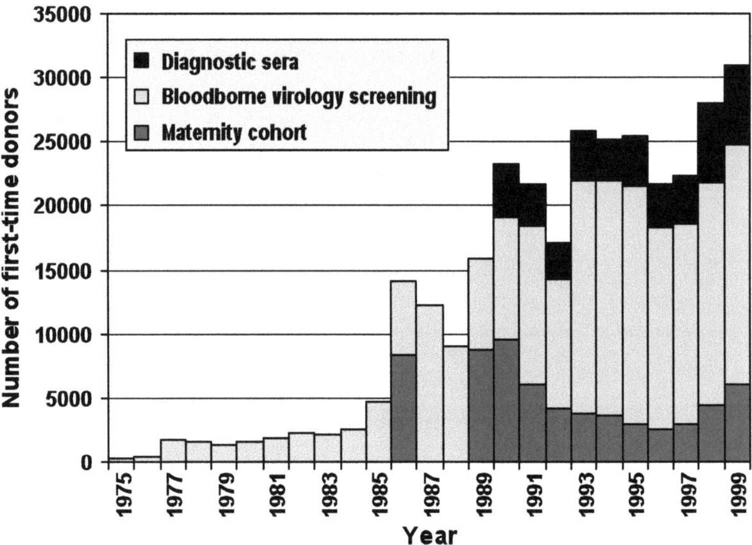

Fig. 2. Annual numbers of first-time donors in the Malmö Microbiology Biobank 1975–1999, by subcohort.

and similar simple classifiers. The numbers should correspond to the known facts about the size of the subjects. Any irregularity in the time series should be documented in the history of cohort formation.

For instance, the distribution of the number of pregnant women in Malmö Maternity Cohort (Fig. 2) reveals that the samples from 1987 and 1988 have been destroyed to save storage space. It has been quite common that such tabulations reveal differences between what has been believed to be the historical coverage of a biobank and what is the actual one. It is better to know this type of discrepancies *before* designing a biobank-based study than *after* the laboratory analyses have been done.

3.2. Are the End-of-Follow-Up Data Complete?

It is crucial to know for how long the cohort members are at risk. Therefore, information on vital status and emigration should also have been obtained for every cohort member. The simplest tabulation to control completeness of follow-up data is to count the annual numbers of deaths (and emigrations) of the cohort members by year of death (or emigration). Figure 3 demonstrates two such trends for real biobank cohorts and two artificial situations demonstrating problems in follow-up of vital status.

Annual numbers of deaths among the 722,500 women in the Finnish Maternity Cohort (FMC) are very small in the 1980s but increase heavily during the present millennium. This is expected because the women who were pregnant in the beginning of the biobank collection in the 1980s now gradually reach ages when the mortality among women starts to increase. The dynamic nature of the FMC (new women join the cohort every year) also increases the number of annual deaths.

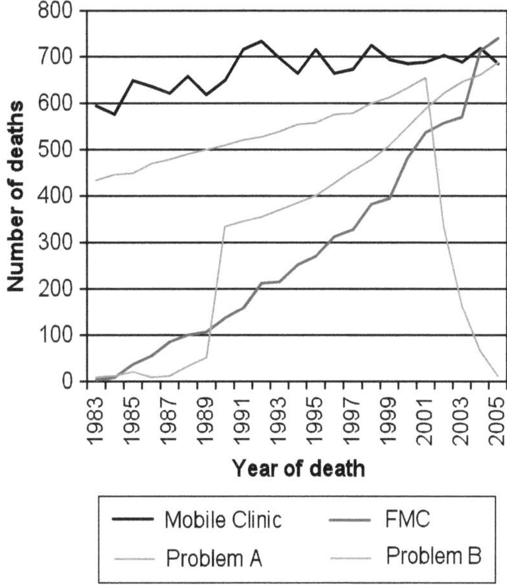

Fig. 3. Annual numbers of deaths among the 722,500 women in the Finnish Maternity Cohort (FMC), and the 50,400 persons in the Finnish Mobile Clinic Health Examination Survey, and two artificial situations demonstrating problems in follow-up of vital status (*see* text).

The trend of the annual numbers of deaths among the 50,400 persons who were 15+ years old when they participated in the Finnish Mobile Clinic Health Examination Survey in 1966–1972 is quite stable over the years. This is a correct trend in a cohort with large amount of old people. The number of persons alive in the cohort simply decreases so quickly that despite the strongly increasing relative mortality rate of the remaining cohort members the absolute numbers of annual deaths start to decrease.

In numerous newly collected study cohorts, we see a mortality trend demonstrated as "Problem A" in Fig. 3: there is a very small mortality in the beginning of follow-up. The reason for this problem is that some old samples (or their records) have been destroyed, if the person had died, to save space. Often this type of deletion is not documented and may (or may not) be revealed as too low mortality rates in the first years of follow-up.

Trend curve "Problem B" in Fig. 3 demonstrates another common problem in follow-up of vital status: the number of deaths decreases in the most recent years. This happens when the systematical follow-up for vital status via national death register files has not been done.

3.3. Are the Person Identification Data Accurate?

Every resident of the Nordic countries has a unique personal identification (PID) code that is used in all main registers and makes computerised linkages accurate and effective (36). The identification data of each biobank cohort member should have been compared with the national Population Register data to check that

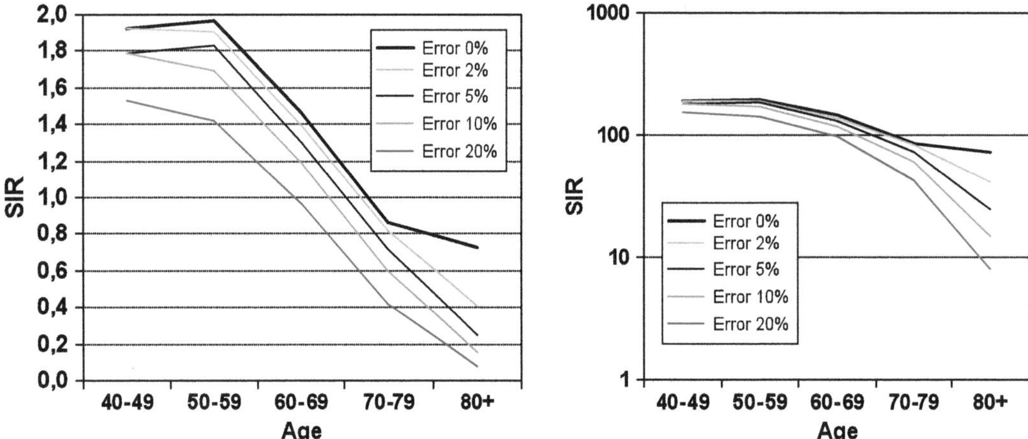

Fig. 4. Example of the effect of error in the identifier (*link key*) to the relative risk estimate. Standardised incidence ratio (SIR) for cancer (all sites combined) during 1953–2005 among 750 male workers of an anthophyllite asbestos mine, by age. The correct SIRs are indicated with "Error 0%" line; the other lines demonstrate situations when part of the cancer cases and deaths of the cohort members are missed because of failure in the person identifier data randomly produced in the cohort.

the personal identifiers are the correct ones and persons really exist in the population. In the same occasion, information on vital status and emigration can be obtained for every cohort member, but this information needs to be updated regularly (see Subheading 3.2).

The bias related to failures in record linkage with vital status and cancer very much increases along with increasing age at follow-up. Because the biobank cohorts are still quite young, the example is taken from another type of cohort, namely workers of an old asbestos mine in Finland (36). Figure 4 illustrates effect of error in the identifier (link key) to the SIR for cancer among male anthophyllite asbestos miners. The true SIR (all ages combined) during 1953–2005 is 1.35 (95% confidence interval 1.17–1.55). If there would be an error in 2% of the identifiers (at random), the observed number of the cases would decrease by 2% but the expected number would increase by 8% because of missing death information and subsequent addition of person-years at risk in the oldest age groups. The SIR related to 2% linkage error would be 1.23 (1.06–1.41), to 5% linkage error 1.06 (0.91–1.22), to 10% linkage error 0.87 (0.75–1.01) and to 20% linkage error 0.62 (0.52–0.72). Hence, a highly significant cancer risk related to asbestos mining would look like a significant protective effect if about 10% of the person IDs would be incorrect. In some of the older biobanks originally collected for non-scientific use, the proportion of incomplete IDs may have been several percentages. If these errors would not have been corrected, they would in long run have had serious effects on the results based on those samples.

The possibility of such linkage errors when the events of a *different* persons are linked together would be connected is not a major issue in the Nordic countries where all linkages nowadays are based on the unique PIDs. Although it is possible that a registered person may get a PID of another person, this event is so rare that it hardly affects conclusions of any study. There are examples of older times when the linkage was done manually based, e.g., on name, date and place of birth, place of residence. The linkage failures were much more common than in the later automatic linkages based on PIDs (37).

3.4. Standardised Incidence Ratios as Tools of Cohort Quality and Representativeness

The SIR is a useful tool to reveal occasions when the cohort members do not represent typical risk situation of the base populations. Although in nested case-control studies, it is not required that the cases and controls are selected from a population with average population risk level, understanding of the baseline risk level is important if the relative risk estimates are generalised to population attributable risk fractions. There are also situations when cohort members have temporally very special conditions that modify their risk level in a way that is not be easy to take into account in the RR analyses (see examples below).

The cancer cases among the serum donors included in the NBSBCCC serum banks have been traced through automatic record linkages with the national Cancer Registries.

In the person-year calculation needed for calculation of expected numbers of cancer cases, the follow-up starts at the date of first serum donation and ends at death, emigration or on the general closing date (depending on the lag of national cancer registration), whichever is first. Because the dates of emigration are not known in the Icelandic biobanks, about 4,000 emigrated persons of the Icelandic biobanks have to be excluded.

The numbers of observed cases and person-years at risk are counted for each calendar year, by gender and five-year age group. Sometimes it is useful to make further stratification according to the time elapsed since the sample donation. The expected numbers of cases for total cancer and for selected specific cancer types in the following examples were calculated by multiplying the number of person-years in each stratum by the corresponding cancer incidence rate in the national population, but sometimes regional cancer incidence rates may be a more informative reference.

The SIR was defined as the ratio of the observed to expected number of cases. The 95% confidence intervals (CI) for the SIR were based on the assumption that the number of observed cases followed a Poisson distribution.

3.4.1. Selection Related to the Indication of the Sampling?

Sometimes factors related to the reason of the serum donation may make the cohort temporally quite different from average of the baseline population it presents. In screenings of random samples

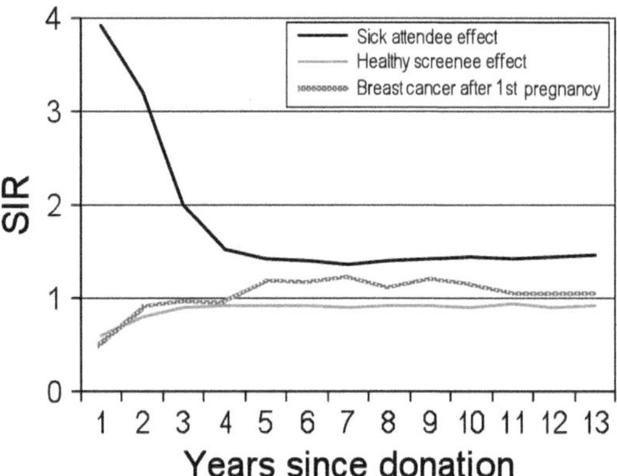

Fig. 5. Examples of biases related to the indication of serum donation that can be studied via trends of the standardszed incidence ratio (SIR) stratified by time elapsed since serum donation.

of population, incidence and mortality of chronic diseases tends to be decreased during the first months or years after the baseline study. This bias, illustrated with the "healthy screenee effect" curve in Fig. 5, is related to the selective participation: those who have severe early symptoms of a disease participate less frequently than the other people.

In the biobanks including samples of symptomatic persons, part of the symptoms may actually reveal to be symptoms of the outcome disease and therefore the SIR of that disease is very high soon after the serum donation. In the situation illustrated by the "sick attendee" curve in Fig. 5, the risk level is stabilised to the normal level of the base population of the cohort after about 5 years. If one would design a study within that cohort, it would be safest to exclude cases diagnosed during the five first years. If these cases would be included, there would be a risk of "reverse causality bias" (see Chapter 5): the hidden disease may have affected the values of biological parameters at baseline.

The third curve in Fig. 5 describes another atypical risk pattern, namely the "dual effect" of the pregnancy to the risk of breast cancer (38). The real-data example is taken from the Finnish Maternity Cohort. The SIR for breast cancer is first low but there is peak of increased risk some years after the pregnancy before the protective effect of the pregnancy starts to decrease the risk.

3.4.2. What Does Cancer Incidence Pattern Reveal of the Biobank Cohort?

The following observed and expected numbers of cancers are based on altogether 1.95 million subjects under follow-up in the 17 biobank cohorts, which were ready to produce person-years at risk calculations. The accumulated number of person-years from the date of first donation until the closing date (1999–2006, depending on the biobank) was 29.3 million (Table 2). The mean length of follow-up of a person was 13.4 years and the longest follow-times almost 40 years. The number of malignant cancer cases diagnosed between sampling and closing date exceeds 130,000.

The above numbers exclude the subjects from the Swedish Institute for Infectious Disease Control biobanks and those donors from other biobanks who donated their first sample after the closing date, altogether more than one million donors.

The specific cancer types selected a priori for the analysis (see Table 3) included cancer sites with known risk factors that reveal deviating risk behaviour among the cohort members, and other common cancer types selected to give a representative picture of the cancer situation among the cohorts.

3.4.2.1. Biobanks Based on Invitation of the General Population

The observed number of malignant neoplasms among persons (6,219) in the prospective cardiovascular *Reykjavik Study* exceeded slightly the expected rates based on Icelandic national rate, similarly in both genders, yielding an SIR of 1.06 (95% CI 1.03–1.08) (Table 3). Men had significantly elevated incidence of cancers of the prostate (SIR 1.13; 1.06–1.19) and kidney (1.23; 1.04–1.43) and significantly low risk of lip cancer (SIR 0.59; 0.35–0.94). Women had significant excess risk of ductal carcinoma of the breasts (SIR 1.17; 1.06–1.27) and leukaemia (1.29; 1.02–1.61). Incidence of malignancies of unknown primary was lower than in the general population of Iceland (SIR 0.82; 0.68–0.98). From the malignancies not included above, basal cell carcinoma of the skin showed elevated incidence (SIR 1.26; 1.18–1.34). The low lip cancer rate indicates a low proportion of high-risk categories, namely farmers and fishermen (39) that was expected because the cohort represents city people. All elevated SIRs are in cancers that are most common among urban populations.

There were 7,754 malignant neoplasms diagnosed between serum donation (1966–1972) and 31 December 2005 among the 50,448 subjects of the *Finnish Mobile Clinic Health Examination Survey* for whom serum sample is available. The SIR for all cancers combined was 0.94 (95% CI 0.92–0.95), similarly in both genders. Incidence of cancers of the genital organs is significantly below the national average: prostate (0.93; 0.88–0.98), penis 0.50 (0.21–0.97), breast (0.90; 0.84–0.95), cervix uteri (0.75; 0.57–0.97) and Fallopian tube (0.33; 0.07–0.97). Only penile cancer showed an SIR above 1.0 (1.98; 1.17–3.12). From the other cancers low SIRs were seen in adenocarcinoma in lungs (0.80; 0.68–0.93), liver cancer (0.78; 0.62–0.95). In males but not in females, there was a low SIR in non-Hodgkin lymphoma (0.81; 0.68–0.95) and cancer with unknown primary site (0.74; 0.58–0.92).

The total number of cancer cases among the *FINRISK* study members who have donated serum in 1992, 1997 or 2002 was 1,104 (SIR 0.97; 0.92–1.03). There was an excess of prostate cancer of localised stage (SIR 1.26; 1.05–1.48) but no excess

Table 3
Numbers of observed (O) and expected (E) cancer cases diagnosed between first serum donation (1,967+) and 31 December 2005 among the 19,257 participants of the cardiovascular Reykjavik Study. Expected numbers based on national population; standardised incidence ratios (SIR = O/E) given with 95% confidence intervals (CI). Statistically significant SIRs are in bold

ICD-7	Cancer site	O	E	SIR	95% CI
140–207	All malignant neoplasms	6,219	5874.45	**1.06**	**1.03–1.08**
140	Lip	20	33.18	**0.60**	**0.37–0.93**
143–144	Oral cavity	14	13.13	1.07	0.58–1.78
145–148	Pharynx	31	34.03	0.91	0.62–1.29
150	Oesophagus	81	90.25	0.90	0.71–1.11
151	Stomach	347	349.52	0.99	0.89–1.10
153	Colon	523	494.65	1.06	0.97–1.14
154	Rectum	184	170.57	1.08	0.93–1.24
155	Primary liver	46	46.19	1.00	0.73–1.32
155.1	Gall-bladder, biliary tract	45	44.29	1.02	0.74–1.35
157	Pancreas	188	180.37	1.04	0.90–1.19
161	Larynx	34	41.55	0.82	0.57–1.14
162–163	Lung	790	757.58	1.04	0.97–1.11
170	Breast	714	635.16	**1.12**	**1.04–1.20**
171	Cervix uteri	41	51.46	0.80	0.57–1.08
172	Corpus uteri	149	139.12	1.07	0.91–1.24
175	Ovary	101	120.06	0.84	0.69–1.01
177	Prostate	1,013	900.18	**1.13**	**1.06–1.19**
180	Kidney	221	201.39	1.10	0.96–1.24
181	Bladder	353	315.92	**1.12**	**1.00–1.23**
190	Melanoma of the skin	83	79.16	1.05	0.84–1.29
191	Non-melanoma skin	206	202.07	1.02	0.88–1.16
193	Brain and nervous system	165	151.71	1.09	0.93–1.25
194	Thyroid	123	107.61	1.14	0.95–1.35
200,202	Non-Hodgkin lymphoma	141	139.51	1.01	0.85–1.18
204	Leukaemia	164	136.20	**1.20**	**1.03–1.39**
199	Unknown site	116	140.77	**0.82**	**0.68–0.98**
Not included above					
Basal cell carcinoma of the skin		824	653.77	**1.26**	**1.18–1.34**

of non-localised prostate cancer. Also diagnosis of basal cell carcinoma of the skin was more common than in the general population (SIR 1.18; 1.05–1.32). Cancer of gallbladder was rare (SIR 0.33; 0.07–0.97).

Persons participating in health examinations in Norway (and allowing use of their sera for anonymous cancer research) in the *Janus biobank* cohort had less incidence of cancer than the general Norwegian population (21,889 cases observed by the end of 2001 vs. 24,086 expected). Significantly decreased SIRs were observed for cancers of the oral cavity (0.84; 95% CI 0.70–1.00), pharynx (0.77; 0.65–0.92), oesophagus (0.83; 0.70–0.97), primary liver (0.77; 0.61–0.95), lung (0.87; 0.83–0.90) and cervix uteri (0.82; 0.75–0.90), i.e., cancers related to high alcohol consumption and generally way of life not directed to healthy habits. None of the SIRs was significantly elevated.

In the *Malmö Diet and Cancer Study* cohort, there were 1,852 cancer cases, while the expected number based on incidence rates of the entire Swedish population was 1,568 (SIR 1.18; 1.13–1.24). This significant excess was mainly attributable to excesses in prostate cancer (84 excess cases, SIR 1.40; 1.25–1.57), breast cancer (59 excess cases, SIR 1.22; 1.09–1.38), skin melanoma (39 excess cases, SIR 1.72; 1.39–2.11) and bladder cancer (30 excess cases, SIR 1.42; 1.16–1.72). There were no significantly decreased SIRs in the cohort. This cancer incidence pattern is typical to a cohort representing population from southernmost Sweden with rates often more similar to the Danish cancer incidence rates than the Swedish average (Fig. 6).

The other invitational Southern Swedish cohort, that of the *Preventive Medicine in Malmö* project, produced more cancers (4,343), but the SIR was similar (1.17; 1.13–1.20). The pattern of cancer sites with increased incidence was partly similar to that of Malmö Diet and Cancer Study – breast cancer (SIR 1.24; 1.13–1.36), bladder cancer (1.46; 1.30–1.63), and skin melanoma (1.33; 1.16–1.53) – but some other cancers also had increased SIRs: lung cancer 1.48 (1.36–1.61), laryngeal cancer 1.41 (1.02–1.89), pharyngeal cancer 1.56 (1.14–2.08) and pancreatic cancer 1.23 (1.02–1.48). Despite the large numbers of cases, none of the 22 primary sites studied separately showed an SIR significantly below unity.

The *Northern Sweden Health and Disease Study* consists of three cohorts randomly selected from the population of given ages in that region. The largest number of cancer cases (2,426) was found among members of the Västerbotten Intervention Program (VIP). The expected number was slightly higher (2,531). The SIR was significantly decreased for lung cancer (0.82; 95% CI 0.68–0.98); otherwise, there were no major aberrations from 1.0.

There were 289 cancer cases in the smaller MONICA cohort as compared to 310 cases expected. The difference is not significant, and none of the site-specific SIRs was significantly different from unity.

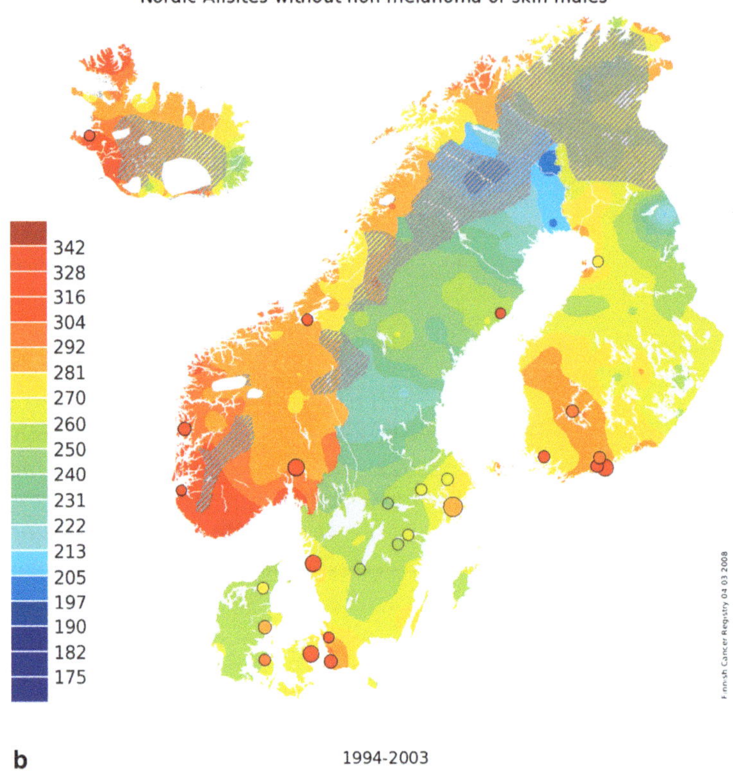

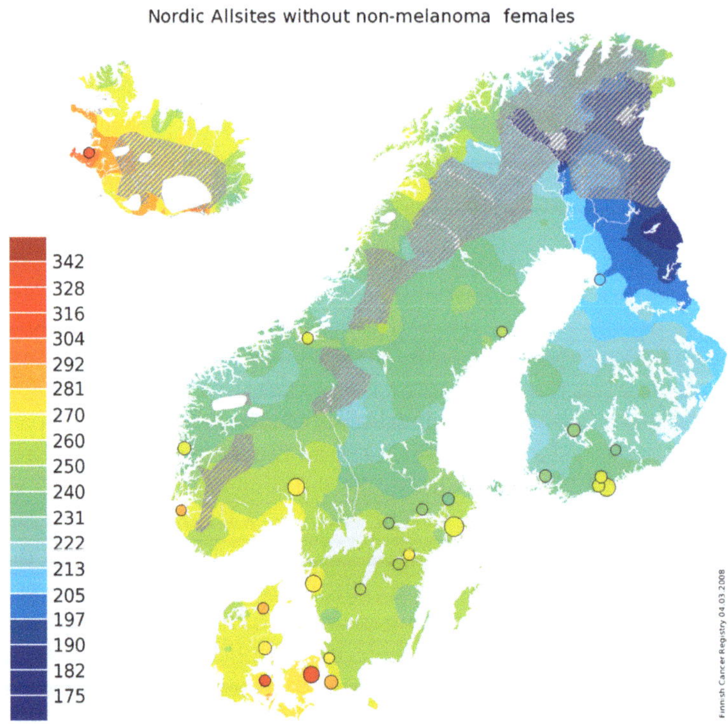

Fig. 6. Spatial variation of age-adjusted incidence rates per 100,000 of cancer (all sites, excluding non-melanoma skin cancer) in the Nordic countries, 1994–2003. For mapping method, *see* (57).

The mammography screening cohort (also part of Northern Sweden Health and Disease Study) showed an SIR of 0.99 (1,159 observed cases vs. 1,174 observed). These women – age range 50–69 years – had a significantly lowered SIR for lung cancer (0.75; 0.56–0.98), while none of the other sites showed an SIR significantly different from unity. Incidence of breast cancer was significantly increased during the first year after mammography and serum sampling date (SIR 1.89; 1.58–2.24), but this excess was compensated by a significantly decreased incidence in the later years (SIR 0.88; 0.78–0.98).

3.4.2.2. Maternity Cohorts

There were 14,973 cancer cases observed after sampling (from 1983 until August 2005) and before 31 December 2005 in the *Finnish Maternity Cohort*, which is the biggest one of the four Nordic biobanks based on the population screening of pregnant women. The expected number based on average Finnish female population was 15,770 and the SIR was 0.95 (95% CI 0.93–0.96).

There were 6,861 cases of breast cancer, equal to the expected number. The SIR for lobular type of breast cancer (15% of breast cancers) was 1.12 (1.05–1.18). The incidence of breast cancer was above the national average after sera drawn in the context of the first pregnancy (SIR 1.08; 1.04–1.11) but the SIR gradually declined along with the subsequent pregnancies and was after fifth pregnancy 0.62 (0.38–0.94). The SIR for endometrial cancer among all pregnant women was 0.64 (0.57–0.70) and decreased after the third pregnancy to only 0.30 (0.15–0.56).

There was an excess of the rare placental choriocarcinoma during the first year after sampling (11 cases; SIR 6.06; 3.03–10.84), which is by definition related to pregnancy. Borderline tumours of the ovary were less frequent than in the population on average (SIR 0.85; 0.76–0.94) and invasive ovarian tumours even more rare (SIR 0.73; 0.67–0.79). The SIR for lung cancer was 0.79 (0.69–0.89), with the strongest decrease in adenocarcinoma (SIR 0.60; 0.48–0.74). The SIR for stomach cancer was 0.88 (0.77–0.99), for soft tissue sarcoma 0.84 (0.69–0.99) and for cancer with unknown primary site 0.82 (0.68–0.96).

In the *Icelandic Maternity Cohort*, there were 1,453 malignant neoplasms observed versus 1,466 expected (SIR 0.99; 0.94–1.04). The SIRs for single cancer sites were similar as those reported above for the Finnish Maternity cohort but none of them reached statistical significance in this ten times smaller data set

Women in the *Malmö Maternity Cohort* (part of Malmö Microbiology biobank) also had overall cancer incidence similar to the national population (493 observed cases vs. 498 expected, SIR 0.99; 0.91–1.08), but there was a tendency for higher lung cancer incidence than the reference population (SIR 1.28; 0.68–2.18). None of the other cancer sites deviated significantly from the expected incidence.

In the *Northern Sweden Maternity Cohort*, there were 1,625 cancer cases observed after sampling and before end of follow-up. The expected number was 1,717 and the SIR 0.95 (0.90–0.99). Significantly decreased SIRs were seen for lung cancer (0.59; 0.40–0.83) and endometrial cancer (0.69; 0.49–0.94).

3.4.2.3. Specific Cohorts with Clearly Defined Enrolment Criteria

Men in the *Helsinki Heart Study* had 3,638 cancer cases, less than expected (SIR 0.92, 95% CI 0.89–0.94). The SIRs were significantly decreased for cancers of the pharynx (SIR 0.55; 0.28–0.99), stomach (0.78; 0.65–0.91), pancreas (0.81; 0.66–0.96), nose (0.14; 0.00–0.77) and unspecified sites (0.69; 0.52–0.90). SIR for lung cancer was below the national average in all main histological types: in squamous cell carcinoma 0.72 (0.61–0.83), adenocarcinoma 0.71 (0.57–0.88), and small cell carcinoma 0.63 (0.49–0.78).

Incidence of non-melanoma skin cancer (SIR 1.37; 95% CI 1.15–1.59) and basal cell carcinoma of the skin (1.24; 1.16–1.31) was significantly above the national average. Also meningiomas of the brain were in excess (SIR 1.59; 95% CI 1.05–2.29).

The incidence pattern of the health-interested volunteers of Helsinki Heart Study is very different from that of the cohort of smoking men in the *Alpha-Tocopherol-Beta-Carotene (ATBC) study* (Fig. 7). The latter cohort has been utilised in studies aiming to confirm whether various diseases are related to smoking or not (40).

In the ATBC cohort, there is an excess risk of cancer in most sites. The observed number of cancers in the end of 2005 was

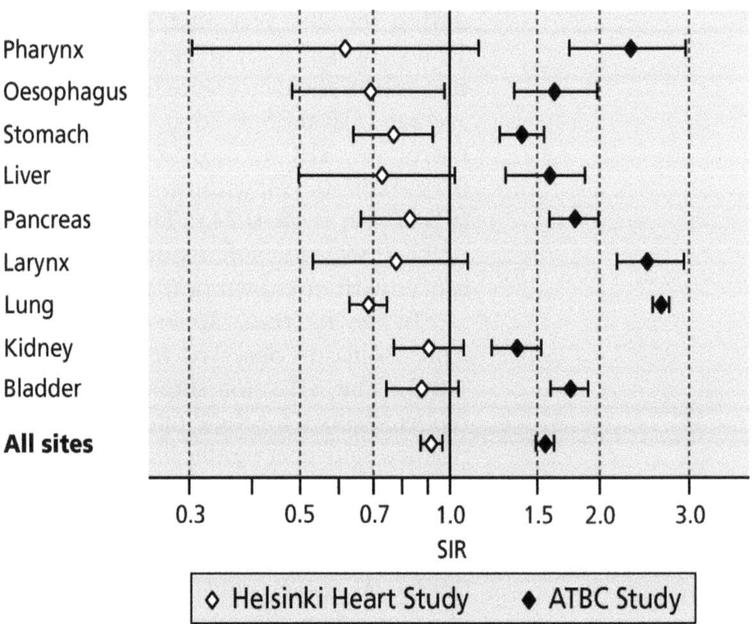

Fig. 7. Standardised incidence ratios (SIR) of selected cancers 1984–2003 among the 19,000 Finnish men in Helsinki Heart Study, and the 29,000 men in Alpha-Tocopherol-Beta-Carotene (ATBC) Study, with 95% confidence interval bars.

9,420, i.e., 3,202 more than the expected number calculated on the basis of incidence rate of average Finnish same-aged men (SIR 1.52; 1.48–1.54). The SIRs were significantly increased for cancers of the tongue (SIR 1.77; 95% CI 1.23–2.46), other oral cavity (1.86; 1.31–2.54), pharynx (2.22; 1.68–2.88), oesophagus (1.64; 1.36–1.95), stomach (1.36; 1.23–1.50), colon (1.14; 1.02–1.25), liver (1.61; 1.36–1.87), pancreas (1.76; 1.58–1.94), larynx (2.50; 2.13–2.89), lung (2.71; 2.62–2.80), prostate (1.11; 1.06–1.15; for non-localised prostate cancers 1.22; 1.10–1.34), kidney (1.36; 1.22–1.50; includes renal pelvis 2.41; 1.69–3.33), bladder (invasive 1.77; 1.64–1.91, and for papilloma 2.22; 1.32–3.51), and unspecified sites (1.79; 1.57–2.03). Excess risk was seen in acute myeloid leukaemia (SIR 1.37; 1.00–1.83) –according to literature that seems to be related to smoking (41) – but not in other types of leukaemia. The only sites with an SIR below 1.0 were the skin melanoma (0.69; 0.56–0.84) and the basal cell carcinoma of the skin (0.89; 0.84–0.94).

The *Red Cross blood donors* in capital Oslo and surrounding areas (the smaller part of Janus biobank) had lower than average overall cancer incidence (2,286 cases observed vs. 2,399 expected; SIR 0.95; 95% CI 0.91–0.99). The SIRs of cancers of the stomach, primary liver and larynx are as low as 0.36–0.46, all significantly decreased. The SIR for lung cancer was 0.76 (95% CI 0.65–0.88). The SIR for breast cancer was significantly elevated (1.29; 1.17–1.42), and so was the SIR for skin melanoma (1.24, 1.07–1.42).

3.4.2.4. Viral Screening and Clinical Testing Biobanks

In the part of the *Malmö Microbiology cohort* including samples submitted for testing because of clinical suspicion of infection with blood-borne viruses (e.g. jaundice or impaired liver function, drug addicts, haemophiliacs and dialysis patients), there were 2,055 cancer cases more than the expected number 4,455 (SIR 1.46; 95% CI 1.43–1.50). All SIRs were above 1.0, except those for breast cancer and endometrial cancer. The highest SIRs were seen for primary liver cancer (5.58; 4.87–6.36), pancreatic cancer (3.28, 2.93–3.67) and gall-bladder cancer (2.52; 1.94–3.22).

The *Malmö Microbiology* subcohort consisting of sera submitted for other virus serology had even higher relative overall cancer risk (SIR 2.08; 95% CI 1.97–2.20; 1,328 cases observed vs. 638 expected). Very high SIRs were seen in primary liver cancer (4.15; 2.63–6.22), pancreatic cancer (2.71, 1.91–3.74), lung cancer (2.95; 2.46–3.53) and cancers of the brain and nervous system (3.05; 2.37–3.88).

3.5. Accuracy of Variables Associated to Persons and Samples

Most samples of the biobanks include variables related to the sample itself, to the sampling occasion, or to the person who donated the sample. If these variables were in major role in the original setting of a study – such as the questionnaire data related to study persons' health habits (for example smoking, diet, physical exercise, body mass index) – they are normally stored and documented systematically. If such data has been asked in a context of clinical practice, these data may be kept non-systematically, possibly on paper format only, or even lost.

A high-quality biobank database should include some variables directly related to the sample that are crucial for nested case-control studies based on the biobank:

1. date of sampling,
2. indication of sampling (in biobanks with mixed origin of sampling),
3. number of freeze–thaw cycles,
4. amount of sample left,
5. indicator of damaged sample.

All of these variables may be used as matching criteria in control selection. If they are missing, the quality of the study will not be as good as it could be. If these factors can only be confirmed after search the samples from the fridge, the logistics of any such study becomes clumsy and laborious.

4. Prospective Cancers: Basis for Nested Case Control Studies

4.1. Numbers of Prospective Cancer Cases

Maybe the most important tool for quality assurance of a big biobank network is a simple tabulation of numbers of persons in the biobank cohorts and numbers of cancer cases. The NBSBCCC network has agreed to collect such data – stratified by biobank, year, gender, age and cancer type – in a centralised database, which will be automatically updated after each new linkage of the records of any biobank and cancer registry data.

Even though the number of new donors to the NBSBCCC has been decreasing in the latest years (Fig. 8), the number of prospective cancer cases increases year by year (Fig. 9). The annual number seems to drop in the very latest years in some of the regions. This is an artefact related to technical reasons. For instance, the biobanks in Malmö have had to wait for a new update because of slow progress in getting permission from privacy issue officials to link cancer data with their cohort; this problem is now solved and very soon the closing date for the Malmö cohorts will be moved from 1999 to 2005. In Norway, the Janus cohort has been linked with cancer data until about 2005, but there have been problems and principle issues to tabulate the numbers for the NBSBCCC quality assurance database.

The numbers of cancer cases diagnosed after serum donation among persons in each serum bank are given in Table 4, for all cancers combined and for 64 subcategories. These numbers are based on the routine linkages between the serum banks and cancer

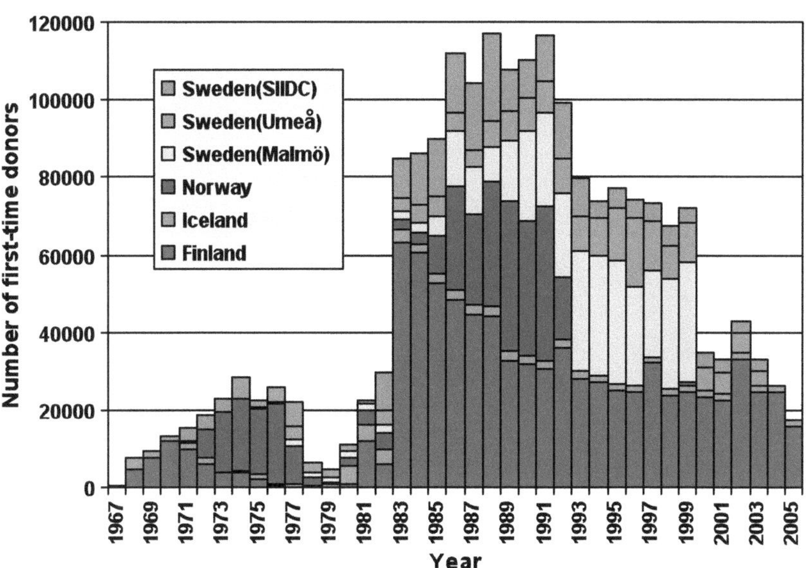

Fig. 8. Annual numbers of first-time donors of the Nordic biobanks as reported to the joint NBSBCCC quality assurance surveillance database by June 2007, by region.

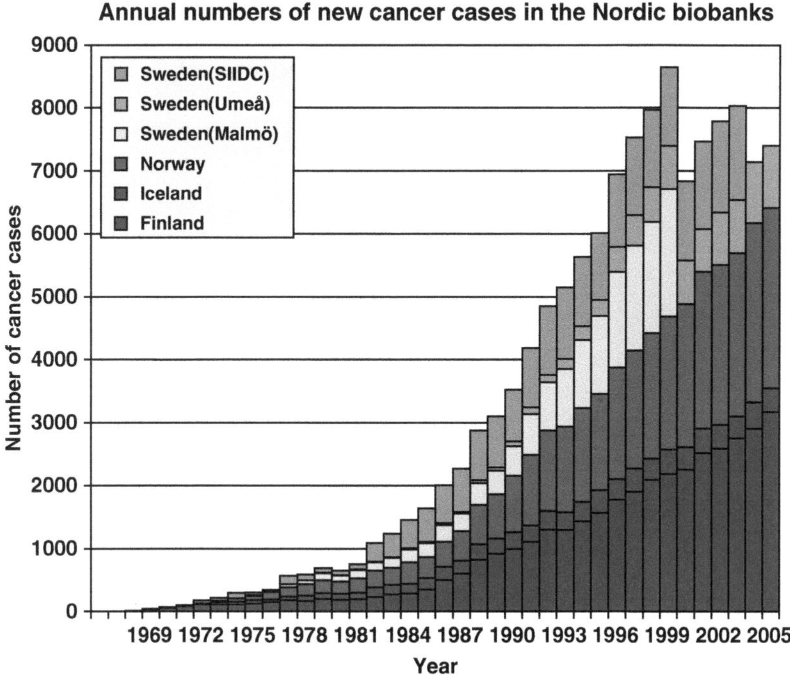

Fig. 9. Annual numbers of registered cancer cases among subjects in the Nordic biobanks diagnosed after serum donation and before 31 Dec 2005, as reported to the joint NBSBCCC quality assurance surveillance database by March 2008 by region.

Table 4

Numbers of cancer cases registered among serum donors after donation to the Nordic Biological Specimen Banks for Cancer Causes and Control (NBSBCCC) according to the latest cancer registry linkages, by cancer site and serum bank. Table constructed February 28 2008

	Cancer site/type	Number of cancer cases																			
		Finland					Iceland		Norway		Sweden										Total nordic
									Janus		Northern Sweden Health and Disease Study					Malmö Microbiology Biobank					
ICD-7 (or internal code)		Maternity cohort (females)	Mobile Clinic	FINRISK	Helsinki Heart Study (males)	ATBC (males)	Maternity cohort (females)	Heart Preventive Clinic	Health examinations	Blood donors	Northern Sweden Maternity Cohort (females)	VIP	MONICA	Mammography (females)	Institute for Infectious Disease Control	Diet and cancer (Malmö)	Preventive medicine (Malmö)	Maternity cohort (females)	Blood-borne virus screening	Clinical viral serology	
Years covered		1983–2005	1967–2005	1992–2005	1980–2005	1985–2005	1980–2005	1967–2005	1973–2006	1973–2006	1975–2005	1985–2005	1986–2005	1995–2005	1971–2003	1991–1999	1974–1999	1985–1999	1969–1999	1990–1999	
140–207	All malignant neoplasms	14,973	8,463	1104	3,638	9,420	1,453	6,219	36,371	2,406	2,192	3,786	476	1,864	21,802	1,852	4,343	493	6,511	1,328	128,694
140	Lip	4	78	4	31	70	1	20	131	7	2	9	1	1	76	7	18	0	32	10	502
141	Tongue	48	41	9	9	35	1	15	154	14	6	6	2	3	54	3	18	2	25	2	447
142	Salivary glands	61	28	2	11	8	6	10	67	4	4	7	1	5	39	3	7	0	18	6	287
143–144	Oral cavity	35	29	5	15	38	5	29	183	8	2	11	1	6	68	10	22	1	26	6	500
145–148	Pharynx	34	30	2	11	57	3	16	194	17	5	14	3	4	115	10	46	1	40	7	609

150	Oesophagus	14	107	12	35	119	1	81	263	20	2	24	2	10	108	18	53	1	51	10	931
151	Stomach	246	516	37	133	369	8	347	995	38	27	92	14	38	392	36	125	6	157	26	3,602
152	Small intestine	33	32	7	13	19	2	27	135	13	8	12	2	6	92	10	23	0	37	5	476
153	Colon	525	445	56	220	366	42	523	3,104	195	67	254	31	161	1,066	105	232	16	311	75	7,794
154	Rectum, rectosigma	275	351	38	160	264	25	184	1,697	118	38	155	23	75	545	85	172	6	205	28	4,444
155	Primary liver	42	89	19	48	148	2	46	141	4	7	26	5	13	359	15	39	0	223	23	1,249
155.1	Gall-bladder, biliary tract	55	109	3	23	57	1	45	172	8	6	9	0	7	132	9	16	0	65	6	723
157	Pancreas	135	354	38	111	368	7	188	828	49	22	100	11	47	476	42	119	3	316	37	3,251
160	Nose	9	14	5	1	12	1	10	63	5	4	9	0	3	17	1	6	0	5	4	169
161	Larynx	2	80	6	37	165	1	34	260	9	1	8	1	1	71	4	44	0	43	1	768
162–163	Lung	236	1,157	108	485	3,125	63	790	3,548	190	62	185	22	93	1,698	137	535	13	523	127	13,097
(162A)	lung, adenoca	85	146	18	86	416	32	269	1,075	60	26	65	7	34	466	45	149	2	160	40	3,181
(162S)	lung, small cell	41	167	16	76	519	10	154	685	28	7	28	7	17	231	27	93	2	80	20	2,208
(162E)	lung, squamous cell	32	354	29	155	945	8	172	801	74	9	17	2	13	406	25	125	2	115	21	3,305
170	Breast	6,861	998	173	4	10	512	714	5,215	413	903	696	75	614	1,265	324	466	165	635	112	20,155
(170D)	breast, ductal	5,389	703	129	3	9	406	485	3,925	192	560	478	53	424	1,047	138	117	67	243	57	14,425
(170L)	breast, lobular	1,031	100	33	1	0	50	48	507	52	99	99	7	100	194	66	50	13	75	12	2,537
171	Cervix uteri, invasive	559	57	4	0	0	134	41	539	54	174	36	5	17	1,795	13	27	56	101	10	3,622
172	Corpus uteri	316	237	39	0	0	33	149	1,176	59	54	152	17	161	329	48	75	2	76	18	2,941
173	Choriocarci-noma	28	1	0	0	0	.	.	2	2	5	0	1	0	.	0	0	0	2	0	41
175.0	Ovary	508	163	30	0	0	41	101	925	59	103	100	7	65	638	35	59	13	103	16	2,966

(continued)

Table 4 (continued)

ICD-7 (or internal code)	Cancer site/type	Number of cancer cases* Finland: Maternity cohort (females)	Finland: Mobile Clinic	Finland: FINRISK	Finland: Helsinki Heart Study (males)	Finland: ATBC (males)	Iceland: Maternity cohort (females)	Norway: Heart Preventive Clinic	Norway (Janus): Health examinations	Norway (Janus): Blood donors	Sweden: Northern Sweden Maternity Cohort (females)	Sweden (Northern Sweden Health and Disease Study): VIP	Sweden (Northern Sweden Health and Disease Study): MONICA	Sweden (Northern Sweden Health and Disease Study): Mammography (females)	Sweden: Institute for Infectious Disease Control	Sweden: Diet and cancer (Malmö)	Sweden: Preventive medicine (Malmö)	Sweden (Malmö Microbiology Biobank): Maternity cohort (females)	Sweden (Malmö Microbiology Biobank): Blood-borne virus screening	Sweden (Malmö Microbiology Biobank): Clinical viral serology	Total nordic
175.1	Tuba	18	3	1	0	0	2	3	43	.	2	2	0	3	11	4	4	0	7	0	103
176.0	Vulva	47	23	4	0	0	4	14	87	6	9	3	0	6	50	5	8	2	6	3	277
176.1	Vagina	11	8	2	0	0	2	3	30	1	6	2	1	2	11	1	0	1	2	0	83
177	Prostate	0	1,076	227	1,127	2,000	0	1,013	4,632	266	0	740	100	0	1,985	292	606	0	694	128	14,886
(177L)	prostate, localised	0	501	133	546	854	.	.	.	.	.	.	.	.	.	.	.	.	.	.	2,034
(177N)	prostate, non-localised	0	214	23	156	366	.	.	.	.	.	.	.	.	.	.	.	.	.	.	759
178	Testis	0	8	2	9	8	0	2	220	30	0	22	3	0	117	2	17	0	70	2	512
(178S)	testis, seminoma	0	5	2	7	6	0	1	150	20	0	16	2	0	117	1	12	0	36	1	376

(178N)	testis, non-seminoma	0	3	0	2	2	0	1	57	3	0	6	1	0	162	1	5	0	34	1	278
179.0	Penis	0	18	2	10	10	0	11	74	.	0	8	2	0	20	1	4	0	8	2	170
180	Kidney	211	296	37	156	331	19	221	1,009	71	22	79	10	45	589	43	141	5	174	41	3,500
(180.1)	renal pelvis	7	22	3	12	36	0	25	98	1	2	8	3	9	57	7	25	0	21	2	338
181	Bladder, ureter, urethra	84	327	25	180	623	19	353	1,757	60	20	149	25	51	728	102	310	5	292	28	5,138
190	Melanoma of the skin	778	209	28	107	95	149	83	2,036	200	138	123	9	43	623	93	208	65	256	33	5,276
191	Non-melanoma skin cancer	140	250	24	143	177	20	206	1,029	83	13	36	9	29	2,792	71	171	8	489	144	5,834
192	Eye	43	22	3	8	9	3	13	106	11	9	17	0	7	82	1	5	1	8	4	352
193	Brain and nervous system	1,149	238	39	114	129	65	165	1,280	95	143	154	23	72	817	58	162	40	242	67	5,052
(193G)	glioma	415	84	13	47	56	22	73	534	49	28	54	11	27	114	1	8	4	30	10	1,580
(193M)	meningeoma	384	65	10	28	29	36	65	430	23	49	52	7	31	222	19	48	8	70	12	1,588
194	Thyroid	1,108	116	14	17	25	127	123	415	27	49	18	1	13	94	10	28	17	60	11	2,273
(194F)	follicular	52	17	1	5	3	10	20	48	4	3	1	0	1	5	2	2	0	5	2	181
(194P)	papillary	1,017	77	9	10	14	113	90	311	11	24	11	1	8	73	4	6	9	23	4	1,815
195.0	Glandula suprarenalis	20	7	0	3	5	1	2	34	5	11	8	0	7	43	3	11	2	24	4	190
195.1	Glandula parathyreiodea	0	1	0	0	0	0	3	.	.	34	11	5	4	232	16	53	8	75	13	455
195.2	Thymus	10	2	2	0	1	0	2	14	.	2	4	0	2	17	1	4	2	10	5	78
195.3	Hypophysis	0	0	0	0	0	8	22	1	.	14	16	0	6	78	9	17	6	40	8	225
195.4	Corpus pineale	4	1	0	0	0	0	0	1	.	.	.	.	.	.	0	0	0	0	0	6

(continued)

Table 4
(continued)

ICD-7 (or internal code)	Cancer site/type	Number of cancer cases*																			
		Finland					Iceland	Norway		Sweden											Total nordic
								Janus				Northern Sweden Health and Disease Study				Malmö Microbiology Biobank					
		Maternity cohort (females)	Mobile Clinic	FINRISK	Helsinki Heart Study (males)	ATBC (males)	Maternity cohort (females)	Heart Preventive Clinic	Health examinations	Blood donors	Northern Sweden Maternity Cohort (females)	VIP	MONICA	Mammography (females)	Institute for Infectious Disease Control	Diet and cancer (Malmö)	Preventive medicine (Malmö)	Maternity cohort (females)	Blood-borne virus screening	Clinical viral serology	
196	Bone	54	14	3	4	4	10	14	50	4	6	4	0	1	43	3	5	1	15	3	238
197	Soft tissue	118	55	6	21	27	10	16	131	18	17	24	1	6	193	14	26	4	46	12	745
199	Other/unknown site	133	190	16	54	230	19	116	760	53	29	94	16	62	755	37	103	3	211	40	2,921
200.202	Non-Hodgkin's lymphoma	457	249	29	167	226	43	141	1,266	72	58	138	17	59	1,429	69	138	13	271	111	4,953
201	Hodgkin's disease	216	41	1	12	14	19	14	112	12	22	10	3	6	340	3	15	9	52	40	941
203	Multiple myeloma	68	106	15	47	60	8	91	464	12	11	80	8	30	308	28	57	1	83	17	1,494
204–207	Leukaemia	243	192	21	84	129	26	164	646	40	40	71	12	34	1,591	42	91	11	158	67	3,662
(204CLL)	chronic lymphocytic	35	76	4	45	52	1	61	223	4	7	32	7	15	253	14	32	1	41	17	920

(204AML)	acute myeloid	109	57	7	17	45	6	46	250	10	13	16	2	8	81	10	28	5	54	23	787
Not included above																					
171C	Cervix, CIN3/ in situ/ dysplasia gravis	5,335	88	23	0	0	.	.	.	.	2406	225	26	53	1,607	0	0	0	0	0	9,763
175B	Ovary, border-line tumour	345	28	7	0	0	47	20	273	7	61	25	5	13	73	0	0	0	0	0	904
181P	Bladder, papilloma	10	11	1	10	18	1	23	.	60	16	140	23	47	3	90	271	4	252	22	1,002
191B	Skin, basal cell carcinoma	2,269	1677	270	920	1,038	254	824	.	.	.	.	.	.	.	.	.	.	.	.	7,252
Total benign/ semimalignant	7,959	1,804	301	930	1,056	302	867	273	67	67	390	54	113		90	271	4	252	22	17,238	

Not applicable
Not registered by the national cancer registry
Classification not available from the national cancer registry

registries (status in June 2007). There were altogether 110,217 cases traditionally counted as real cancers, and 15,428 basal cell carcinomas of the skin, precancerous cervical lesions, borderline tumours of the ovaries and cancers of low malignant potential of the bladder. The registration of the latter outcome categories varies over the Nordic cancer registries, and the same is true for in situ cancers and several other cancer-like lesions not included in Table 4.

After update of the biobanks in Malmö and taking into account the cancer cases for Janus biobank missing from the above tabulation, the numbers of subsequent cancer cases exceed 30,000 in Sweden, Finland and Norway (27,000), giving a balanced three-country setting in the future studies. The Icelandic number (7,700) is smaller but very large as compared to the small population size in Iceland (less than 300,000).

4.2. Simple Cross-Tabulations: A Powerful Quality Assurance Tool

There are several ways how simple tabulations of numbers of persons and prospective cancer cases may improve the quality of network activity.

When designing a case-control study nested in the biobank cohorts, it is good to know how many eligible cases there will be to be sure that the study power will be satisfactory. The number of cancer-free individuals makes it easy to select matching criteria in such a way that required number of eligible controls will be found but there will not be unnecessary large variation in matching criteria such as storage time of the sample or age of the individual.

Once the cases and controls have been selected in each participating biobank, it is always good to check whether the numbers match with those to be found from the NBSBCCC tabulation. If the numbers do not match, there are two possibilities:

1. There has been an error in the case-control selection. In this case, the error can be corrected before sending the samples to laboratory analyses.
2. There are good explanations for the drop of the case number, such as additional exclusion criteria whose prevalence was wrongly estimated. If this happens, the design may be modified to replace the missed cases from other biobanks or by extending the period of case recruitment.

This gives the principle investigator of any network-based study tools to control for accuracy of the study materials, and also eliminate attempts of fraud, such as fabrication of data. Although the scientific moral in the NBSBCCC network has been high and fraud would never been expected in this research society, there are examples from the latest years from other research groups that makes it important to be able to demonstrate that even such extreme possibilities can be controlled. Those research institutes whose studies have never been linked to any scientific miscarriage are most eager to offer materials to any quality assurance operation that would improve possibilities to external evaluation of the accuracy of the data, such as the simple tabulations described above.

4.3. Future Numbers of Prospective Cancer Cases

The tabulations of numbers of prospective cancer cases indicate that the annual number of new prospective cancer cases in the NBSBCCC biobank cohorts will be about 10,000 cancer cases in the next few years. The basic data readily collected to the joint database allow predictions based on age-period-cohort analyses, and a project to predict numbers of cancer cases up to the year 2015 has been started. In short, this requires estimation of future incidence rates of selected cancers, and prediction of future person-years at risk.

Even when the total number of prospective cancer cases is huge, there are rare cancer types, special subcategories and rare exposures where the current study power may not be satisfactory. Therefore, in a situation when we anyway have to set priorities on what to do first, an optimally coordinated research cluster should postpone underpowered studies to the future. Predictions on the future numbers of case help in long-term planning of study schedules.

5. Discussion

5.1. Strengths of Biobank-Based Study Designs

The infrastructure described allows multi-national and multi-disciplinary networking for comprehensive prospective epidemiological studies nested in several biological specimen banks. There are several strengths in studies based on samples readily collected in biobanks to the alternative situation that there is no biobank, i.e., samples from cases and controls have to be collected after the disease of the case has been diagnosed:

(a) Use of biobank data offers proper time order of exposure data collection and outcome and decreases the possibility of "*reverse causality bias*", i.e. the mixing up of cause and effect. For instance, herpes viruses are frequently reactivated by severe diseases, such as cancer, and may indeed induce cellular genes related to cellular proliferation (4, 21). If the virus is measured from a sample taken at the time of cancer diagnosis, it is difficult to assess whether associations between reactivatable viruses and cancer are causal or mere secondary associations with opportunistic infectious agents. In the prospective design, we have been able to show that cancer reactivates herpes simplex virus type 2 and not vice versa (21, 42, 43).

(b) A related type of bias is the *differential measurement bias*, i.e., situations where the fact that the patient has disease influences measurements. Even existent (pre)cancer may influence both antibody levels and cellular immunity because of the immune dysfunctions seen in cancer (11, 25, 27). Also, it may be easier to obtain cancer tissue than control tissue. When measurement biases are related to case status, their effect is particularly unpredictable. Studies using samples taken from individuals long time before the cancer diagnosis suffer only from misclassification bias that is non-differential with regard to case status, which may result in a conservative and readily quantifiable bias.

(c) Many exposures are associated with *non-attendance* in retrospective case-control studies, biasing results. In biobank-based studies, there may be baseline selection in the formulation of the study base (that makes the study base different as compared to the population from which it was originally drawn), but after that all samples from the study base are available for testing, and there is no selection related to later case-control status.

(d) Studies based on readily collected biosamples are *time-effective* and – if the biobank is used in many studies –*cost-effective*. The classical prospective cohort study, where samples are not stored but analysed immediately after sampling, requires very long follow-up, often decades. Study hypotheses and measurement assays may be outdated when the outcomes are finally obtained. The establishment and maintenance of population-based biological specimen banks is costly, but when such banks are established they can be used for a variety of prospective studies on the aetiology of several reasonably common diseases, e.g., association of HPV infections with various human cancers (2, 3, 6–9, 16, 19). The marginal cost for a prospective study can be reduced to the level where also rather unlikely, innovative hypotheses (that may result in breakthroughs) can be reliably evaluated., e.g., the role of *Chlamydia trachomatis* in cervical cancer causation (14, 17, 44). Since biological specimen banks are already established, the time required for completion of a reliable prospective study with decade-long follow-up of a recently emerged epidemiological problem is short.

(e) The Nordic biological sample banks contain a very high proportion of *serial samples*; the mean number of samples per person is two to three (Table 2). For instance, the maternity cohorts includes complete sets of serial samples related to pregnancies of majority of the parous women (13), and some specific research cohorts may include very tight set of samples, e.g. there are up to 28 samples from part of the Helsinki Heart Study subjects. For studies of chronic diseases, such as cancer, that develop over a very long time span, a considerably more reliable and complete assessment of the importance of various exposures can be obtained by studying multiple serial measurements of the same person (10, 13). Furthermore, the unavoidable variability of measurements in a single sample will cause a systematic underestimation of the importance of a risk factor (regression dilution towards the mean) (45), which can be corrected for using serial measurements. Serial samples can also be used to pin point the time-point of exposure (13). There are many examples of disease causation by an exposure that occurs only if the exposure occurs at a certain time-point. Poliomyelitis as a result of delayed exposure to poliovirus is a well-known example. Only if samples taken at many different time-points preceding development of disease have been stored, one can attempt to study time-point of exposure by biochemical and molecular assays.

Biobank samples may sometimes offer *objective measurements* for variables that are hard to be accurately registered via questionnaire surveys. The extent of misclassification of self-reported exposures can be considerable, especially for sensitive questions, such as addictions. Even a very modest amount of misclassification may lead to very misleading conclusions. There now exist an increasing arsenal of biochemical measurements that can be used for objective measurement of exposures in stored biological specimens, e.g. serum cotinine measurements for assessment of smoking habits (46). The accuracy of extensive questionnaires on environmental exposures, diet and life style that most of the NBSBCCC banks contain may be validated with biochemical measurements in a relatively small amount of samples derived from the biobank.

5.2. Stability and Validity of Old Samples

A potential weakness of studies based on historical biobank samples is the stability and validity of the old samples. The oldest samples in the Nordic biobanks are more than 30 years old and many are stored at –25°C. Validations of the Janus biobank have shown that most of the substances commonly analysed in epidemiological studies, for instance proteins (in particular antibodies), organic acids, carbohydrates, trace metals, inorganic salts and polyunsaturated fatty acids are stabile when they are stored at –25°C. However, not all enzymes and vitamins are stabile under these conditions (46).

Genotyping from archival serum and plasma samples is, following the development of efficient whole genome amplification methods, a fairly routine method also from very old samples stored at –25°C (47). However, investigators contemplating amplification-based methods such as PCR should be aware that in the 1960s and 1970s disposable pipettes and tips may not always have been used in all biobanks.

Possible deterioration of the oldest sera is commonly outweighed by consideration of increased statistical power, reduced reverse causality biases with longer follow-up and possibility to detect causative exposures that occur many years before diagnosis of disease and may not be detectable in samples taken at or close to diagnosis.

5.3. Follow-Up Procedures

Initial calculations of SIRs in some of the biobanks did not include *follow-up for vital status*, which produced erroneous, markedly lowered SIRs in older ages. As demonstrated above for cohort analyses, the problem with missing data on vital status slowly becomes a serious problem also in case-control settings: a control subject that is registered as being alive may actually have had died before the respective case is diagnosed with cancer. For the quality assurance tabulations presented in this paper, all NBSBCCC biobanks were linked with national population registers to get dates of death up-to-date, and the procedure will from now on become a regular routine procedure.

Follow-up for emigration has not been considered very important because its magnitude has been rather small. However, in younger cohorts of modern Europeans emigration really has an effect. For instance, almost 4,000 women (6%) of the Icelandic maternity cohort had emigrated after serum sampling. Because the Icelandic registration system did not give the dates of emigrations, there was no information on how long the persons had been at reach of Icelandic follow-up possibilities, and all emigrated persons have to be excluded from all studies. In this type of situation, additional effort in seeking the missing dates of emigration would return several thousands of readily collected and carefully stored samples back to useful study materials and might be worth doing.

Incorrect PIDs is another source of errors on cancer risk estimates as demonstrated for cohort analyses in Fig. 4. The practice to check all PIDs against the population registries was not in routine use by all biobanks before the NBSBCCC quality assurance evaluation, but the procedure will from now on become a regular routine procedure.

The data quality requirement for the standardised incidence ratio calculation was a good way to improve accuracy of identifiers and completeness of follow-up for vital status, which is crucial in case-control studies for picking up controls that really are at risk of getting the cancer. Lack of follow-up for vital status and presence of some incorrect identifiers are likely to have caused minor errors in control selection in previous studies (Table 1): controls might have died or got cancer which was not known to the researchers. This type of errors would have reduced the risk estimates towards unity, i.e., any excess risks published so far are rather under- than overestimates of the true risk. Computerised record linkage procedures based on the unique PIDs are unambiguous (48). Therefore, linking failures do not bias cancer risk estimates.

5.4. Follow-Up for Cancer Incidence

The nationwide Nordic cancer registries have been in operation since 1950s and have virtually complete coverage for cancer incidence (49). The tabulation of observed numbers of cancer cases given in Table 4 demonstrate that the cancer registries are able to produce data also by cancer classifications based on variables other than the topography alone (such as subtypes of leukaemia, histology and stage-specific categories) and tabulations of certain precancerous lesions. These specific categories are often useful for focused hypotheses testing. The data collection procedures prepared for the NBSBCC must be made to be able to design nested case-control studies, as knowledge of the number of cases is required to estimate statistical power. The predicted numbers for the years to come, further help in deciding the optimal time to start a given study.

The numbers of cancer cases diagnosed after sample donation and accumulated to Table 4 are based on the latest linkages between the biobanks and cancer registries: 10,000–20,000 of newly diagnosed cases are missing due to the normal delay of cancer registration and about 10,000 are missing because some biobanks are not linked with cancer registry very often. In some countries, each linkage for a specific research purpose requires a new ethical permission.

5.5. Cancer Incidence Rates in Cohorts in Relation to National Cancer Incidence Rates

None of the biobank cohorts had exactly the incidence pattern of the national general population. Some of them were known to deviate from the general population by enrolment design. For instance, the maternity cohorts included only pregnant women who are known to have lower risk of cancers of breast, corpus uteri and ovary than nulliparous women. Information on parity and age at first pregnancy is available from the databases and can be taken into account when designing studies on diseases related to reproductive parameters.

Studies on samples taken during pregnancy are not necessarily generalisable to non-pregnant women. On the other hand, these samples offer a unique possibility to study the effect of in utero exposures to the health of the children (25, 28). The large Nordic Maternity cohorts are the main source of prospective cancer cases diagnosed in ages before the age of 50 (Fig. 10).

The most extreme example of an a priori known selection was the ATBC cohort which included only smoking men, who have a more than twofold excess incidence of numerous cancer types than the average male population (Fig. 7). Clinical biobanks also deviated from population averages due to the clinical diagnostics selection process, the impact of which could not have been estimated in advance.

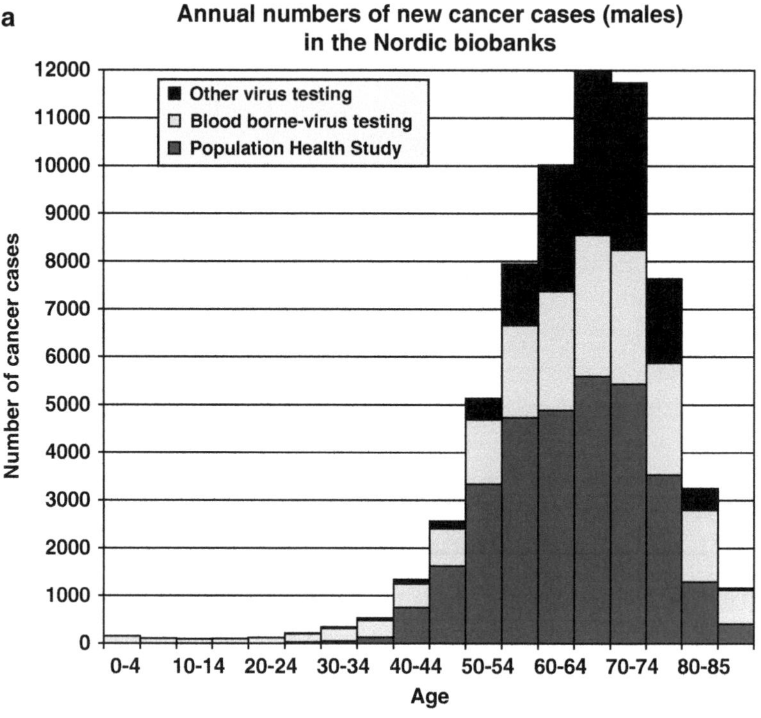

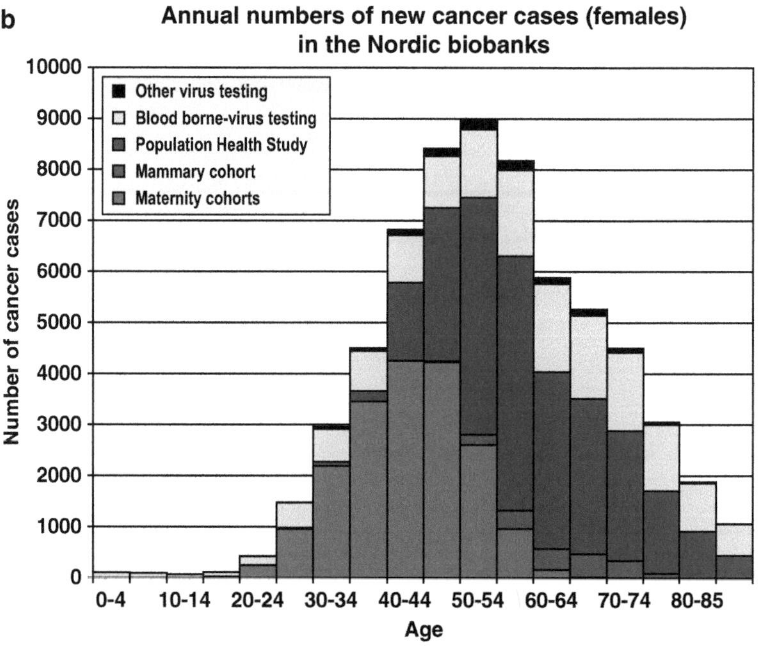

Fig. 10. Numbers of registered cancer cases among subjects in the Nordic biobanks diagnosed after serum donation, by sex, age and type of biobank. The numbers refer to cancer update status in March 2008, when coverage was complete only until 1999–2006.

The overall cancer incidence among men increases and among women decreases towards the lower socio-economic position (37, 39, 50). Typical cancers associated with low socio-economic

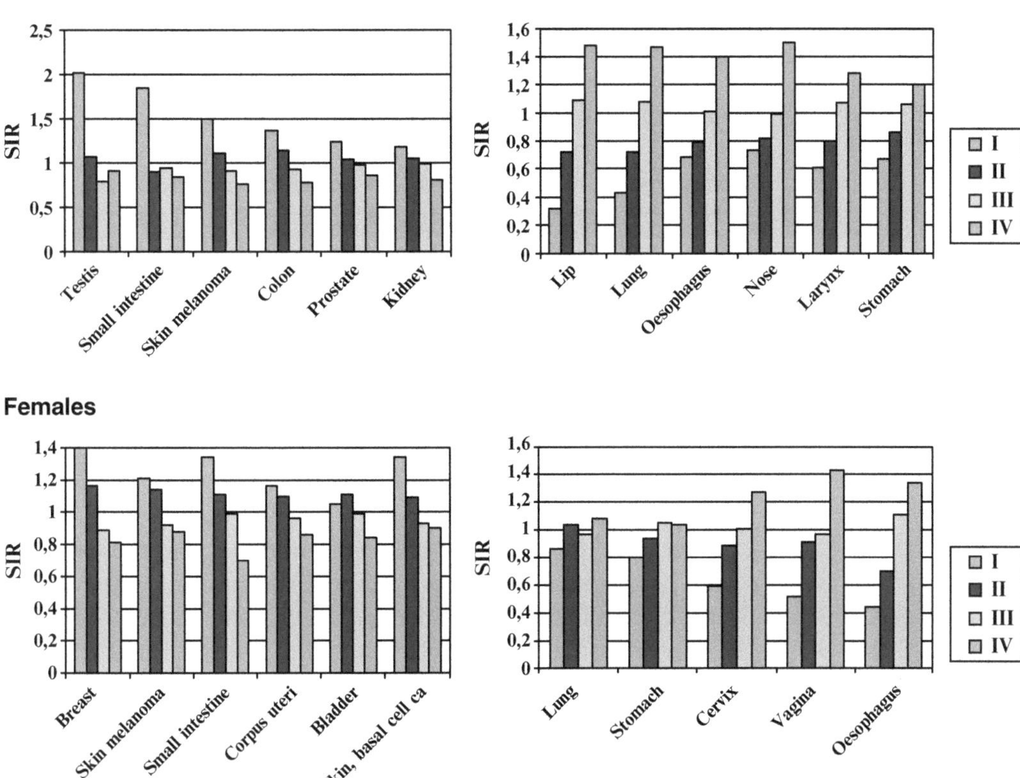

Fig. 11. Standardised incidence ratio (SIR) of selected cancers according to social class in Finland, 1971–1995. Social classes: I = managers, higher administrative; II = lower administrative/clerical; III = skilled/specialised blue-collar; IV = labourers (for details, *see* 51). Cancers of the *left* are related to high and on the *right* to low socioeconomic position.

status or educational level are cancers of the lip, oesophagus, stomach, larynx and nose, and multiple myeloma in both sexes, cancers of cervix uteri and vagina in women and lung cancer in men (Fig. 11). Cancers of the colon, breast, testis and soft tissue, and skin melanoma (especially in the trunk and limbs) are most common in high social strata. A person who knows the variation of cancer incidence over socio-economic or health habit strata can estimate from the cancer pattern whether a cohort is representative of the general population in terms of these factors.

Most biobank cohorts showed slightly lower than average cancer risk. The biobanks that were based on population registry-based invitations presumably contain a *representativity bias* related to better participation rate among health-conscious subjects. Participation rate seems not to be a especially strong indicator of this selection; e.g., the cancer pattern for the Malmö Diet and Cancer Study, with participation rate of only 40%, was rather typical for the entire population in Southern Sweden, and similar to the population samples with higher participation rates, suggesting that selection is commonly related to a never-attending non-health-conscious population.

Some serum banks contain clearly *discernible subcohorts* with obviously different cancer incidence patterns. In nested case-control studies, it is therefore recommended to consistently match for such subcohorts. Malmö Microbiology Biobank is the best example of a biobank technically collected in same place by the same organisation, but that contains clearly discernible subcohorts enrolled for different reasons. As described in this paper, these subcohorts have clearly different background cancer risks. Matching for subcohort in case-control selection is important to maintain validity in the rate ratio estimation.

The fact that symptoms related to the outcome disease of the study may increase the likelihood for sampling will increase the likelihood to encounter reverse causality biases (mix-up of cause and effect). In Malmö Microbiology Biobank, the SIRs for liver, gallbladder and pancreatic cancer were extremely high during the first year after serum sampling. Symptoms from these cancers (such as jaundice) are likely to cause testing for hepatitis viruses. While the risk for gallbladder and pancreatic cancers were not elevated after the first year after sampling, the risk for liver cancer remained elevated, presumably reflecting a true etiologic link (such as infections with hepatitis B and C viruses being causes of liver cancer). When using clinical biobanks for prospective studies, we therefore suggest excluding samples that do have shorter follow-up between sampling and diagnosis of the endpoint disease than the length of the "sick attendee effect" as demonstrated in Fig. 4.

In the cohort collected in association of mammography screenings in Northern Sweden, there was an almost twofold incidence of breast cancer during the first year after sampling. Mammography screening is indeed expected to find non-symptomatic breast cancer cases that will have a diagnosis date shortly after the screening visit. The *cohort formation principle* therefore produces an atypical collection of breast cancers in terms of timing of diagnosis and stage distribution that must be considered if these cases are used, e.g., in studies on natural latency times.

While calculation of observed and expected rates is very helpful for characterising cohorts and estimating generalisability, it should be pointed out that the main focus of biobank-based studies is more on studies of new aetiologies than on generalising to total cancer occurrence in national populations. When cases and controls are selected from the same prospectively followed cohort (strictly defined using personal identifiers and enrolment date) representing relatively homogeneous baseline population there is internal validity and possibility to make valid aetiologic inferences regardless of the degree of population representativeness of the cohort.

5.6. Recommended Study Design for Biobank-Based Studies

The nested case-control design and the case-cohort design are commonly used in molecular epidemiological studies within cohorts. They are methods of sampling from an assembled cohort study (51).

In the nested case-control design, for each case controls are randomly sampled from those eligible to be controls. In the classical case-cohort design, a simple random sample of the cohort, a subcohort, is used as a comparison group for all cases in the cohort. In the stratified case-cohort design, the subcohort is selected applying stratified random sampling.

The controls for nested case-control studies are appropriately selected applying incidence-density sampling. For each case controls are randomly sampled from all control candidates alive and free of cancer at the time of case's diagnosis. A subject is eligible to be selected as a control for more than one case and a case can serve as a control for cases with earlier date of diagnosis (52). This is called sampling with replacement. The odds ratio will then be an estimate of the incidence rate ratio in the source population between those exposed and not exposed. This holds true regardless of a disease rarity assumption, provided that the control sampling is independent of the exposure given the factors used in matching (53). In sampling with replacement, there is a small probability of multiple use of the same sample, and therefore more complicated statistical pseudo-likelihood approach is usually not necessary.

In case-cohort studies, the subcohort is selected without regard to disease status. The subcohort provides information about the person-time experience in the random sample. The case-cohort design allows direct estimation of risk ratio.

Among the advantages of the nested case-control design is that there is no need to follow up the controls beyond case's diagnosis. Effects of analytic batch, storage time and freeze–thaw cycles can be removed by matching (54). The major advantage of the case-cohort design is that the subcohort can be used for several diseases and for extended follow-up.

Among the drawbacks of the nested case-control design is that the controls are not a representative sample of the cohort and thus cannot necessarily be used as controls for future cases. Control for batch and storage effects and freeze–thaw cycles is cumbersome in case-cohort design compared to nested case-control design. Batch effect will cause bias when subsequent case series are studied in case-cohort design (55).

The case-cohort design might be preferable if the biomarkers would not suffer from storage length, batch effects and freeze–thaw cycles. The nested case-control design provides tools for dealing with such issues in principle, and is therefore more appropriate design for the NBSBCCC studies. Hence, *the optimal design is the nested case-control design applying incidence-density sampling with replacement.*

5.6.1. Matching in Nested Case-Control Design

Matching is restriction on selection of control series. The goal of matching is to balance the ratio of cases to controls within matched sets, and to make controls' distributions of the potentially confounding matching variables more like those of cases'.

The network of Nordic biobanks has attempted to use *uniform control selection algorithms* in all biobanks participating in a given joint study. For each cancer case of interest, typically one to four control donors of same sex are randomly selected among persons who were alive at the time of case's diagnosis have donated a sample around the same time as the case and were born within two years of the case's date of birth. As pointed out above, in the case of heterogeneous biobanks, matching for subcohort (e.g. Malmö Maternity Cohort and Blood-borne virus screening within Malmö Microbiology Bank) is essential. Rather exact matching for sampling date has been considered important, because different length of storage time in the bank can have profound influence on some biological markers. For some markers, seasonal variation is so large that it is also therefore important to select the control samples from same time of the year as the sample of the case. In NBSBCCC studies typically only a difference of 1–2 months in sampling date is accepted. As freezing and thawing can affect a number of biomarkers, it is also highly recommended to match on the number of freeze-thawing cycles a sample has been subjected to. The biobanks have not necessarily recorded the numbers of freeze–thaw cycles. The effect freeze–thaw cycles should be in any case prevented by sufficient aliquoting or other suitable methods, for example the straws in the EPIC study (55). Samples of the matched set are typically pipeted in random order on same panel to minimise the effects of analytic batch.

While matching is a means of reducing bias due to confounders, matching on variables intermediate in the causal pathway between exposure and disease will bias estimates (56). This is also true for matching on variables affected by exposure and disease. Therefore, matching on other variables than those mentioned above is generally not allowed in NBSBCCC studies. Matching may increase the random error, e.g., matching on a non-confounder associated with exposure but not disease reduces efficiency. Hence, matching for only a limited number of variables, typically sex, age, storage time and subcohort, is preferable.

There are certain practices in control selection that are bound to specific features of the unique sample materials. First, because most biobank databases do not include variables indicating how many times a sample has been used as a control and how much serum is left, it is often necessary to pick up one or two *extra control candidates* that will be used if the actual controls are missing or do not contain enough materials. Second, persons who have been diagnosed with other cancers have in some studies in some biobanks not been accepted as controls (to save these valuable samples), although formally they would be eligible at least until the date of cancer diagnosis of the respective case. This causes only a negligible theoretical error, because the pool of eligible controls for each case normally includes hundreds of subjects.

To assure protection of integrity and ensuring equal analyses of cases and controls, the samples must be blinded before they are sent to the analysing laboratory. After the laboratory analyses are ready, the researchers receive the code key that tells them which samples are cancer cases and which are controls.

5.7. Quality Assurance

Since 2001, a system of Quality Assurance (QA) for Good Biobanking Practice is used on a routine basis at the Medical Biobank in Umeå. Quality control and auditing by an external expert or organisation is performed at regular intervals. QA is a process that aims at measuring, evaluating and continuously re-evaluating the quality and when required, improving the quality. The QA work should have a plan of activity and schedule for work, and all employees of a biobank should be involved in the QA system. The QA system is supposed to guarantee that the biological samples, questionnaires and data have the quality that corresponds to the intended use. Database systems that document historical storage conditions, aliquoting history, number of thawings/freezings and amounts available are highly recommended.

The QA system should include procedures for how the completeness and accuracy of the attached database (non-material part of the biobank) should be maintained, kept up-to-date and how the pitfalls of selection and follow-up biases should be traced. Many biobanks have no instruments to make basic person-year at risk calculation from their cohorts or other means to control the coverage and population representativeness of their data. We suggest that calculation of cancer incidences and SIRs should be included as a basic QA practice of essential importance in biobanking QA, which should be asked for in reviews of biobank-based studies.

Many clinical biobanks do not give high priority to such check-ups of registered data that are absolutely necessary for epidemiological follow-up studies. The system described in this paper, where the data management of clinical biobanks was entrusted to cancer registries or experienced epidemiological biobanks, is likely to be essential for valid use of clinical biobanks for epidemiological studies.

6. Conclusions

The high internal validity of internal comparisons within a defined biobank cohort make prospective biobank-based study designs preferable for aetiological studies. Limited population-representativeness implies that generalisation of results to entire national populations should be made with caution. Because the described biobanks are committed to work towards joint Quality Assurance standards, including defined accessibility to external requests for samples and as the biobanks together contain a huge numbers of prospectively occurring cases of cancer, the Nordic biobank cohorts provide a solid basis for prospective studies on cancer causes and control.

In practical terms, each biobank cohort should at least once be checked using the best quality assurance methods traditionally used for many other types of study cohorts, including calculations of standardised incidence ratios and correction of any erroneous data. After that, regular simple cross-tabulations such as those described in this chapter may well be enough to keep the quality high. A real quality biobank also takes care of the future of its materials; future predictions of outcome events belong to this vision.

Acknowledgement

The publisher of the main source article of this chapter (34), Acta Oncologica (http://www.informaworld.com) kindly granted the permission to copy parts of that article. I am also pleased to acknowledge my co-authors of that article, all of whom made a great contribution to that text: Aage Andersen and Randi Gislefoss from the Cancer Registry of Norway, Institute of Population-Based Cancer Research, Oslo, Norway; Göran Berglund from Malmö Diet and Cancer Study, Lund University, Malmö, Sweden; Joakim Dillner and Anders Widell from the Department of Medical Microbiology, Lund University, University Hospital at Malmö, Sweden; Vilmundur Guđnason from the Icelandic Heart Association, Kópavogur, Iceland; Göran Hallmans from the Department of Public Health and Clinical Medicine, Nutritional Research, Umeå University, Umeå, Sweden; Egil Jellum from the Institute of Clinical Biochemistry, Rikshospitalet University Hospital, Oslo, Norway; Pekka Jousilahti, Paul Knekt, Pentti Koskela, Matti Lehtinen, Leena Tenkanen and Jarmo Virtamo from the National Institute for Health and Welfare, Helsinki and Oulu, Finland; Pentti Kyyrönen, and Tapio Luostarinen from the Finnish Cancer Registry, Institute for Statistical and Epidemiological Cancer Research, Helsinki, Finland; Per Lenner from the Cancer Registry of Northern Sweden and Department of Radiation Sciences, Umeå University Hospital, Umeå, Sweden; Arthur Löve from the Department of Medical Virology, Landspitali-University Hospital, University of Iceland, Reykjavik, Iceland; Helga Ögmundsdóttir from the Molecular and Cell Biology Laboratory, Icelandic Cancer Society, Reykjavik, Iceland; Pär Stattin from the Department of Urology, Umeå University

Hospital, Umeå, Sweden; Laufey Tryggvadóttir from the Icelandic Cancer Registry, Reykjavik, Iceland; and Göran Wadell from the Department of Virology, University of Umeå, Sweden. Several network researchers – in addition to those listed above – gave valuable comments in topics related to this study discussed in numerous joint network meetings. Special thanks are due to Tapio Luostarinen, key statistician in the biobank-related analyses in the Finnish Cancer Registry, who drafted virtually the entire text for section 5.6.; and Timo Hakulinen (Finnish Cancer Registry) and Esa Läärä (University of Oulu, Finland) who participated actively in finalisation of the paragraphs related to the principles of theoretical statistics related to this chapter. Joakim Dillner and Matti Lehtinen as present and former coordinators of the network had an essential impact in pinpointing examples of the general strengths and weaknesses experienced in practical research projects during the years of network operation.

Jan Ivar Martinsen at the Cancer Registry of Norway, Björn Tavelin in Umeå, and Håkan Krzeszowski and Henrik Månsson in Malmö made a great effort in creating the O/E calculation procedures for biobank cohorts in Norway and Sweden. Gudrí ur Olafsdóttír took care of quality control of the data related to the Icelandic serum cohorts. Anna Törner kindly offered the numbers of cancer cases from the test linkage from the Swedish Institute of Infectious Disease Control. Kari Pasanen from the University of Kuopio prepared the informative map on biobanks' coverage areas and Toni Patama from the same institute the wonderful maps on cancer incidence in the Nordic countries.

The study was supported by the Nordic Council of Ministers longitudinal epidemiology programme, by the European Union fifth framework Concerted Action on Evaluation of the Role of Infections in Cancer and by the sixth framework Network of Excellence on Cancer Control using Population-based Registries and Biobanks. The map production was supported by the Nordic Cancer Union.

References

1. Dillner, J., Knekt, P., Schiller, J.T. and Hakulinen, T.: Prospective seroepidemiological evidence that Human Papillomavirus type 16 infection is a risk factor for oesophageal squamous cell carcinoma. Br. Med. J. 311: 1346 (1995)
2. Lehtinen, M., Dillner, J., Knekt, P., Luostarinen, T., Aromaa, A., Kirnbauer, R., Koskela, P., Paavonen, J., Peto, R., Schiller, J.T. and Hakama, M.: Serologically diagnosed infection with human papillomavirus type 16 and risk for subsequent development of cervical carcinoma: nested case-control study. Br. Med. J. 312: 537–539 (1996)
3. Lehtinen, T., Luostarinen, T., Dillner, J., Aromaa, A., Hakama, M., Hakulinen, T., Knekt, P., Leinikki, P., Lumio, J., Lähdeaho, M.-L., Maatela, J., Teppo, L. and Lehtinen, M.: Serum p53 accumulation and altered antibody responses to Epstein-Barr virus proteins precede diagnosis of haemopoietic malignancies of lymphoid origin. Br. J. Haematol. 93: 104–110 (1996)
4. Dillner, J., Kallings, I., Brihmer, C., Sikström, B., Koskela, P., Lehtinen, M., Schiller, J.T.,

Sapp, M. and Mårdh, P.A.: Seropositivities to Human Papillomavirus type 16, 18, or 33 capsids and to *Chlamydia trachomatis* are markers of sexual behavior. J. Infect. Dis. 173: 1394–1398 (1996)

5. Bjørge, T., Dillner, J., Anttila, T., Engeland, A., Hakulinen, T., Jellum, E., Lehtinen, M., Luostarinen, T., Paavonen, J., Pukkala, E., Sapp, M., Schiller, J., Youngman, L. and Thoresen, S.: Prospective seroepidemiological study of role of human papillomavirus in non-cervical anogenital cancers. Br. Med. J. 315: 646–649 (1997)
6. Dillner, J., Lehtinen, M., Bjørge, T., Luostarinen, T., Youngman, L., Jellum, E., Koskela, P., Gislefoss, R.E., Hallmans, G., Paavonen, J., Sapp, M., Schiller, J.T., Hakulinen, T., Thoresen, S. and Hakama, M.: Prospective seroepidemiologic study of human papillomavirus infection as a risk factor for invasive cervical cancer. J. Natl. Cancer Inst. 89: 1293–1299 (1997)
7. Bjørge, T., Hakulinen, T., Engeland, A., Jellum, E., Koskela, P., Lehtinen, M., Luostarinen, T., Paavonen, J., Sapp, M., Schiller, J., Thoresen, S., Wang, Z., Youngman, L. and Dillner, J.: A prospective, seroepidemiological study of the role of Human Papillomavirus in esophageal cancer in Norway. Cancer Res. 57: 3989–3992 (1997)
8. Dillner, J., Knekt, P., Boman, J., Lehtinen, M., af Geijersstam, V., Sapp, M., Schiller, J., Maatela, J. and Aromaa, A.: Sero-epidemiological association between human-papillomavirus infection and risk of prostate cancer. Int. J. Cancer. 75: 564–567 (1998)
9. af Geijersstam, V., Kibur, M., Wang, Z., Koskela, P., Pukkala, E., Schiller, J., Lehtinen, M. and Dillner, J.: Stability over time of serum antibody levels to Human Papillomavirus type 16. J. Infect. Dis. 177: 1710–1714 (1998)
10. Luostarinen, T., af Geijersstam, V., Bjørge, T., Eklund, C., Hakama, M., Hakulinen, T., Jellum, E., Koskela, P., Paavonen, J., Pukkala, E., Schiller, J.T., Thoresen, S., Youngman, L.D., Dillner, J. and Lehtinen, M.: No excess risk of cervical carcinoma in women seropositive for both HPV16 and HPV6/11. Int. J. Cancer. 80: 818–822 (1999)
11. Lehtinen, M., Luostarinen, T., Youngman, L.D., Anttila, T., Dillner, J., Hakulinen, T., Koskela, P., Lenner, P. and Hallmans, G.: Low levels of serum vitamins A and E in blood and subsequent risk for cervical cancer: interaction with HPV seropositivity. Nutr. Cancer. 34: 229–234 (1999)
12. Kibur, M., af Geijerstamm, V., Pukkala, E., Koskela, P., Luostarinen, T., Paavonen, J., Schiller, J., Wang, Z., Dillner, J. and Lehtinen, M.: Attack rates of Human Papillomavirus type 16 and cervical neoplasia in primiparous women and field trial designs for HPV16 vaccination. Sex. Transm. Infect. 76: 13–17 (2000)
13. Koskela, P., Anttila, T., Bjørge, T., Brunsvig, A., Dillner, J., Hakama, M., Hakulinen, T., Jellum, E., Lehtinen, M., Lenner, P., Luostarinen, T., Pukkala, E., Saikku, P., Thoresen, S., Youngman, L. and Paavonen, J.: *Chlamydia trachomatis* infection as a risk factor for invasive cervical cancer. Int. J. Cancer. 85: 35–39 (2000)
14. Sigstad, E., Lie, A.K., Luostarinen, T., Dillner, J., Jellum, E., Lehtinen, M., Thoresen, S. and Abeler, V.: A prospective study of the relationship between prediagnostic Human Papillomavirus seropositivity and HPV DNA in subsequent cervical carcinomas. Br. J. Cancer. 87: 175–180 (2002)
15. Mork, J., Lie, A. K., Glattre, E., Hallmans, G., Jellum, E., Koskela, P., Møller, B., Pukkala, E., Schiller, J.T., Youngman, L., Lehtinen, M. and Dillner, J.: Human Papillomavirus infection as a risk factor for squamous-cell carcinoma of the head and neck. N. Engl. J. Med. 344: 1125–1131 (2001)
16. Anttila, T., Saikku, P., Koskela, P., Bloigu, A., Dillner, J., Ikäheimo, I., Jellum, E., Lehtinen, M., Lenner, P., Hakulinen, T., Närvänen, A., Pukkala, E., Thoresen, S., Youngman, L. and Paavonen, J.: Serotypes of *Chlamydia trachomatis* and risk for development of cervical squamous cell carcinoma. JAMA. 285: 47–51 (2001)
17. Stattin, P., Adlercreutz, H., Tenkanen, L., Jellum, E., Lumme, S., Hallmans, G., Harvei, S., Teppo, L., Stumpf, K., Luostarinen, T., Lehtinen, M., Dillner, J. and Hakama, M.: Circulating enterolactone and prostate cancer risk: a Nordic nested case-control study. Int. J. Cancer. 99: 124–129 (2002)
18. Bjørge, T., Engeland, A., Luostarinen, T., Mork, J., Gislefoss, R.E., Jellum, E., Koskela, P., Lehtinen, M., Pukkala, E., Thoresen, S.Ø. and Dillner, J.: Human Papillomavirus infection as a risk factor for anal and perianal skin cancer in a prospective study. Br. J. Cancer. 87: 61–64 (2002)
19. Stattin, P., Lumme, S., Tenkanen, L., Alfthan, H., Jellum, E., Hallmans, G., Thoresen, S., Hakulinen, T., Luostarinen, T., Lehtinen, M., Dillner, J., Stenman, U-H. and Hakama, M.: High levels of circulating testosterone are not associated with in creased prostate cancer risk: a pooled prospective study. Int. J. Cancer. 108: 418–424 (2004)
20. Lehtinen, M., Koskela, P., Jellum, E., Bloigu, A., Anttila, T., Hallmans, G., Luukkaala, T., Thoresen, S., Youngman, L., Dillner, J. and Hakama, M.: Herpes simplex virus and risk of

cervical cancer: a longitudinal nested case-control study in the Nordic countries. Am. J. Epidemiol. 156: 687–692 (2002)

21. Lehtinen, M., Pawlita, M., Zumbach, K., Lie, K., Hakama, M., Jellum, E., Koskela, P., Luostarinen, T., Paavonen, J., Pukkala, E., Sigstad, E., Thoresen, S. and Dillner, J.: Evaluation of antibody response to human papillomavirus early proteins in women in whom cervical cancer developed 1 to 20 years later. Am. J. Obstet. Gynecol. 188: 49–55 (2003)
22. Kapeu, A.S., Luostarinen, T., Jellum, E., Dillner, J., Hakama, M., Koskela, P., Lenner, P., Löve, A., Mahlamaki, E., Thoresen, S., Tryggvadóttir, L., Wadell, G., Youngman, L., Lehtinen, M.: Is smoking an independent risk factor for invasive cervical cancer? A nested case-control study within Nordic biobanks. Am J Epidemiol. 169: 480–8 (2009)
23. Paavonen, J., Karunakaran, K.P., Noguchi, Y., Anttila, T., Bloigu, A., Dillner, J., Hallmans, G., Hakulinen, T., Jellum, E., Koskela, P., Lehtinen, M., Thoresen, S., Lam, H., Shen, C. and Brunham, R.C.: Serum antibody response to the heat shock protein 60 of *Chlamydia trachomatis* in women with developing cervical cancer. Am. J. Obstet. Gynecol. 189: 1287–1292 (2003)
24. Lehtinen, M., Koskela, P., Ögmundsdóttir, H.M., Bloigu, A., Dillner, J., Gudnadóttir, M., Hakulinen T., Kjartansdóttir, A., Kvarnung, M., Pukkala, E., Tulinius, H. and Lehtinen, T.: Maternal Herpesvirus infections and risk of acute lymphoblastic leukemia in the offspring. Am. J. Epidemiol. 158: 207–213 (2003)
25. Tuohimaa, P., Tenkanen, L., Ahonen, M., Lumme, S., Jellum, E., Hallmans, G., Stattin, P., Harvei, S., Hakulinen, T., Luostarinen, T., Dillner, J., Lehtinen, M. and Hakama, M.: Both high and low levels of blood vitamin D are associated with a higher prostate cancer risk: a longitudinal, nested case-control study in the Nordic countries. Int. J. Cancer. 108: 104–108 (2004)
26. Luostarinen, T., Lehtinen, M., Bjørge, T., Abeler, V., Hakama, M., Hallmans, G., Jellum, E., Koskela, P., Lenner, P., Lie, A. K., Paavonen, J., Pukkala, E., Saikku, P., Sigstad, E., Thoresen, S., Youngman, L.D., Dillner, J. and Hakulinen, T.: Joint effects of different Human Papillomaviruses and *Chlamydia trachomatis* infections on risk of squamous cell carcinoma of the cervix uteri. Eur. J. Cancer. 40: 1058–1065 (2004)
27. Lehtinen, M., Ögmundsdóttir, H.M., Bloigu, A., Gudnadóttir, M., Hakulinen, T., Hemminki, E., Kjartansdóttir, A., Paavonen, J., Pukkala, E., Tulinius, H., Lehtinen, T. and Koskela, P.: Associations between three types of maternal bacterial infection and risk of leukemia in the offspring. Am. J. Epidemiol. 162: 662–667 (2005)
28. Anttila, T., Tenkanen, L., Lumme, S., Leinonen, M., Gislefoss, R.E., Hallmans, G., Thoresen, S., Hakulinen, T., Luostarinen, T., Stattin, P., Saikku, P., Dillner, J., Lehtinen, M. and Hakama, M.: Chlamydial antibodies and risk of prostate cancer. Cancer Epidemiol. Biomarkers Prev. 14: 385–389 (2005)
29. Stolt, A., Kjellin, M., Sasnauskas, K., Luostarinen, T., Koskela, P., Lehtinen, M. and Dillner, J.: Maternal Human Polyomavirus infection and risk of neuroblastoma in the child. Int. J. Cancer. 113: 393–396 (2005)
30. Hakama, M., Luostarinen, T., Hallmans, G., Jellum, E., Koskela, P., Lehtinen, M., Thoresen, S., Youngman, L. and Hakulinen, T.: Joint effect on HPV16 with *Chlamydia trachomatis* and smoking on risk of cervical cancer: antagonism or misclassification (Nordic countries). Cancer Causes Control. 11: 783–790 (2000)
31. Tedeschi, R., Bidoli, E., Agren, Å., Wadell, G., De Paoli, P. and Dillner, J.: Epidemiology of Kaposi's sarcoma herpesvirus (HHV8) in Västerbotten county, Sweden. J. Med. Virol. 78: 372–378 (2006)
32. Tedeschi, R., Luostarinen, T., De Paoli, P. Gislefoss, R.E., Tenkanen, L., Virtamo, J., Koskela, P., Hallmans, G., Lehtinen, M. and Dillner, J.: Joint Nordic prospective study on human herpesvirus 8 and multiple myeloma risk. Br. J. Cancer 93: 834–837 (2005)
33. Korodi, Z., Dillner, J., Jellum, E., Lumme, S., Hallmans, G., Thoresen, S., Hakulinen, T., Stattin, P., Luostarinen, T., Lehtinen, M. and Hakama, M.: Human papillomavirus 16, 18, and 33 infections and risk of prostate cancer: a Nordic nested case-control study. Cancer Epidemiol. Biomarkers Prev. 14: 2952–2955 (2005)
34. Pukkala, E., Andersen, A., Berglund, G., Gislefoss, R., Gudnason, V., Hallmans, G., Jellum, E., Jousilahti, P., Knekt, P., Koskela, P., Kyyronen, P. P., Lenner, P., Luostarinen, T., Love, A., Ogmundsdottir, H., Stattin, P., Tenkanen, L., Tryggvadottir, L., Virtamo, J., Wadell, G., Widell, A., Lehtinen, M. and Dillner, J.: Nordic biological specimen banks as basis for studies of cancer causes and control – more than 2 million sample donors, 25 million person years and 100,000 prospective cancers. Acta Oncol. 46: 286–307 (2007)
35. ATBC (The Alpha-Tocopherol B-CCPSG): The effect of vitamin E and beta carotene on the incidence of lung cancer and other cancers in male smokers. N. Engl. J. Med. 330: 1029–1035 (1994)

36. Meurman, L.O., Pukkala, E. and Hakama, M.: Incidence of cancer among anthophyllite asbestos miners in Finland. Occup. Environ. Med. 51: 421–425 (1994)
37. Pukkala, E. (1995) Cancer risk by social class and occupation. A survey of 109,000 cancer cases among Finns of working age. Contributions to Epidemiology and Biostatistics. Basel: Karger; 7
38. Liu, Q., Wuu, J., Lambe, M., Hsieh, S.F., Ekbom, A. and Hsieh, C.C.: Transient increase in breast cancer risk after giving birth: postpartum period with the highest risk (Sweden). Cancer Causes Control. 13: 4, 299–305 (2002)
39. Andersen, A., Barlow, L., Engeland, A., Kjaerheim, K., Lynge, E. and Pukkala, E.: Work-related cancer in the Nordic countries. Scand. J. Work Environ. Health. 25, Suppl. 2 (1999)
40. Malila, N., Virtanen, M.J., Jarmo Virtamo, J., Albanes, D. and Pukkala, E.: Cancer incidence in a cohort of Finnish male smokers. Eur. J. Cancer Prev. 15: 103–107
41. Dreyer, L., Winther, J.F., Pukkala, E. and Andersen, A.: Tobacco smoking. APMIS, 105, Suppl. 76: 9–47 (1997)
42. Lehtinen, M., Leminen, A., Kuoppala, T., Tiikkainen, M., Lehtinen, T., Lehtovirta, P., Punnonen, R., Vesterinen, E. and Paavonen, J.: Pre- and post-treatment serum antibody responses to HPV16 E2 and HSV-2 ICP8 proteins in women with cervical carcinoma. J. Med. Virol. 37: 180–186 (1992)
43. Lehtinen, M., Hakama, M., Knekt, P., Heinonen, P.K., Lehtinen, T., Paavonen, J., Teppo, L. and Leinikki, P.: Serum antibodies to the HSV-2 specified major DNA-binding protein are elevated before the diagnosis of cervical cancer. J. Med. Virol. 27: 131–136 (1989)
44. Wallin, K.L., Wiklund, F., Luostarinen, T., Hallmans, G., Anttila, T., Koskela, P., Lehtinen, M., Paavonen, J., Stendahl, U., Wadell, G. and Dillner, J.: *Chlamydia trachomatis* infection: a risk factor in cervical cancer development – a population based prospective study. Int. J. Cancer. 101: 371–374 (2002)
45. Clarce, R., Shipley, M., Lewington, S., Youngman, L., Collins, R., Marmot, M. and Peto, R. Underestimation of risk associations due to regressios dilution in long-term follow-up of prospective studies. Am. J. Epidemiol. 150: 341–353 (1999)
46. Parish, S., Collins, R., Peto, R., Youngman, L., Barton, J., Jayne, K., Clarke, R., Appleby, P., Lyon, V., Cederholm-Williams, S., Marshall, J. and Sleight, P.: Cigarette smoking, tar yields, and non-fatal myocardial infarction: 14,000 cases and 32,000 controls in the United Kingdom. The International Studies of Infarct Survival (ISIS) Collaborators. Br. Med. J. 311: 471–477 (1995)
47. Jellum, E., Andersen, A., Lund-Larsen, P., Theodorsen, L. and Orjasaeter, H.: Experiences of the Janus Serum Bank in Norway. Environ. Health Perspect. 103, Suppl. 3: 85–88 (1995)
48. Pukkala, E. Use of record linkage in small-area studies. In: Elliot, P., Guzick, J., English, D., Stern, R., eds. (1992) *Geographical and environmental epidemiology*. Oxford: Oxford University Press: 125–131
49. Teppo, L., Pukkala, E. and Lehtonen, M.: Data quality and quality control of a population-based cancer registry. Acta Oncol. 33: 365–369 (1994)
50. Pukkala, E. and Weiderpass, E.: Time trends in socio-economic differences in incidence rates of cancers of the breast and female genital organs (Finland 1971–1995). Int. J. Cancer. 81: 56–61 (1999)
51. Langholz, B. Entries: case-cohort study and case-control sudy, nested. In: Armitage, P., Colton, T., eds. (1999) *Encyclopedia of biostatistics*. Chichester: John Wiley & Sons: 497–503 & 514–519
52. Greenland, R. and Thomas, D.C.: On the need for the rare disease assumption in case-control studies. Am. J. Epidemiol. 116: 547–553 (1982)
53. Rothman, K.J. and Greenland, S. (1998) *Modern Epidemiology 2nd ed.* Philadelphia: Lippincott-Raven: 95–96
54. Rundle, A.G., Vineis, P. and Ahsan, H.: Design options for molecular epidemiology research within cohort studies. Cancer Epidemiol. Biomark. Prev. 14: 1899–1907 (2005)
55. Riboli, E., Hunt, K.J., Slimani, N., Ferrari, P., Norat, T., Fahey, M., et al.: European Prospective Investigation into Cancer and Nutrition (EPIC): study populations and data collection. Public Health Nutr. 5: 1113–1124 (2002)
56. Wacholder, S., Silverman, D.T., McLaughlin, J.K. and Mandel, J.S.: Selection of controls in case-control studies. III. Design options. Am. J. Epidemiol. 135: 1042–1050 (1992)
57. Pukkala, E., Söderman, B., Okeanov, A., Storm, H., Rahu, M., Hakulinen, T., Becker, N., Stabenow, R., Bjarnadottir, K., Stengrevics, A., Gurevicius, R., Glattre, E., Zatonski, W., Men, T. and Barlow, L. (2001) Cancer atlas of Northern Europe. *Cancer Society of Finland* Publication No. 62, Helsinki (69 pages + CD)

Chapter 4

Biobanks Collected for Routine Healthcare Purposes: Build-Up and Use for Epidemiologic Research

Joakim Dillner and Kristin Andersson

Abstract

The routine health services collect large amount of samples for biobanking, particularly in clinical laboratory medicine, mainly for clinical diagnostic purposes. These samples provide a large-scale and clinically relevant biobanking infrastructure that can be used for research if these conditions apply. There must be a system for database management that can obtain data on clinical endpoints, vital status, and additional required information via registry linkages. There must be an appropriate ethical system for handling consent for research use. There should be an active effort to optimize the usefulness of clinical biobanks also for research use. Major steps in this direction include measures to stop the ongoing discarding of old samples, reformatting to minimize pick-up times, external quality assurance and formal accreditation of biobanks, building of a dedicated high-quality database that is regularly used for registry linkages, and considerations on whether usefulness and accessibility for research can be optimized by extended saving or pre-treatment of samples. Systematic clinical biobanking could become a major asset for clinical research and public health if biobanking is considered as a routine part of everyday clinical practice, and the science of biobanking is considered an essential part of the science of laboratory medicine.

Key words: Healthcare biobank, Registry linkage, Quality assurance

1. Introduction

Biobanks can be classified into two major groups based on their purpose of storage: those with samples collected mainly for research purposes and those with samples collected within the healthcare system mainly for healthcare purposes. However, medical research is an important and commonly integrated part of the healthcare services, and usefulness for medical research is commonly specified as an important secondary purpose for the collection of health services-based biobanks. When research is not the main purpose, there are important practical and ethical issues

Joakim Dillner (ed.), *Methods in Biobanking*, Methods in Molecular Biology, vol. 675,
DOI 10.1007/978-1-59745-423-0_4, © Springer Science+Business Media, LLC 2011

that have to be considered both in the buildup and use of the biobank.

In this chapter, we will review the major healthcare-based biobanks, their usefulness for research purposes, issues that can promote usefulness, and how to create an infrastructure suitable for governing and improving the usefulness of healthcare-based biobanks, including ethical considerations.

2. Characteristics of Clinical Biobanks in Relation to Research Use

A biobank created within the healthcare system, with the main aim of being used for healthcare purposes, usually only has limited information about the sample donor attached to the sample. Typically, the information may be restricted to the information provided on the referral sheet, such as the age, sex, name and personal identifiers, and sometimes a brief summary of the clinical condition that resulted in the request for analysis. For usefulness in research, it must be possible to link the biobank database to comprehensive health data registries, such as cancer registries, death registries, birth registries, and so on, where more information on the subjects can be retrieved (e.g., clinical phenotypes and exposure information) and where endpoints (such as diagnosis of a certain disease or death from a certain cause) can be obtained for case ascertainment.

As for all biobanks, it is important that the attached database is of good quality and is well documented, as inadequate information on identities and/or failure to update the vital status will make data obtained from registry linkages inaccurate (see Chapter 5, by Pukkala for a more detailed description).

Older biobanks that have stored samples for very long time spans are especially valuable, because of the possibility to do a long follow-up of the subjects. This is essential for several reasons: (1) Statistical power. With more follow-up time, more cases of disease will have occurred in the biobank, which may make it possible to perform epidemiologic studies with adequate statistical power on rare diseases also. (2) Clinical outcome. For all studies investigating the ability to predict the clinical course and whether the treatment given had any effect, data on major clinical outcomes such as, e.g., 10-year survival rates are essential. (3) Life-course studies and window of susceptibility. There is increasing evidence suggesting that the causes of major chronic diseases such as cancer have occurred many decades before the diagnosis, sometimes even in utero. It may not be possible to study the causes of the disease unless samples are saved from the time point when the causal exposures occurred.

However, although old samples may open new possibilities for use, they also have drawbacks. The longer the samples have been stored, the higher the demands for controlling the quality of the samples. Historical data on storage and handling may be hard to find, as appropriate Quality Assurance (QA) programs have typically not been used for more than a few decades. Different analytes in the specimens may have different degrees of sensitivity for conditions of storage and handling. Matching for storage time and handling (e.g., number of freeze-thaw cycles) is recommended for all analytes that are not perfectly stable, but such matching is only a partial remedy as rates of decay are typically not uniform, but vary for each individual sample.

Biobanks created for healthcare purposes collect samples for different reasons, and it is important to be aware of these reasons. Population-based screening programs are particularly interesting from an epidemiologic point of view, as these target the entire population and usually have very high attendance rates, making such biobanks optimally generalizable to the general populations. Samples left over from diagnostic testing are generalizable in particular to the real-life situation in clinical laboratory medicine, and such biobanks are optimally relevant for research on and development of new diagnostics in clinical laboratory medicine. The indication why the samples are sent for diagnostics is important to be aware of and to consider when discussing generalizabilty of biobanks based on diagnostic tested samples.

The total amount of left-over samples that are stored in clinical laboratory medicine is immense. In Sweden alone, we have estimated that about two to three million samples are stored every year, out of a population of about nine million inhabitants. Thus, we estimate that most subjects already have multiple samples stored in various health services-based biobanks. Although it is commonly perceived that lack of sufficiently high numbers of samples is the major bottleneck of the research in molecular medicine today, very large amounts of samples do indeed exist and bottlenecks are more related to the fact that the clinical biobanks have not been designed to work as a scientific infrastructure. A major purpose of this chapter is to describe the tasks that would be required to turn the health services-based biobanks into a more effective scientific infrastructure.

2.1. The Purposes of Health Services-Based Biobanks

A discussion on the scientific use of health services-based biobanks needs to start with an understanding of the main clinical purposes of why health services-based biobanks have been established at all. Major purposes include the following:

1. *Diagnostics that require comparison of new samples with previous samples from the same person.* The basic idea is that the existence of a change compared to the previous situation is

more informative than the result of analysis of a single sample. Examples include assessment if there is a treatment response in cancer, if a new tumor in a patient with previous cancer is a new cancer or a relapse with the previous disease (crucial information for design of treatment), and the serologic diagnosis of infectious diseases that commonly rely on sero-conversion (change in antibody status) to decide if antibody positivity reflects past or present infection.

2. *Enabling request for additional diagnostic analyses.* Particularly in pediatrics (when it may be difficult to obtain a new blood sample) or in diagnostics based on biopsies from non-renewable tissue, saving of left-over samples may be critical for diagnosis. For example, the clinical course of a disease may necessitate additional laboratory analyses – and if it is difficult or impossible to obtain new specimens, the clinical biobank is essential.
3. *Choice of treatment.* There are several examples where cancer treatment may depend on the availability of original tumor, e.g., treatment that is dependent on the expression of Her2neu in breast cancer. As there is a rapid development of new treatments, patients who do not have the original tumor saved may not be eligible for receiving such treatments. Many comprehensive cancer centers today consider comprehensive biobanking of cancer tissue as a state-of-the-art requirement in clinical oncology.

   Another example is the design of antiviral treatments (e.g., for hepatitis C) that may depend on virus kinetics of serial samples.
4. *Responsibility and documentation requirements.* If there is doubt about whether an analysis was accurately performed, a repeat analysis of the sample should be possible. This is important for the legal safety of both the patient and the laboratory.
5. *Quality control.* Most quality control systems require that a repeat analysis of the same sample should give the same results, even if performed days, weeks, or months later.
6. *Clinical development.* Clinical development work can mean many different tasks for the continuous improvement of the way of working. The most common use of the clinical biobank in clinical development work is in the assessment that new diagnostic tests perform as well as the old ones. Typically, comprehensive series of samples submitted for clinical diagnosis for a particular test are retrieved and re-assayed with a new assay that is intended to have higher accuracy and/or lower cost than the previously used test. The clinical laboratory medicine departments that have focused on the systematic buildup of well-characterized sample collection of

such diagnostic request series are attractive partners for collaborative research projects with the diagnostics development industry.

7. *Infectious disease protection*. There are many examples where biobanks have assisted in the location of the source of outbreaks of serious epidemics. For example, with the analysis of the evolution of the virus nucleotide sequence in serial samples from an infected person (quasi-species development), the time-point of infection can be ascertained – sometimes with an accuracy even pinpointing the exact date of infection. Another example is the comparison of nucleotide sequences from samples taken during prolonged time intervals, where it may not be obvious that they originate from the same source (e.g., food-borne infections in frozen foodstuffs).
8. *Education*. Particularly in clinical pathology laboratories, it is essential that the new personnel in the diagnostic laboratories can be trained in diagnostic skills using old samples.

The many and vital clinical uses of clinical biobanks have important implications for ethical and organizational issues, as elaborated elsewhere. For example, the fact that discarding of the sample may result in health hazards for the patient resulted in most guidelines indicating that if it is unknown if the patient has consented to storage or not, the sample should not be discarded until the consent status of the patient becomes known. Another major implication is that the release of a sample for research purposes requires an assessment of whether this is possible without jeopardizing the medical needs of the patients and the needs of the diagnostics laboratory. Obviously, the situation may be different for different types of samples and may vary between laboratories, necessitating that this assessment must be made by an expert with medical responsibility for the diagnostics.

2.2. Some Examples of Healthcare-Based Biobanks

2.2.1. Microbiology Biobanks

Microbiology Biobanks comprise left-over clinical samples sent to a clinical microbiology laboratory for routine diagnostics. A majority of these samples are serum samples that have been submitted for diagnosis of blood-borne viral infections (for example, hepatitis viruses and HIV) (1). Most microbiological laboratories store these samples for 6 years or more; nowadays some store indefinitely. The Southern Sweden Microbiology Biobank is an example of a biobank that resulted as a consequence of a clinical microbiological laboratory stopping the discarding of the oldest stored samples. About 1.3 million samples from 550,000 individuals (about 60% of the entire catchment area population) are stored today.

2.2.2. Maternity Cohorts: Maternity Care Serologic Screening Biobanks

Many countries target pregnant women with nation-wide screening programs for rubella immunity. Many countries also offer screening for hepatitis B, HIV, and syphilis. These serum samples are typically taken during the first trimester, during weeks 12–14 of pregnancy. Assessment of primary infection with rubella is a good example of the clinical usefulness of a clinical biobank, as a positive result for rubella IgM is firmly diagnostic of a primary rubella infection only if it represents a seroconversion (i.e., it was not present in a previous, stored sample).

2.2.2.1. Finnish Maternity Cohort

The Finnish Maternity Cohort is a biobank based on samples from serology screening of pregnant women in Finland. Since 1983, 98% of all pregnant women attending maternity clinics in Finland have participated in serologic screening of syphilis, HIV, hepatitis B, and rubella immunity during the first trimester, and donated serum samples to the Finnish Maternity Cohort, which is stored by the National Public Health Institute in Finland. Altogether 750,000 women are included in the cohort and many of them have donated samples from more than one pregnancy. Today about 1.5 million serum samples are stored at -25°C. This is a biobank that has been widely used for research.

2.2.2.2. Icelandic Maternity Cohort

Also, Iceland has a nationwide maternity cohort biobank, where serum samples form all pregnant women in the country has been stored since 1980. About 50,000 women are included in the biobank, with samples originating from over 90,000 pregnancies.

2.2.2.3. Maternity Cohorts in Sweden

The maternity care screening program is the same for the entire Sweden, but since analysis of the samples is not nationally centralized, each laboratory decides by itself if the samples should be discarded after the mandatory 10-year storage time. In the Southern and Northern parts of Sweden, the samples are stored indefinitely. The Swedish Institute for Infectious Disease Control has launched an initiative to save valuable Maternity Cohort samples that otherwise would be discarded, but this has not yet reached a nationwide coverage.

The Southern Sweden Maternity Cohort is a part of the Southern Sweden Microbiology Biobank (see Subheading 2.2.1). The biobank contains serum samples from 130,000 women and 190,000 pregnancies. From 2005 onwards, both the serum and the clotted part of the blood sample are stored. The blood clots are useful as a good source of high-quality DNA.

The Northern Sweden Maternity Cohort has stored serum samples from maternity screening since 1975 onwards. It contains samples from over 86,000 women and from over 118,000 pregnancies in the three northernmost counties of Sweden.

2.2.3. Clinical Pathology Biobanks

The clinical pathology biobanks comprise paraffin blocks of surgical and autopsy tissue samples and corresponding histologic slides as well as cytologic material consisting of slides of vaginal smears, fine needle aspiration biopsies, and exfoliative cytologic material.

Most clinical pathology biobanks have been in operation for almost a century, and many still store samples that are more than 50 years old.

2.2.3.1. Tissue Array Biobanks

Retrieval and sectioning of these blocks are both laborious and may jeopardize the clinical usefulness of the specimen. The requirement for systematic tissue array construction for an effective accessibility of the clinical biobanks with formalin-fixed paraffin-embedded material (FFPE) is described in the accompanying Chapter 22.

2.2.3.2. Frozen Tissue Biobanks

Many comprehensive cancer centers have organized separate biobanking of fresh frozen tissue, in order to obtain maximal usefulness for a variety of analysis technologies targeting DNA, RNA, and proteins, as elaborated in the accompanying Chapter 16. The ground-breaking work of the group behind the European Union consortium TUBAFROST (tumor biobanking of fresh-frozen tissue) is a good example of how the production of scientific evidence-based standard operating procedures has had a major impact on the usefulness and international harmonization of an important biobanking resource.

2.2.3.3. The Clinical Cytology Biobanks

In most parts of the world, where organized cervical screening is being offered, the cervical smears are stored for at least 10 years and sometimes indefinitely. The ongoing switch to liquid-based cytology, where samples are taken in a methanol-based fixative instead of being smeared on slides, has opened a new possibility for generation of a useful clinical biobank, as these samples contain high amounts of high-quality DNA, RNA, and protein (see Chapter 15).

2.2.3.4. Mammography Biobanking

This approach is mentioned here, as it (similar to the cervical cytology and maternity cohort biobanks) also exploits the existence of a well-attended population-based routine screening program for generation of a biobank. Mammography biobanking has no clinical use today and is best classified as a research biobank. The Northern Sweden Mammography Cohort (part of the Northern Sweden Health and Disease Study Cohort) has been launched because of the belief that the ongoing major efforts toward discovery of new biomarkers for early detection will most likely result in discovery of breast cancer biomarkers whose usefulness for screening will need to be evaluated using samples from a real-life breast cancer screening program with comprehensive follow-up data on long-term incidence and mortality from the disease. The biobank has been built using blood samples and questionnaires from women aged 50–69 years attending the population-based mammography screening every second year. Samples have been stored since 1995 and include about 48,000 samples from 27,500 women.

3. An Infrastructure Governing the Appropriate Use of Health Services-Based Biobanks

Samples stored in a healthcare biobank are originally stored for the benefit of the patient and the diagnostics laboratory. Samples have been stored for a century, but informed consent from sample donors also for diagnostics of left-over samples is either a rather new practice or not yet used even today. As a consequence, a majority of the samples in healthcare biobanks do not have an expressed consent from the sample donor – since this was not an issue at the time of the sampling. Should samples collected without informed consent be barred from use? It can be argued that it would be a waste of resources important for health to not use existing collections for the benefit of both the patient and the mankind. Integrity of the individuals who donated the samples must be protected – an important issue for both scientists and ethical committees. Researchers using biobank samples are not interested in individual results, but in the overall results in a population and health perspective. When using samples from a biobank that was collected without obtaining consent, the ethical committee has the responsibility to protect the donor's interest and integrity (see Chapters 1 and 2). Requiring that the researchers should go back and ask every donor for consent is not realistic, and the older the biobank is, the more unrealistic it gets, since many of the donors would have moved, died, or become ill since they donated the sample.

A solution to this problem that was proposed in the Netherlands is the use of the Opt-out form of consent. The donors should be given the opportunity to contact the biobank at any time point and – without giving a reason – withdraw their participation in the biobank. To respect and follow the expressed wish of the donors is most important for any researcher and an absolute requirement in the building of the public trust in the biobanks.

When creating new healthcare-based biobanks, it is important to obtain consent from the sample donors, not only to both respect autonomy and build the public trust, but also to ensure that the samples can actually be used for health-related research. Legislation in different countries have different demands, but since it is not possible to know today what the samples might be used for tomorrow, the need for a broad consent is obvious if we wish to make biobanks useful for health-related research (2).

A solution to these issues was proposed by the Swedish National Biobanking Program in 2002 (http://www.biobanks.se). It involves the following parts: (1) standardized patient information asking for a broad consent for biobanking and future medical research (with approval of each new project by an ethical committee). Today, one single, standardized, broad consent is collected throughout the healthcare system in Sweden (about 2–3 million consents/year). (2) Referral sheet-based documentation. The physician ordering the diagnostic tests marks the consent status of the patient on the referral sheet. (3) A comprehensive Opt-out system where any patient wishing to withdraw consent can easily do so – for all samples stored in all healthcare-based biobanks. Today, this is managed by regional biobank registries; it is hoped that these will in the near future be replaced by a national biobank registry. A national opt-out registry for biobanking has been in operation since 2004 in Denmark and has been very popular as an effective solution to the issue.

4. Creation of an Infrastructure with Improved Usefulness of a Prospectively Collected Healthcare Biobank

Transforming a clinical health services-based biobank into a dedicated and effective research infrastructure requires an understanding of both the needs for efficiency and high-quality registry linkages in epidemiologic research, and the strengths and weaknesses of clinical laboratory medicine. The transformation work of the Southern Sweden Microbiology biobank is taken as an example.

1. *Stop discarding old samples.* As for many clinical microbiology laboratories, the samples were stored for 10 years – but then thrown out! This was stopped in 1999 and the biobank has grown rapidly in importance as a research infrastructure ever since. The usefulness of the samples for the clinical diagnostics of the patient who donated the sample decreases with time. By contrast, the research usefulness increases with longer follow-up times. It is, therefore, entirely logical that the usefulness for research should be upgraded from a secondary reason for storage to a prime reason when the samples have been stored for some years.
2. *Reformat the storage to minimize pick-up time.* Before 1999, the biobank consisted of a series of –20°C household freezers, with samples stored in a complicated and not logical order. The average pick-up time for one sample could be 5 min, which is acceptable for clinical diagnostics but unacceptable for epidemiologic studies where typically many thousands of samples are analyzed. The samples were transferred to ordered, standardized box/crate system in a dedicated freezing hall. About 650,000 samples were manually transferred by available clinical diagnostics department personnel in "spare moments." This was completed after 3 years and the pick-up time per sample is now about 0.5 min. Today, there are straightforward technical possibilities to reformat existing

biobanks into completely automated retrieval systems, and work to create efficient pick-up times; therefore, existing biobanks should consider this option.

3. *Implement a formal Quality Assurance system – with all procedures and handling of samples traceable and standardized.* This is not a problem for biobanks in clinical laboratory medicine, as they are usually very experienced in quality assurance. The Southern Sweden Microbiology Biobank is part of a clinical laboratory that is formally accredited according to ISO 15189, and the accreditation was extended to include the biobank in 2004. The availability of an experienced Quality Assurance Officer dedicated to quality work is essential. In our instance, the existence of an Internet-based electronic Quality Manual System was useful. In the future, it is highly likely that the professional organizations in clinical laboratory medicine will be organizing external quality assurance systems for biobanks.
4. *Building a high-quality biobank database.* In 1999, the microbiological laboratory used a routine clinical laboratory LIMS (laboratory information management system) for storing data on biobank samples, with the result that the biobank was essentially useless for the epidemiologic sciences.

   As an example, linkage of the biobank database with the population registry found no less than 26,983 subject identities that were incorrect (linkages with population registry finding non-existent subjects). Subjects with incorrect personal identifiers can result in severe biases as registry linkages to assess vital status and health will find that these subjects never die and never get sick. Similarly, there were no less than 1956 subjects with change of identity (sex change, witness protection, etc.). To make the biobank useful, it is necessary to build a separate high-quality database, checked to have only real identifiers and a relational function in case of identity change.

   Data on follow-up status (vital status (death) and emigration) should be added to the database regularly. If the assumption is made that subjects who did not develop disease did remain healthy – when they may have been non-diseased because of being dead – the studies are severely biased, as elaborated in the accompanying Chapter 5.

   If it is allowed to also add family data, clinical data, and exposure data to the database by registry linkages (e.g., with multigeneration registry/medical birth registry/population registry), the biobank will be immensely useful. Such linkages are virtually always allowed for a specific study with a specific ethical permission, whereas the practice of giving permissions for linkages for more general, planning purposes by the biobank itself has in our experience been rather unpredictable. Time and efficiency are essential for epidemiologic research and the biobank should strive as far as possible to obtain the required data for being able to plan optimize the scientific use of the biobanks as well as minimize the time required for a specific study.

5. *Explore extended saving of samples.* It should be explored whether the routine practices used in clinical laboratory medicine can be changed for optimization of the research usefulness of the biobank, without disrupting the routine work.

One example of such extended saving is the saving of blood clots for the Southern Sweden Maternity Cohort. Before 2005, only serum samples were stored. These contain only limited amounts of DNA, and the pellets of the serum tubes – with blood clots containing large amounts of high-quality DNA – were thrown into the waste, in spite of a population-based enrolment program with collection of consent for biobanking being in place. Since 2005, the blood clots also have been saved.

Another example is the liquid-based cervical cytology samples that contain large amounts of high-quality DNA/RNA/protein and also derive from a population-based screening program with broad consent for biobanking already being collected. With the switch from smears to liquid samples, the cytology laboratories stopped saving the samples (for reasons of lack of space – tubes take more space than slides). Since 2007, the liquid-based cervical cytology samples are now saved.

For both these examples, a high-quality, formally accredited, clinical, population-based biobanking system targeting (a) all pregnant women or (b) all women aged 23–65 years was possible with the existing personnel and QC system of the clinical diagnostics laboratories.

5. Strengths and Weaknesses of Clinical Laboratory Medicine for Biobanking

The deposit component ("sample in") in clinical biobanking is very strong. All procedures are accredited and Quality Assured, and are typically governed by electronic Quality Manual Systems. A large critical mass of experienced staff, robotic systems for handling samples (centrifugation and aliquoting), and entry of personal identifiers/clinical data via electronic referral sheet systems are other strengths. Saving a few 100,000 extra samples is a barely noticeable extra effort, requiring very limited external funding.

The routine screening programs typically have very high attendance (maternity care >95%; cervical screening >70% over 3 years and >90% over 10 years). The clinical biobanks based on these programs are, therefore, highly population-representative biobanks (not biased by low attendance rates).

The study base definition component is typically weak. Usually, clinical laboratories have no experience in data cleaning and data management. Sometimes even the simplest registry linkages (to get the accessory data; case ascertainment; sample picking lists) cannot be managed. Building this capacity at clinical laboratory biobanks is not so easy and in our experience, the easiest way to overcome this weakness is to send the database for cleaning and registry linkages at experienced centers, such as Cancer Registries or major experienced research biobanks.

The retrieval component ("sample out") is usually weak. The clinical biobanks are organized to pick out one sample from a given patient. There are no personnel to organize withdrawals of thousands of samples. This bottleneck is rather easily solved if there is external funding. An example of this is the development of the Malmö Biobank Consortium that during 1969–2002 made seven Scientific Retrievals. During 2002–2009, the Swedish National Biobanking Program and the EU FP6 Network of Excellence on biobanking CCPRB subsidized retrieval personnel, increasing the output to 263 retrievals for research and >100 publications.

6. Opening the Gold Mine of Biobanks in Clinical Laboratory Medicine

Even the very large microbiology/cytology/pathology biobanks constitute only about 2% of all samples that are handled in clinical laboratory medicine. The other 98% of samples that are handled are thrown out – a vast unused potential.

We could harness the capacity of clinical laboratory medicine for large-scale and optimally clinically relevant biobanking infrastructure if we:

1. Recognize that a basic biobanking infrastructure is an essential component of modern clinical medical care. Sending a sample for biobanking should have the same *rules*, *logistics*, and *funding* as sending samples for clinical diagnosis.

   Epidemiologic studies on healthcare-based biobanks are mainly aiming at improved health for the population. It could be accomplished by finding new diagnostic markers, evaluating risk for disease, etc. The studies are designed to look at risks in the population or a specific cohort, but never in the individual.

2. Institute an efficient infrastructure that provides an appropriate ethical basis.

As mentioned above, we think this should constitute (1) standardized, broad consent; (2) referral sheet-based documentation of consent status; and (3) an easy and effective mechanism for withdrawal of consent, preferably nationwide opt-out registries.

3. Have education and scientific visibility: The number of scientists who really know how to exploit the clinical biobanking system optimally is too small, resulting in underuse of this unique infrastructural resource. Similarly, the staff building the biobanks needs to have competent knowledge on what the samples will eventually be used for. The biobanks will need a critical mass of well-educated scientific staff willing to devote their career to biobanking. The fact that biospecimen & biorepositories research has only recently been recognized as a science has slowed down the development of the entire field of biobanking. Development based solely on opinion and administrative considerations is less efficient than the use of scientific and evidence-based progress in the building and exploitation of biobanks. It is particularly important that studies in biobanking methodology are performed using scientific methodology and that their results are routinely published in the scientific literature. Laboratory medicine is by definition the science involved in handling and analysis of biospecimens. If the resources and experiences of the scientific community in laboratory medicine can be committed to building an efficient biobanking infrastructure, we may be coming closer toward the fulfillment of the promises of molecular medicine in providing new and better diagnostics for new and better prevention as well as new and better treatment.

References

1. Pukkala, E., Andersen, A., Berglund, G., Gislefoss, R., Gudnason, V., Hallmans, G., Jellum, E., Jousilahti, P., Knekt, P., Koskela, P., Kyyronen, P. P., Lenner, P., Luostarinen, T., Love, A., Ogmundsdottir, H., Stattin, P., Tenkanen, L., Tryggvadottir, L., Virtamo, J., Wadell, G., Widell, A., Lehtinen, M., and Dillner, J. (2007) Nordic biological specimen banks as basis for studies of cancer causes and control – more than 2 million sample donors, 25 million person years and 100,000 prospective cancers. *Acta Oncol* **46**, 286–307.
2. Hansson, M. G., Dillner, J., Bartram, C. R., Carlson, J. A., and Helgesson, G. (2006) Should donors be allowed to give broad consent to future biobank research? *Lancet Oncol* 7, 266–269.

Chapter 5

Biobanks and Registers in Epidemiologic Research on Cancer

Eero Pukkala

Abstract

The Nordic countries have a long tradition of register-based epidemiologic studies. Numerous population-based specialized registers offer high-quality data from individuals, and the extensive use of register data further improves the quality of the registers. Unique personal identity codes given to every resident and used in all registers guarantee easy and accurate record linkage. A legislation that makes the use of the existing data possible for purposes that benefit both registered individuals and the society – instead of forcing researchers to use their energy in repeated questionnaire studies, disturbing individuals' privacy and leading to response and recall biases – is a prerequisite for effective epidemiologic research.

Biobanks can be considered an additional type of registers. They may offer data from individuals that cannot be reliably collected via questionnaire surveys. In turn, other types of registers are crucial in biobank-based studies (1) in defining for how long the persons in biobank cohorts are at risk of getting the diseases, (2) to get information on cofactors that may modify the relative risk measured by the biomarkers, and (3) to get information on the long-term outcome events.

This chapter describes the possibilities of register use mainly in Finland – a typical representative of the Nordic "paradise of register-based epidemiological research" – in research of cancer etiology. The ongoing Nordic research project *Changing work life and cancer risk in the Nordic countries* (NOCCA) will be described as an example of a massive register use, including both direct linkages on an individual level and indirect group level linkages.

Key words: Registers, Biobanks, Record linkage, Causal factors, Risk determinant, Censoring event, Outcome event, Cancer incidence

1. Introduction

Record linkage is the combination of data items, often from different files, for a certain unit of observation. The data may originally have been collected for some other purpose, without knowledge of the future uses to which the data might be put. In epidemiology, record linkage is usually used to connect data

Joakim Dillner (ed.), *Methods in Biobanking*, Methods in Molecular Biology, vol. 675,
DOI 10.1007/978-1-59745-423-0_5, © Springer Science+Business Media, LLC 2011

for a particular individual. It is often used for causal research and is applied when the data on causes (treatment, exposure, etc.) are to be related to the effect (survival/risk of the disease).

In the Nordic countries, there are a large number of registers, both old manual and newer computerized, containing individual health data over the entire life span of the individuals. In fact, the first data on individual health are recorded before birth, i.e., those obtained during pregnancy. Further data on the health of individuals are recorded at the time of birth, when individuals use the national health system, and at time of death. For some diseases, there are special registers; in Finland, congenital malformations, cancers, certain heart diseases, tuberculosis, and some other diseases are registered for the whole country. A registration system for causes of death has been in operation for hundreds of years. There are also countrywide records on some intervention procedures, such as mass screening for cancer of the cervix or breast. All hospital visits, with codes of treatment for individual diseases, are registered centrally.

A system that would provide the life-long history of the health and health-related events of any individual whenever needed for generally accepted and scientifically important studies would be optimal for etiologic research (and also helpful in best diagnostics and treatment selection for diseased persons). In practice, this kind of system would do better to build on linkable specialized registries rather than use a huge all-in-one database. First, in a centralized system, it might not be possible to have all the expertise needed to maintain data quality, which is usually the case with specialized registers. Second, the privacy of the registered individuals is protected better if only the data actually required for each specific study are put together.

2. Prerequisites of Good Register-Based Research

The Nordic countries (Denmark, Finland, Iceland, Norway, and Sweden) have a long tradition of high-quality epidemiologic research based on existing registers. This effective research methodology requires certain base elements discussed below.

2.1. Idea

A sound a priori defined hypothesis is the most important requirement of a high-quality register-based research. Registers offer an endless collection of variables that may attract people to run analyses without any idea of biologically or otherwise plausible mechanisms. Multiple testing situations – often giving millions of risk estimates – always produce significant observations. One should not do such analyses, and especially not interpret every significant observation as a scientific truth.

2.2. Exposure Data

Exposure measurement is normally the critical variable in an etiologic research. The extent of misclassification of self-reported exposures can be considerable, especially for sensitive questions such as addictions. A non-differentiate measurement error dilutes relative risk estimates toward unity and a differentiate error may indicate a risk in situations when there is no risk. Therefore, the accuracy of the risk estimate should always be carefully verified.

If there is a marker of the exposure that can be measured from the historical biological biosample, the biobanks offer objective means to validate register-based exposure variable in a sample of exposed persons. For example, cotinine can be used to control quality of smoking data (1).

2.3. Linkage Key

The unique personal identity code (PIC) given to every resident of the Nordic countries since the 1960s and now used as main key in virtually all registers including data on individuals offers a powerful tool to make accurate record linkages. However, even a small proportion of erroneous PICs decrease the relative risk estimates of any outcome event far below its true value.

Sometimes the risk variables are rather linked to the environment than to each single individual, or there is data of a risk factor measured only for groups of people. In that kind of situations, the variable used as linkage key is geographic region or group indicator.

2.4. Data Indicating When the Persons Were At Risk

It is extremely important to know when the persons in the cohort stop being at risk of getting the outcome event. Even a rather small fraction of missing end-of-follow-up data may markedly decrease the risk estimates in studies with long follow-up times (2).

2.5. Data on Confounders

In the epidemiologic articles, it is often mentioned that there was no data of factors that are known to be related to risk of the outcome event. The authors, reviewers, and readers of the article have no means to evaluate whether the prevalence of these cofactors is not correlated to the values of the main risk factor of the study or not. If they are, the result of the study is biased. In the Nordic countries, information on such cofactors actually often exists in registers, if not for every resident, at least for a large sample of people. The researchers may not search for such data, because (1) they are not aware of all existing data sources, (2) they find the required permission bureaucracy too cumbersome, (3) the study will be accepted for publication even without that extra work, or (4) they like to keep their study directly comparable with studies done outside the Nordic region (with no access to multi-register data).

2.6. Permissions

It is important that data collected to registers are never used for purposes that may violate the privacy of the registered individuals. It is also wrong if such data are NOT used in research that would bring gain to the individuals, their family members, or to the entire society. Therefore, a legislation to forbid misuse and promote acceptable use of the register data is needed. Any scientific research project must be evaluated by external review committees in terms of ethical acceptability of the research topic and the process how the study will be performed. If the study is ethically sound, a formal written permission is needed to document that the research also fulfills the formal legal requirements.

Even if the permission procedure may sometimes be nerve-wracking and slow, there are hardly any cases in Finland where permission would not have been finally permitted to run a register-based epidemiologic study on an ethically acceptable and scientifically relevant issue.

2.7. Funding

Because of the exceptionally good data infrastructure in the Nordic countries and overwhelming experience of utilization of those recourses, research teams of the Nordic countries are repeatedly asked to perform studies that are of high scientific interest, but these cannot be done with the same accuracy and/or efficacy outside the Nordic region. This brings external funding also, which in addition to the good Nordic research funding sources normally covers the costs of the studies.

2.8. Epidemiologic Skills

Since availability of data or research funding is not a problem, the limiting factor of effective use of all existing data starts to be the lack of experienced researchers educated to utilize the possibilities offered by the Nordic registries. Therefore, education programs have been developed that especially highlight the unique possibilities of the register-based data of the Nordic countries that are not included in the international textbooks or epidemiology courses. Special emphasis is put to create the creative thinking and courage to try new approaches in the research. Two examples of specific courses that add to the standard educational programs of the Nordic universities are described below.

2.8.1. Pregraduate Education

Nordic Summer School in Cancer Epidemiology has been operating on a biennial cycle since 1991, organized by the Association of the Nordic Cancer Registries and sponsored by the Nordic Cancer Union (http://www.ancr.nu/summerschool). The course is designed for pregraduate students of medicine, biology, sociology, statistics, and related fields to attract talents to epidemiology before they get fixed to other disciplines.

2.8.2. Postgraduate Education

The EU Network of Excellence "Cancer Control using Population based Registries and Biobanks (CCPRB)" organizes courses on registry linkage studies as part of the Spreading of Excellence (SoE) Doctoral Program in Public Health of Tampere School of Public Health in Finland (http://www.cancerbiobank.org/SoE%20Announcement-2007.pdf). The main objective is to educate the students – who in this program mainly come from countries far away from the Nordic region – to understand and utilize the data network of biobanks and other registers in their doctoral theses and hopefully in their later research activity.

3. Finnish Health-Related Registers

The Finnish main registers related to research on health and welfare have been nicely described in the publication by Mika Gissler and Jari Haukka (3). In the following, some parts of their text have been selected to give understanding of the magnitude of the registers in Finland. The same special issue "Epidemiological registries – access, possibilities and limitations" of the Norwegian Journal of Epidemiology (http://www.ub.ntnu.no:80/journals/norepid/2004-1.html) where the Finnish register infrastructure is described also includes extensive description of Norwegian registers, in Norwegian (4). Many of the data sources described below are being used in epidemiologic studies on cancer etiology (Fig. 1).

3.1. Long Tradition of Maintaining Registers

Record keeping in general has a long tradition in Finland. Population registers have a long history in Finland, with population information having been registered since the 1530s (http://www.vaestorekisterikeskus.fi/vrk/home.nsf/pages/C06B93B4C73B0447C2257244002D3488). The registration of vital statistics, including, for example, births, deaths and marriages was initiated

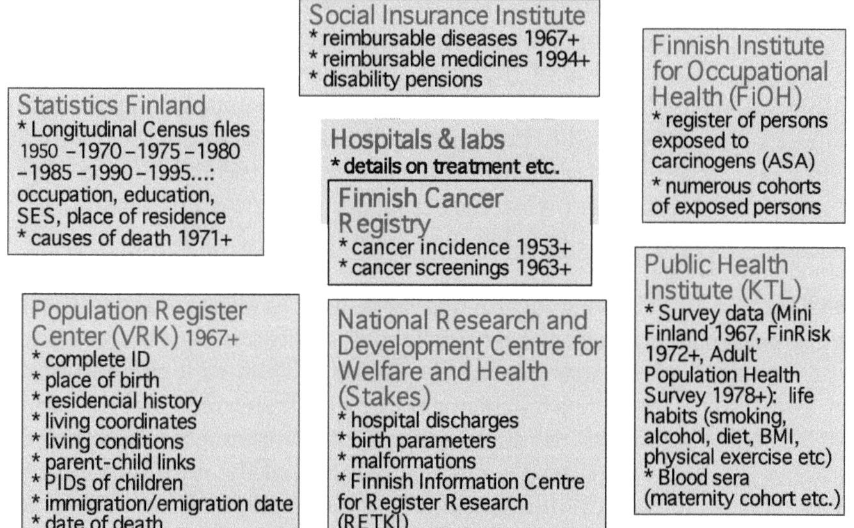

Fig. 1. Typical computerized and linkable register data sources used in studies related to etiology of cancer in Finland. *Blue background* indicates registers that cover entire population; data with orange background are available only for sample of population.

as early as 1749. The first nationwide, computerized disease register, the Finnish Cancer Register, was started in 1952 (Table 1). Cancer registries, the main source of outcome events in cancer-related research, are described in detail in Subheading 7.1. The Mass Screening Register was started as part of the Finnish Cancer Registry in mid-1960s primarily to register women invited to organized Pap smear screening for precancerous lesions of the cervix uteri, and later extended to breast cancer screening (mammography) in 1987 and colorectal cancer screening (fecal test) in 2003.

The different hospital discharge registrations were merged to a new Hospital Discharge Register covering all public hospitals (1967). In 1994, the Hospital Discharge Register (renamed as the Care Register) was widened to cover all social institutions, such as elderly care homes and institutions for the mentally disabled.

The Finnish Central Population Register was created in the 1960s. Currently, the register covers information on all Finnish citizens and permanent residents of Finland. Their residential histories and family relations exist in the central register since the 1970s.

In addition to the specific health and social welfare registers, many other registers are important for epidemiologic research. Statistics Finland compiles the computerized Cause-of-Death Register, which can be linked with other registers since 1971. Statistics Finland also annually gathers Population Census data. Finland – along with Denmark – nowadays is one of the rare countries that base its Census on already compiled register information instead of collecting similar information from all citizens by postal questionnaires and/or interviews. The data for the Population Census are gathered from approximately 30 different registers and administrative files.

One of the main prerequisites for the utilization of register data is good data quality, i.e., all events are included in the database, and the registered data are in accordance with the reality. This has been shown to be true for several Finnish administrative registers in studies comparing register information with patient records or other information from the primary source (5–9).

3.2. Finnish Data Protection Legislation

The first health registers were compiled under the legislation covering the data collecting institution, but there was no separate legislation on health registers (3). Such legislation was passed in the Finnish Parliament in 1987. It ensured citizens' right to privacy despite the increased use of computerized registries containing sensitive data, and also recognized the need to collect health and medical information. These statutes, which are still in force, gave health authorities the right to gather and register relevant information on individual level, including PICs, and obliged both public and private healthcare personnel to provide these data for them.

Table 1
The nation-wide health and social welfare registers in Finland. Table modified from (3)

Register	Established in computer format	Current register keeper
Healthy registers		
Cancer Register	1953	STAKES[a]
Central Register on Healthcare Personnel	1955	NAMA
Register of Congenital Malformation	1963	STAKES
Register on Occupational Diseases	1964	FIOH
Register on Adverse Drug Reactions	1966	NAM
Hospital Discharge Register (health care institutions)	1967	STAKES[b]
Mass Screening Register (cervical and breast cancer)	1968	STAKES[a]
Cause-of-Death Register	1969	Statistics Finland
Register on Induced Abortions and Sterilisations	1977	STAKES
Register on Persons Exposed to Cancer-hazardous Material	1979	FIOH
Implant Register on Orthopaedic Endoprostheses	1980	NAM
Drug Surveillance Register	1982	NAM
Register on Visual Impairments	1983	STAKES[c]
Medical Birth Register	1987	STAKES
Register on Infectious Diseases	1989	NPHI[d]
Register on Dental Implants	1994	NAM
Social welfare registers		
Register on Pensions	1962	FCP
Registers on Social Benefits under the National Sickness Insurance[e]	1967	SII
Register on Social Assistance	1985	STAKES
Child Welfare Register	1991	STAKES
Discharge Register (social institutions)	1994	STAKES

FCP Finnish Centre for Pensions, *FIOH* Finnish Institute of Occupational Health, *NAM* National Agency for Medicine, *NAMA* National Authority for Medicolegal Affairs, THL National Institute for Health and Welfare, *SII* Social Insurance Institution, *THL* National Institute for Health and Welfare (the predecessors of which were the National Board of Health and the National Board of Social Welfare, National Public Health Institute, and National Research and Development Centre for Welfare and Health)

[a] Technical management has been given to the Finnish Cancer Societies

[b] The data collection also included previous hospital discharge registers, which included care in tuberculosis sanatorium (since 1956), psychiatric hospitals (since 1957), and general hospitals (since 1960). Complete identification numbers are available since 1969

[c] Technical management has been given to the Finnish Federation of the Visually Impaired

[d] Register on New Cases of Tuberculosis and Sexually Transmitted Diseases started in 1958

[e] The data collection includes information on special reimbursements of medicine since 1964; on national pensions (guaranteed minimum pension), family pensions, and child disability allowances since 1970; on sickness allowances and reimbursed visits in private healthcare services since 1971; on rehabilitation since 1978; on conscripts' allowances and basic unemployment allowances since 1985; on family allowances and child care subsidies since 1993; on maternal grants, labor market subsidies and housing allowances since 1994; on prescribed medicine and reimbursed interventions in private healthcare services since 1996; and on students' allowances since 1997

The legislation listed all the health registers that national authorities may maintain (Table 1).

Finland revised its legislation on the protection of personal life to meet the EU requirements in 1999. According to the Personal Data Act, health and social information can only be gathered by informed consent from the client or patient, with the exception of data collected for statistics and research in history or science. The legislation also clearly states that the Finnish nationwide health and social welfare registers cannot be used in decision making about a registered individual.

Previously collected health information may be used though in research without informed consent if the data are large or the collection of such informed consents is not feasible. Recent experience indicates that individuals are very positive to allow the use of their data in scientific research. Out of the entire birth cohort of 1985–1986 in Northern Finland (10), about 2% did not give the permission to use their data for scientific research carried out at the local university, and almost 5% refused the delivery of their data for collaborating units. Almost everyone who participated in the clinical examination gave their permission to use the compiled data in research.

Biological samples can be combined with register data, but a statement from an ethical board is mandatory. In cases where researchers wish to contact registered persons with health-related events, e.g., for interviews or postal questionnaires, the first contact can only be done through the physician in the healthcare institution where the patient or client was treated.

3.3. Information for Register Research

As in the other Nordic countries, the significant possibilities for register-based research have been noted in Finland. To promote the use of administrative registers in scientific research, the Finnish Information Centre for Register Research was initiated at STAKES in August 2003. The aim of the center (http://retki.stakes.fi/EN/index.htm) is to promote the use of national administrative registers in research, especially in health and social sciences by

- supporting planning and implementation of register-based research,
- improving the capabilities for using register data among researchers,
- increasing cooperation between different registers, and
- improving practices on the utilization of register data.

The center has created a network of contact persons in the register-keeping organizations and introduced an Internet portal presenting the existing registers, data protection legislation and practices, and methods in register-based research. Possible future tasks include training for students and researchers in register-based research, assistance in the process of retrieving authorization for data access, financing of register-based research, data linkages and analyses, and data archival. In 2009, NPIH and STAKES were merged. The current name is the Institute for Health and Welfare (THL).

4. Registers of Exposures and Health Habits

For epidemiologic research, health data alone – even if all possible elements of health data could be linked together – are usually not sufficient. Data sets containing information about risk factors or at least risk determinants are also required. In all Nordic countries, extensive socio-demographic data for all citizens are available from population-based registers. However, some essential data, such as those on diets and smoking, exist only for a part of the population, and researchers may be forced to use averages calculated for subgroups of the population instead of the individual values of the variables. For diseases with a long delay between cause and onset, the most relevant risk factor data may only be available from old manual files or from biobanks including samples taken from healthy population a long time ago.

The most typical – and normally most cost- and time-effective – example of a study based on record linkages in Finland is a cohort study on disease risk. It often starts from a file originally collected for some other purposes (e.g., the pay-roll register of a company and a list of persons exposed to certain chemicals, drugs, radiation, etc.). In this kind of study, the quality of the registers to be linked, especially the accuracy of the key variables, is critical. Occasionally, combining unlinked records is less harmful than not combining records that belong together, since the latter type of error causes a systematic bias in results. If the key variable in any of the linked files is erroneous, no data on death will be found. This increases the number of person-years available, especially in the oldest age groups. Because the incidence of many diseases increases strongly with age, even a small addition of person-years due to failure in record linkage may cause a relatively large artifactual addition to the expected number of cases.

The second record linkage needed for a cohort study is that between the cohort and a disease registry. For members of the cohort with invalid key variables, no observed cases will be found. Thus, where there is incomplete record linkage, the risk estimates calculated as ratios of too few numbers of observed cases and too large numbers of expected cases are systematically too low. This may change the result of the analysis totally, as demonstrated in other chapters of this book (2).

4.1. Population Information System

In Finland, population information has been recorded since the sixteenth century when King Gustavus Vasa of Sweden launched administrative reforms and increased the efficiency of tax collection and military recruitment. The church also introduced its own population register. As early as 1628, the Bishop of Turku ordered that vicars must keep records of births, marriages, and deaths. The order to maintain parish registers was extended to cover the entire country in the 1660s. Finland's first population statistics dates back to 1750.

The computer-based population register was introduced in 1971 (http://www.vaestorekisterikeskus.fi/vrk/home.nsf/en/populationinformationsystem). The Population Register Centre maintains the Population Information System in cooperation with local register offices. The Finnish Population Information System serves a variety of societal functions, including election arrangements, taxation, compilation of statistics, and research. It contains basic information about Finnish citizens and foreign citizens residing permanently in Finland and is the most-used basic register in Finland.

Personal data recorded in the system include name, PIC, address, citizenship and native language, family relations and date of birth, emigration, and death (if applicable). Using building and apartment codes, persons registered in the Population Information System can be linked with the center coordinates of buildings and, using identifiers, buildings can be linked with other national base registers used in Finland. This enables the utilization of the Population Information System in various geographic data applications.

4.2. Statistics Finland

Statistic Finland (http://www.stat.fi) produces statistics of numerous health-related events. For instance, regular population statistics gives the size and structure of the permanent resident population and related changes, such as births, deaths, marriages, migration, employment, families, and household-dwelling units. They also produce population projections by area. Their data on population censuses have been extremely useful as cancer risk determinants (see Subheading 8).

4.2.1. Population Censuses

The full set of Population Census data has been produced in 1950, 1960, 1970, 1975, 1980, 1985, 1995, and 2000. A longitudinal data file spanning from 1970 to 2000 and containing information on all persons resident in Finland in 1970, 1975, 1980, 1985, 1990, 1995, and 2000 has been produced from the census data on individual persons. The file also contains data by families and household-dwelling units. In addition, data on a sample of 400,000 persons from the 1950 Population Census have been attached to the file.

4.3. National Public Health Institute

In the area of public health and chronic disease prevention, the main research areas of the Institute for Health and Welfare

(THL; http://www.thl.fi/en_US/web/en/research/statistics) are cardiovascular diseases, diabetes, health behavior, and nutrition and health promotion. Concerning these areas, e.g., following research programs, projects, and registers are run by the THL.

1. The National FINRISK Study – famous from the North Karelia Project from 1972 – is a large survey on risk factors of chronic diseases carried out every 5 years. The study also collects biological samples that since 1992 are systematically stored and available for important international biobank studies, e.g., in NBSBCCC network (11).
2. MOnica Risk, Genetics, Archiving, and Monograph (MORGAM) is a multinational study to explore the relationship between the development of cardiovascular diseases and their classic and genetic risk factors.
3. National Health Behavior Monitoring Systems include four different health behavior surveys. Especially data from the Finnish Adult Health Behaviour Survey that has collected massive health habit data from random samples of 5,000 Finnish adults every year since 1978 (12) have been used extensively in studies on cancer etiology.
4. The Finnish National Nutrition Surveillance System was launched in February 1995 to collect and distribute data on the status of nutrition in Finland, e.g., to health policy makers, officials, and researches.
5. The Department of Infectious Disease Epidemiology provides information about the occurrence of infectious diseases, epidemics, and suspected epidemics, as well as the related risk factors. This information has a significant impact on decisions at both the societal (health policy and health protection) and the individual level (health habits, behavior, and risk-taking).

4.4. Social Insurance Institution

The national Social Insurance Institution in Finland (Kela) runs research on healthcare issues concentrating on health insurance, rehabilitation, and health promotion (http://www.kela.fi/in/internet/english.nsf/NET/030407124955HJ?OpenDocument). Based on register data, they look at the accumulation of benefit recipiency, how much the insured have to pay in copayments, and the socioeconomic backgrounds of benefit recipients. Kela also boosts research efforts by targeting the funding and register data available to Kela on external research that focuses on questions holding particular relevance to Kela. By participating in international research cooperation, they seek to discover new approaches to conducting research and to organising the provision of health and welfare services.

Kela has since the 1960s kept a register of persons who have got special reimbursements for medicine expenses because of certain chronic diseases such as diabetes, asthma, cardiovascular diseases, rheumatic diseases, and about 50 other chronic diseases. This register has been used both as exposure definition and to identify outcome events that do not lead to hospitalization. An example of the former setting is an old study to test the hypothesis on whether the activated killer cell activity due to asthma could also protect against leukemia (13): there was some indication on that effect, and the study would be worth repeating. In a study on health effects of persons who had lived in houses built on a former dump area, the selection of outcome diseases was extended from cancer to other chronic diseases extracted from the Kela's register: a significant 50% excess was observed for cancer and asthma, and the houses were demolished (14).

A very promising rather new resource for important epidemiologic studies is the register of purchases of all medicaments prescribed to any resident in Finland since about 1994. For instance, every regimen of hormonal (replacement) therapy bought by Finnish women has been registered, and it is now possible to study cancer risk related to various types of hormonal therapies (15).

4.5. Finnish Institute for Occupational Health

The Finnish Institute of Occupational Health (FIOH) has carried out research on occupational hazards and health, and disseminated information since the late 1940s (http://www.ttl.fi/NR/rdonlyres/B7381E28-7F20-4033-81D6-B2323F61BF79/0/SF002.pdf). FIOH maintains several registers and databases on exposures for surveillance, hazard control, epidemiology, and risk assessment purposes (16).

4.5.1. Occupational Biomonitoring Database

Biomonitoring samples have been analyzed at FIOH since the early 1960s. Old data are kept only in manual records but samples from 1991 onwards have been computerized. The general aim of the registry is to monitor nationwide occupational exposure to chemicals on an individual basis, and to improve interpretation of the biomonitoring results by providing exposure histories for individuals, enterprises, and industrial sectors. Intended users of the data include occupational health professionals, employers, researchers, and policy makers. However, only an authorized laboratory personnel has direct access to the data.

More than 60 different determinations of chemicals or metabolites in body fluids are made in automatic analyses of about 8,000 samples annually. The most common determinations are presented in Table 2.

Specimens come from occupational healthcare units all over Finland. Since submission of the specimens is not obligatory (except for lead-exposed workers), and because there are other laboratories in the country that also perform some of these analyses, the register does not have 100% coverage. Still, these records

Table 2
Most common chemical exposures biologically monitored by Finnish Institute of Occupational Health in 1998 ((48), modified). Service measurements only

Chemical agent (and vehicle)	Number of measurements	Mean concentration (μmol/L)	Maximum concentration (μmol/L)
Chromium (in urine)	962	0.08	1.4
Lead (in blood)	849	0.6	4.8
Aluminum (in urine)	525	2.0	28
Nickel (in urine)	376	0.16	3.4
Cadmium (in urine)	372	0.01	0.47
Cadmium (in blood)	286	0.007	0.044
Styrene (mandelic and phenylglyoxyl acid in urine)	246	1,400	16,000
Carbon disulfide (2-tiotiazolidine-4-carboxylic acid in urine)	242	400[a]	4,200[a]
Polychlorinated biphenyls (in serum)	202	0.7[b]	10.0[b]
Cobalt (in urine)	200	0.38	7.88

[a] in μmol/mol creatinine
[b] in μg/L

have offered a good basis, for instance, to construct a cohort of workers exposed to lead for epidemiologic studies on cancer risks (17).

4.5.2. Register of Employees Exposed to Carcinogens

The International Labour Office (ILO) in 1977 recommended the recording of systems to be constructed for the monitoring of occupational exposure to carcinogens. This recommendation prompted the Finnish Register of Employees Exposed to Carcinogens (ASA Register; Finnish abbreviation) in 1979. Employers were obliged to provide data on the use of carcinogens and to notify exposed workers annually to be entered into a database maintained by FIOH.

The ultimate aim of registration was to promote the prevention of occupational cancer in Finland. Obligatory registration was expected to stimulate identification, assessment, and elimination of carcinogenic exposures at workplaces, resulting consequently in decreased risk of occupational cancer among notified workers. In addition, ASA Register was designed as a surveillance system that can be used to follow up cancer risks of exposed workers at the national level by linking its data with cancer data from the Finnish Cancer Register.

Table 3
Annual average number of exposed workers notified to the Finnish Register of Employees Exposed to Carcinogens (ASA), by period (18)

Carcinogen	1979–1984	1990–1994	2000–2004
Environmental tobacco smoke (ETS)	–	–	11,774[a]
Chromium(VI) compounds	3,838	6,799	7,291
Nickel	2,995	5,768	6,371
Asbestos	1,766	3,631	1,696
Benzene	561	1,321	1,628
PAH	685	1,223	1,551
Chloroform	–	978[a]	1,400
Acrylamide	–	145[a]	1,033
Arsenic	457	898	944
Cadmium	610	1,007	917
Other carcinogens	2,130	5,911	7,048
All exposures	13,042	27,593	39,300
All exposed workers	8,495	15,924	25,109

[a]Chloroform, acrylamide, and environmental tobacco smoke were added to the list of carcinogens in 1986, 1993, and 2001, respectively

About 15,000 workers from 1,500 work departments were notified annually to ASA Register from the mid-1980s to 2000. The addition of environmental tobacco smoke (ETS) in the list of carcinogens in 2001 increased the number of annual notified workers to over 25,000 (Table 3). The most common exposures in ASA are those to chromium (VI) compounds, nickel and its inorganic compounds, asbestos, benzene, PAHs, and chloroform. ASA does not include any information on the level of exposure, smoking, or other lifestyle factors of the notified workers.

According to a questionnaire-based survey on the effects of ASA at workplaces (18), the ASA notification process had directly prompted measures to reduce exposure and decreased the exposure of 600 notified workers each year (0.7% of the employed labor force in Finland). During a rather short follow-up, the only significantly increased incidence rate in the ASA cohort was of mesothelioma, probably due to exposure to asbestos (18).

4.6. Radiation and Nuclear Safety Authority

Radiation and Nuclear Safety Authority Finland (STUK; http://www.stuk.fi/en_GB/) maintains, e.g., dose register, register of occupational radon exposure, exposure to internal radiation (radiation workers and Finnish population), chromosome analysis for dose assessment, natural radioactivity in drinking water, and residential indoor radon databases. Many of these data sources have been successfully utilized in epidemiologic cancer research.

4.7. Biobanks

There are numerous biobanks in Finland that can be used in measuring such health-related factors that do not readily exist in the other registers. The following ones – all kept by the National Public Health Institute of Finland – belong to the NBSBCCC network (11):

1. Finnish Maternity Cohort includes more than 1.5 million sera collected during the first trimester of pregnancy for screening of a number of congenital infections.
2. Helsinki Heart Study has sera collected from 19,000 men aged 40–55 years during 1980–1982 for a trial related to regulation of cholesterol level (19).
3. Alpha-Tocopherol, Beta-Carotene (ATBC) Cancer Prevention Study has samples taken in 1985–1988 from 43,000 smoking men. The main aim of the study was to evaluate whether daily supplementation with alpha-tocopherol or beta-carotene will reduce the incidence of lung cancer and other cancers (20).
4. The Mobile Clinic Health Examination Survey was carried out by the Social Insurance Institution during 1966–1972. Blood samples from 40,000 individuals have been stored.
5. The National FINRISK Study has been conducted in Finland every 5 years since 1972. At the beginning, the study was done only in eastern Finland as part of the North Karelia Project, but now it is extended to several other areas. Random samples of 22,900 members aged 25–64 years (since 1997 up to 74 years) are systematically available since 1992.

In addition to these biobanks, biological samples from individuals have been collected and stored by numerous institutes (including hospitals). Access to these samples is more difficult due to discrete storing systems and legislation that strongly controls the use of the samples for purposes other than those for which they were originally collected.

5. Linkage Procedures

5.1. Direct Linkage on Individual Level

The best key to link together observations related to the same individual from different sources is to use the PIC. There may be many people with exactly the same name, but there are no two persons with exactly the same PIC. The code remains unchanged throughout a person's life. The Finnish PIC is issued on the basis of a birth certificate to Finnish citizens. Parents of newborn children need not take any measures to obtain the PIC for their children, as the hospital provides details of all births to the Population Information System. A foreign citizen whose residence in Finland exceeds one year is also issued a PIC. Persons staying in Finland on a temporary basis can also be issued the code.

PICs were introduced in Finland in the 1960s and were given to every resident before the end of the 1960s. For instance, the Finnish Cancer Registry has a rule that every cancer patient who did not die before 1967 must have the PIC. For persons who died before 1967, a manual record linkage can been done based on name, date of birth, and place of residence.

If the linked files do not include PIC but include name and at least one of the data items: date (year) of birth, place of birth, or place of residence, the Population Register Center of Finland is still able to run a computerized record linkage, with the percentage of correct matches varying from 50 to 100% (depending on the amount and accuracy of the key variables). The price of the record linkage without PICs is roughly 50-fold higher than that of the linkage based on PICs.

The routine system of the Finnish Cancer Registry provides a comparison of the accuracy of old manual and modern computerized PIC-based record linkage. Until 1974, the follow-up for annual death files was performed manually by comparing the alphabetical list of persons who died during the year (about 40,000 names) with an alphabetical list of cancer registry patients not known to have died (80,000 names). The maiden names were taken into account as well. The comparison was made by the secretaries at the Finnish Cancer Registry, who were known to be most thorough. From 1975, the linkage has been done automatically, using person-number as a key. A linkage of the whole cancer registry against the population central register later provided the means to evaluate the accuracy of the original linkages. Figure 2 shows the proportion of deaths missed in the original manual and in the automatic record linkage. Manual record linkage did not succeed for about 50–100 cases annually (out of some 10,000) even though Finnish names are ideal for this kind of record linkage; names are always written exactly as they are pronounced, so that no system like the soundex system in the English-speaking world is needed (cf. (21)). The small proportion of mismatches since 1975 (0.05%) represents typically those with cancer notification missing at the time of the original record linkage.

For those concerned about the privacy of the individuals registered, it is worth remembering that in PIC-based linkages, the names or other informative identification data of the persons in the linked files are never revealed, not even to the researchers performing the study. This is not possible with manual record linkage. Only if an error occurs is it necessary to find out the identity of some of the study subjects, a fact that further underlines the need for high-quality data files to be linked.

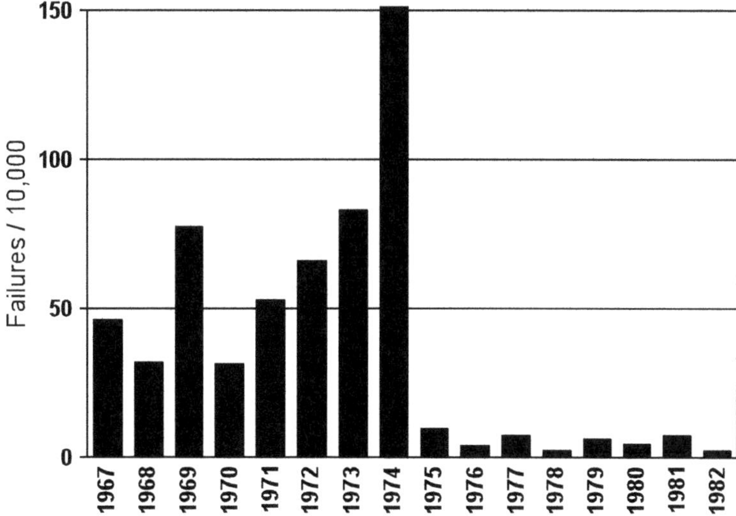

Fig. 2. Failures in record linkage between Finnish cancer registry data and death certificate data carefully carried out manually using names and other variables (1967–1974) and electronically using the person identity code (1975 onwards; edited from ref. 50).

5.2. Indirect Linkage

Sometimes there are no data of some important risk factors or cofactors for every study subject, but such data may be available for a population sample, e.g., from health surveys or from biobanks. From those data, it is possible to create model-based risk estimates on group level. If for instance everybody who have been working in a given occupation has similar occupational exposure to carcinogens, it is enough to know the occupational titles for every study subject and use them to link occupational exposures to individuals. Similarly, geographic coordinates of the residence may link the individuals to factors in their physical living environment, such as distance from a point source of exposure, amount of natural radon radiation, strength of electromagnetic fields due to power lines, etc.

Sometimes the link key may be a mixture of numerous variables. The more homogeneous the categories of exposed people are, the better the risk estimates give this kind of indirect linkage. Often this is not the case but the group level estimates are averages of very differently exposed persons, and in multi-exposure studies, it would be crucial to understand distributions of each of the exposure within the stratum, and also clustering of them into same individuals. Therefore, it would be best if all variables would be known from the very same individuals, i.e., all surveys should use same population samples whenever possible.

Infections are a typical example of a case where linkage via geographic coordinates is a natural choice. The ongoing study by one of the students of the EU Network of Excellence Doctoral Program in Public Health (see Subheading 2.8.2), Felipe Castro, offers an example of the possible use of biobank data to create full-population model of frequency of health-related variables based on biological samples. He has based part of his study (22) on geo-referenced data on cervical cancer-associated HLA antigens and cervical cancer incidence (Fig. 3). Genetic susceptibility to the persistent infection and cervical cancer is associated with HLA types and may determine whether a woman will be protected against infection and cancer or not. For instance, HLA-antigen DR2 is supposed to increase the susceptibility. Only the cervical cancer incidence data can be taken directly from whole-population register. The prevalence of high-risk HPV infections was estimated from a random sample of 8,000 women extracted from the Finnish Maternity Cohort (23, 24), and that of the HLA types on a sample of 19,745 donors from the Finnish Bone Marrow Donor Registry of the Finnish Red Cross (Fig. 4).

6. End of Follow-Up Events

6.1. Death

The simplest event that removes a person from being at risk of an outcome event is death: this happens only once per person, and the time of the event is clearly defined. The dates of death are normally taken from the National Population Register. If cause of death is needed, the source of information is Statistics Finland.

Even with the well-defined causes of death, there are situations when the researchers need to think whether an outcome occurred before end of follow-up or not. For instance, cancer may be diagnosed several weeks after death in a pathologic examination from a sample taken during autopsy. In studies on cancer etiology, this type of cancer is normally counted as an outcome event diagnosed at the time of death. In studies of cancer patient survival, this cancer with "negative survival time" in normally excluded.

6.2. Emigration

If a person moves out of the region of follow-up, it should be the end of person-time calculation. In modern Europe, people move back and forth, which has made the issue of migration more important as it has been before. In the Icelandic Maternity Cohort, 6% of the cohort members cannot be used in studies because they have moved out of the country, but the date of emigration is not registered (11).

In studies coordinated by the Finnish Cancer Registry, the follow-up is normally stopped at first emigration, even if the person returns to Finland. Although this practice causes loss of person-years at risk and observed cancer cases, but on the contrary may add to the quality of the study by elimination of one source of selection bias: some persons may come back to Finland because

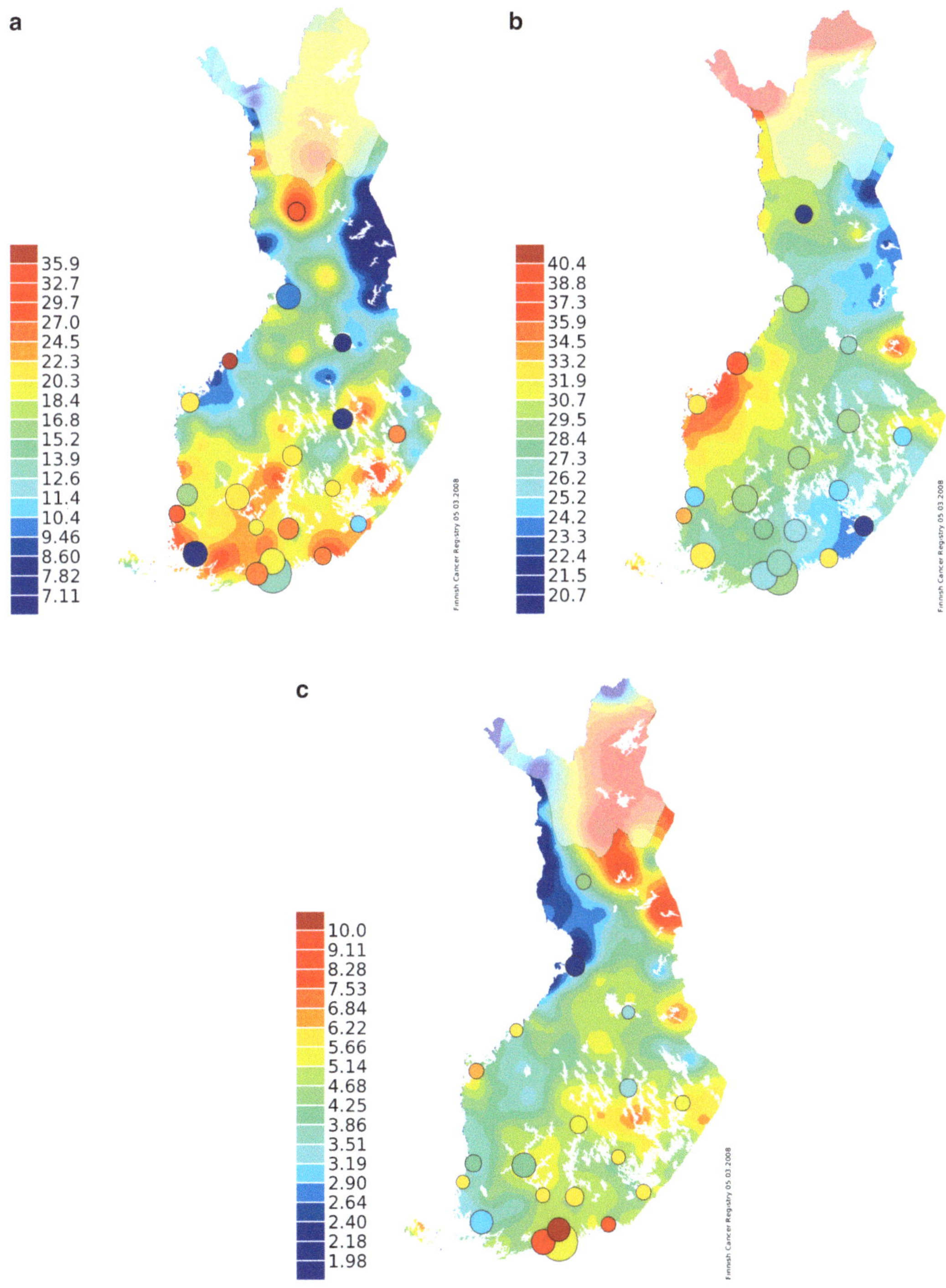

Fig. 3. (**a**) Prevalence (%) of HPV16 virus infections in 1983–1988 among 14–31-year-old women, (**b**) HLA antigen frequencies in young Finnish population generated from a sample of 19,745 donors from the Finnish Bone Marrow Donor Registry in 1992–2004, and (**c**) incidence rate (per 100,000) of cervical cancer in 1995–2005 among 15–49-year-old women in Finland. The larger cities are circled with the size of the circle indicating the size of the city; the rest of the municipality-specific observations presented as population-weighted floating averages.

they have symptoms of cancer and feel safer to get diagnosed and treated in the Finnish high-level medical system.

6.3. Other Events

In specific situations, there may be other events that alter the likelihood of being at risk of the outcome event. One obvious situation is removal of the organ: e.g., every third to fourth post-menopausal woman has undergone hysterectomy (25). These women are not at risk of getting diseases of corpus uteri, and many of them also have cervix or ovaries removed.

There are also examples of half-risk subjects. For instance, in the studies of breast cancer risk in the other breast after removal of one breast, the expected risk level should rather be half the breast cancer risk in the population than the full risk. In other words, the expected risk should rather be calculated per organ-years than per person-years.

6.4. Common Closing Date

If there is no event that would remove a person from being at risk, the calculation of risk time should end on a common closing date. This technical date is related (1) to the date of latest linkage of the end-of-follow-up events and (2) to the delay of registration of the end-of-follow-up events in the register in question. For example, the delay in registration of dates of death in the Finnish Population Register is about 2 weeks. If the study cohort was linked with Population Register on 17 July 2007, then the closing date for follow-up of vital status can be set to 30 June 2007.

It would be best if there would be positive identification of the alive status, i.e., every study subject would have been linked with Population Register and it would be confirmed that this person existed in the register, and had not emigrated or died. Especially in long follow-ups, there are numerous possibilities for a person to miss the end-of-follow-up event, i.e., those without information of an end-of-follow-up event are not necessarily alive and at risk.

7. Cancer Data

7.1. Cancer Registries

7.1.1. Finnish Cancer Registry

Cancer registration in Finland started in 1952 and has since been compulsory. All physicians, all hospitals, and other institutions in the country must send a notification (http://www.cancerregistry.fi/eng/registration/lomakekleng.pdf) to the Finnish Cancer Registry (FCR) of all cancer cases that come to their attention. Pathological, cytological, and hematological laboratories send the respective laboratory notification (http://www.cancerregistry.fi/eng/registration/lomakelbeng.pdf). Vast majority of notifications from the pathological laboratories and smaller part of hospital notifications are currently sent in electronic format according to the instructions created by the FCR. The automatic reporting contains the same information as that in the manual reporting forms, including in the free texts detailed descriptions of the tumor site and histology. In addition, Statistics Finland annually sends a computerized file on death certificates if a malignant disease is mentioned. If only laboratory and/or death certificate information is available, or if the data on the primary site of the tumor or date of diagnosis are incomplete or controversial, requests for further information are sent to the hospitals and physicians.

The following diseases are reported to the Registry:

- all malignant neoplasms, such as carcinomas (also basaliomas), sarcomas, malignant lymphomas, leukemias, multiple myeloma, gliomas, melanoma, etc.;
- carcinoid tumors, pheochromocytomas, thymomas, ameloblastomas, and chordomas;
- carcinoma in situ lesions (except those of the skin); and
- CIN III, dysplasia gravis, and CIL III of the cervix uteri; histologically benign tumors of the central nervous system and meninges, transitional cell papillomas of the urinary tract, and ovarian tumors with borderline malignancy.

PIC is the key in all practical registration procedures: e.g., in combining notifications for one patient received at different times and from different sources. For example, duplicate registration can thus be effectively avoided. The Registry file is annually matched, through computerized record linkage (based on PICs), with the Cause of Death Register located at Statistics Finland, so that the dates and causes of death (also non-cancerous causes, both underlying and contributory causes of death) can be added to the records in the Registry.

The Registry file is also regularly linked with Central Population Register, where the complete name, vital status, possible date of death or emigration, and the official place of residence are obtained.

The cancer notifications submitted to the FCR are immediately stored in the database at the Registry. Thorough visual and automatic checking procedures are carried out, both at data entry and coding. The computer directly announces illegal codes and code combinations, as well as illogical order of dates (of diagnosis, start of treatment, and death). Specific checks are carried out when needed, e.g., completeness of the Registry has been evaluated in comparison with the national Hospital Discharge Registry (6).

Final coding of cancer data is done by qualified secretaries and supervised by the Registry physician. Until May 2008, the Registry has followed a slightly modified version of the ICD-7 nomenclature from 1955 for coding the primary site of cancer and the codes of the American Cancer Society from 1951 for morphology. Now all old codes have been converted to modern ICD-O-3 codes, and only new cases are coded according to ICD-O-3.

Each cancer considered as an independent new primary lesion is registered separately. All independent cancer processes are coded as separate entities. The Registry files contain more than one million cancer cases diagnosed since 1953. In addition, there are a number of cases diagnosed prior to 1953 in persons who died from cancer or got a new primary cancer since 1 January 1953. More than 26,000 new cases of cancer are currently registered each year, plus some 4,500 basal cell carcinomas of the skin and smaller amounts of some other lesions, which in the published statistics are usually excluded from the total numbers of cancers (http://www.cancerregistry.fi/eng/statistics/).

The following *coded* items usually meet the needs of producing statistics and doing analytical research: PIC; municipality of residence; primary site; month and year of diagnosis; basis of diagnosis; stage (localized, regional metastases, and distant metastases); malignancy; histology/cell type; treatment: (surgery, radiotherapy, chemotherapy, hormones, other); specific codes for curative/palliative surgery or radiotherapy; date of death or emigration; and cause of death.

In addition to the items listed above, for instance, names of the notifying hospitals or laboratories, specimen numbers, tumor grade, TNM classification, site of metastases, details of the treatment, or cause for not being treated remain in the computerized database of the FCR and can be used, e.g., for searching the histological slides for re-evaluation.

The FCR has emphasized the importance of data protection and personal privacy years before it became an issue of debate in the Western societies. No violations of the individuals' privacy have occurred during the 57 years of cancer registration in Finland. Data on the individual level can be delivered to researchers working outside the Registry only through permission given by the National Research and Development Centre for Welfare and Health (within the Ministry of Health and Welfare). Every year close to hundred scientific papers are published based on the data and know-how of the FCR.

The newest cancer statistics (incidence, mortality, and prevalence) can always be found at the home pages of the FCR (http://www.cancerregistry.fi/eng/statistics/).

7.1.2. Other Nordic Cancer Registries

The other cancer registries linked to the NBSBCCC network (Iceland, Norway, and Sweden) are in many aspects very similar to the FCR, but there are some differences worth remembering.

The Icelandic Cancer Registry (http://www.cancerregistry.is/krabbameinsskra/indexen.jsp) was established in 1954 and covers cancer incidence since 1955. Cancer registration is still voluntary but there is a new proposal in preparation that would make cancer registration in Iceland compulsory. From the cancer-related data items, stage is not registered in Iceland. A special strength of the Icelandic register system is the systematic link to the family members of each cancer patient.

The Cancer Registry of Norway (http://www.kreftregisteret.no/frame.htm?english.htm) includes data on incident cancer cases since 1953. Basal cell carcinomas of the skin are not registered but there are data, e.g., on all precancerous lesions of cervical cancer.

The Cancer Registry of Sweden has data since 1958, but normally the first year used in statistics is 1960 (26). In Sweden, there are six regional cancer registries that work close to the main hospitals and have, therefore, good links to the hospitals' data. These regional registers send readily coded data to the Swedish central register (http://www.socialstyrelsen.se/en/about/epc/Cancer+Registry.htm) that has a rather technical role of simply combining the contents of the files. The Swedish cancer registration system does not use one source of information that is used elsewhere, namely, death certificate information. This causes a miss of about 4% of all cancer cases, and as much as 18% of leukemia (27).

Swedish central cancer register did not include information on the stage of cancers until recently. Basal cell carcinomas of the skin were not registered until recently.

7.2. Other Registries with Cancer Data

7.2.1. Hospital Care Register

The Hospital Care Registry (sometimes called Hospital Discharge Register or Inpatient Register) should include data on all hospital visits of cancer patients. Cancer diagnosis should be mentioned as an indication of the hospital visit, or as a side diagnosis, if the main reason for hospital visit was something else. The Nordic hospital care registers have been occasionally linked with Cancer Registry data to learn if they could be used in improving the completeness of cancer registries. The conclusion has always been that the data from hospital care registers cannot be utilized as such because they include many cancer codes that do not prove to be cancers at all. One of the reasons for the errors is the imprecise coding made by the lay physicians filling in the hospital discharge form. The other reason is classification of cancer suspicions as cancers; there is no mechanism to cancel the wrong suspicion.

However, this additional data source might be useful in improving the registration of certain malignancies (chronic lymphatic leukemia, multiple myeloma, and old-age meningeomas of the brain).

In general, the hospital care registers do not compete in quality with the specialized cancer registries and should not be used as source of outcome events in cancer research. Still there have been projects to change national cancer registration systems

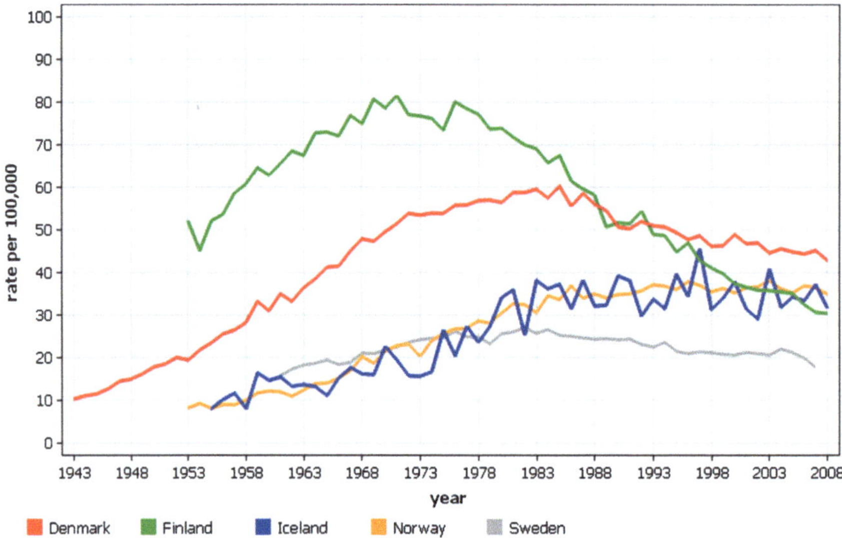

Fig. 4. Age-adjusted incidence of lung cancer among males in the Nordic countries in the years of cancer register coverage (spring 2010). A standard output of NORDCAN tabulation system (26).

similar to hospital discharge registers. For instance, the specialized cancer registry that had been working since 1943 in connection with the Danish Cancer Society was recently moved as one of the automatic registers of the Danish National Board of Health. In the 2007 release of joint Nordic database NORDCAN, the newest year with complete data for Denmark is 2001, while all the other countries have statistics up to 2005 (Fig. 4).

7.2.2. Cause of Death Register

The Cause of Death Register of Statistics Finland includes coded data on underlying cause of death and other diagnoses mentioned in the death certificate. There are several reasons why cause-of-death registry data are normally not comparable with cancer registry data as a measure of cancer frequency in epidemiologic study:

1. Coding of the cancer diagnosis is based on rather limited data and cannot compete with the cancer registry data that are based on multi-source information. Even the primary site may be wrong, and cause-of-death registration does not classify finer subtypes of cancer such as morphological categories. Comparison of the number of cancer deaths in the official mortality statistics (Statistics Finland) and reclassified numbers produced by the FCR indicates high comparability for most sites, but some essential differences for some other sites that are typical sites of metastases. For example, in 2003, there were 205 liver cancer deaths among males in the official mortality statistics but only 126 in the statistics refined by the FCR (28).
2. The competing mortality may be related to the etiologic factor of interest. For instance, cancer patient survival of

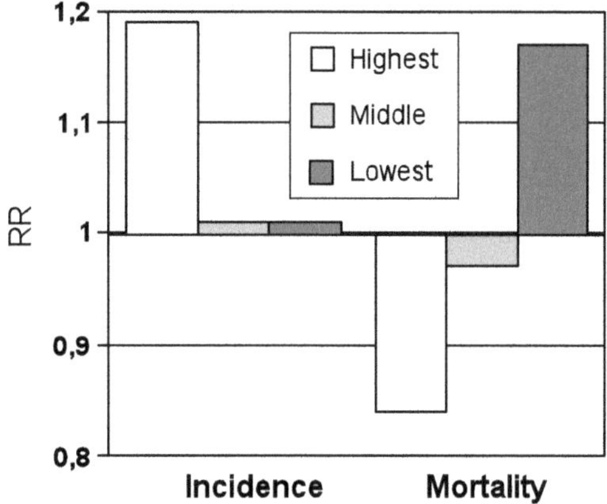

Fig. 5. Socioeconomic variation in rectal cancer among working-aged, economically active, Finnish men measured as standardized incidence or mortality rate. Reference rate: entire Finnish population. Modified from ref. 39.

almost any cancer is essentially better among persons with high socio-economic position than among patients from low socio-economic position (29). This may totally change the character of the outcome. For example, rectal cancer is a disease of the highest social class if measured with incidence but a disease of lower socio-economic status if measured with mortality rate (Fig. 5).

Sometimes cancer mortality may be a more meaningful measure of trend of cancer frequency than incidence. For instance, the incidence of prostate cancer incidence may be misinterpreted, because the strong correlation with the frequency of PSA tests is not understood. Cancer mortality trend describes better the true change in the frequency of prostate cancer in the population. Still, even in this case, the stage-specific incidence rate (excluding localized stage) would probably be the best estimate of true change in prostate cancer risk.

7.2.3. Clinical and Laboratory Data

The cancer registry data are summaries of very detailed data stored in patient records in numerous hospitals. If there is a strong demand to get access to these data, this is possible after getting permissions from the Ministry of Health and Welfare and the head physician of each hospital. The process is laborious but normally successful. It has been used, e.g., in studies of late effects of cancer treatments where details of treatment should be known in much finer precision than what has been recorded in cancer registry database (30–33). Even in these studies, the selection of cases and controls has been based on cancer registry data, and the list of hospitals where the patients have been treated has been extracted from the Cancer Registry.

FCR data also include links to pathological and cytological samples taken from each cancer case. These links have been utilized in numerous studies ((34, 35)) to derive a biological sample from the cancer tissue. The success rate has been high, even for samples taken decades ago.

8. Example of a Multi-register Study: NOCCA

8.1. Background

Assessment of occupational causes of cancer remains an important area of research. The effects of the past exposure will continue to appear in the population over several decades, and new occupation-related risks may have appeared in recent years. It has been estimated that the lack of physical activity will soon cause more cancers in Finland than all known carcinogenic work-related agents together (36).

The on-going study on the effects of the changing work life to cancer risk in the Nordic countries, Nordic Occupational Cancer (NOCCA; http://astra.cancer.fi/NOCCA), is an example of a study based on existing *registries* and population surveys (Fig. 1), which represent a unique research opportunity in the Nordic countries. The project also brings together world-leading *expertise* in industrial hygiene, epidemiology, and biostatistics.

All Nordic residents will be characterized with estimated cumulative life-long exposure to about 30 occupational work-related agents. The large number of cancer cases, about three million, also makes it possible to study rare combinations of cancer and exposure that have never been really studied. Contrasts in exposure levels in five countries are essentially larger than those in any single country.

The pooled database from the Nordic countries presents several features that make it a unique resource for research on occupational cancer:

1. it covers all persons who have reached working ages in five countries;
2. the follow-up after occupational exposures is several decades;
3. data on occupation (basis for exposure estimate) and cancer data are almost complete and of high quality; and
4. data on potential confounders such as smoking, parity, and obesity can be obtained.

This extent and quality of analysis cannot be reached in any other part of the world. Many of the results to be achieved will be novel findings or have importance in confirmation of earlier findings from earlier small studies.

8.2. Data

The study cohort consists of the entire national populations who were 30–64 years old during any computerized population census. In Finland, mainly census data 1970, 1980, and 1990 are used, and the follow-up in the update in 2007 goes up to 31 December 2005 (Fig. 6). In Norway and Sweden also, occupations in 1960 are available in computerized files, while Denmark offers data from 1970 and Iceland from 1981 census only.

The following variables will be collected from registries for each individual. Some of these variables will only be used in specific studies and not all of them can be achieved from all countries.

8.2.1. Baseline Variables

For each cohort member, following socio-demographic variables will be achieved:

- date of birth and gender (both included in the PIC);
- date of immigration (to calculate start of exposure) and country of birth;
- census information from each census: occupation, industry, whether self-employed or not, full-time employment or not, education, and income.

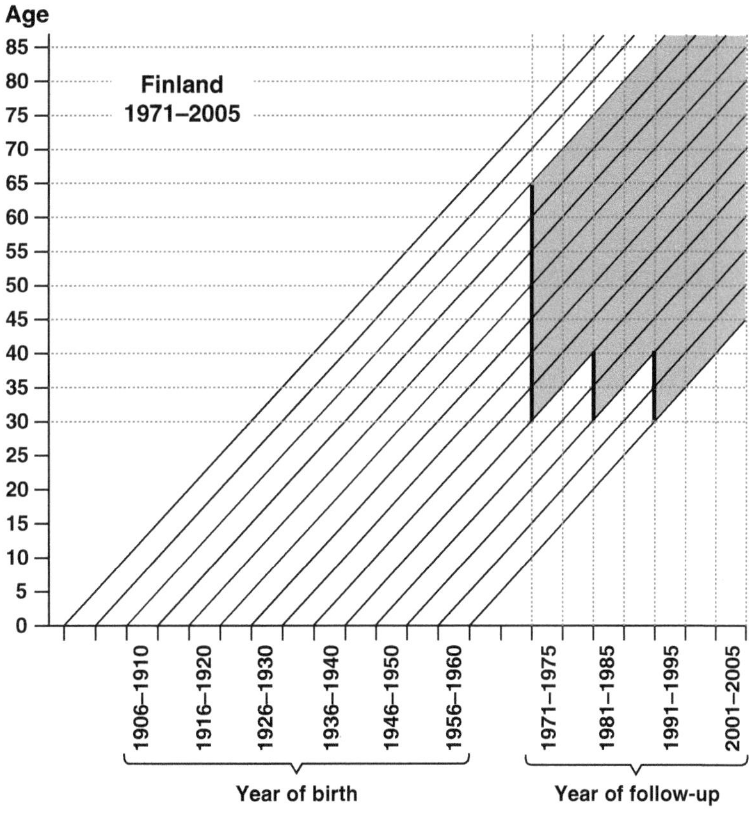

Fig. 6. Setting of NOCCA study, Finland. The thick vertical lines indicate the beginning of follow-up in the occupational cancer risk for each birth cohort of the entire population.

8.2.2. End of Follow-Up Variables

The following data are needed for accurate calculation of person-years at risk:

- date of first emigration; follow-up will end at first emigration even if a person is coming back to country later,
- date of death,
- date of hysterectomy will be used in studies on cancer of the corpus uteri; it changes the relative risk estimates if the prevalence of hysterectomy varies over occupational categories.

8.2.3. Cancer Incidence

In the update done in 2007, all Nordic Cancer Registries had their incidence data ready up to the end of 2003–2005. The large number of cancer cases opens unique possibilities to study less-frequent neoplasms that have hardly ever been studied in the context of occupational exposures. The list of cancer categories will also be expanded to specific histologic subtypes if necessary due to diverging etiology. For example, the suspected occupational risk factors of squamous cell carcinoma of the esophagus are very different from those of adenocarcinoma of the same organ.

We also plan to do stage-specific analyses to separate the roles of varying level of diagnostic activity levels between occupations (typically reflected in non-symptomatic, localized cancers) and real difference in risk. To be able to do all these analyses, following Cancer Registry data items will be achieved for each primary cancer of the individual: date of diagnosis; topography (primary site); morphology (histology); behavior (malignancy); and stage (localized, regional, or distant).

8.2.4. Application of a Job-Exposure Matrix

Exposure to known and suspected carcinogens and other work-related hazards such as work stress, shift work, lack of physical activity, and reduced/postponed parity due to career planning can be estimated via the application of a job-exposure matrix that converts occupational histories known on the level of occupational titles and industries taken from the census data to quantitative estimates of cumulative exposure. Nordic job-exposure matrix has similar structure as possible comprehensive Finnish Job Exposure Matrix (FINJEM) (Fig. 7) that was prepared in the 1990s for the data from Finland (37), and now covers almost 100 occupation-related factors and allows quantitative cumulative exposure estimation and precise timing of relative exposure and lag (38).

8.2.5. Nordic Data on Confounders

Some information on non-occupational risk factors or risk determinants of cancers is available for the entire population and can be directly linked to the individual cohort members. Socioeconomic differentials in reproductive behavior account for some of the socioeconomic variation in the risk of female cancers (30–41). Therefore, information on parity will be obtained from the Swedish Multi-Generation Register and from population registries and

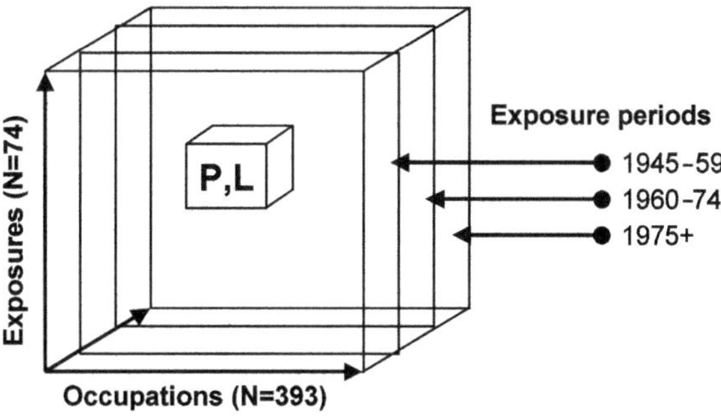

Fig. 7. Dimensions of the Finnish Job Exposure Matrix (FINJEM). Variables P (proportion of exposed persons) and L (mean level of exposure among those exposed, e.g., in ppm) are defined for each stratum.

medical birth registers available in most Nordic countries. The specific risk variables (e.g., age at first birth, age at last birth, and number of children) are formulated from the raw data involving years of birth of all children. It is important to understand that the parity data are completely registered only for a restricted time window, e.g., in Finnish and Norwegian women born after the mid-1930s.

Some others cofactors – including the important cancer risk factors such as tobacco smoking, alcohol drinking, and body mass index – are available only for population sub-samples. When these sub-samples are linked with same census variables as the basic cohort (see Subheading 8.2.1), we will have averaged information on these variables for each occupational, educational, and socio-economic category, by birth cohort. Furthermore, there will be information on the distribution of the values of these variables within each stratum, which makes it possible to use sophisticated hierarchical modeling to reduce the potential risk of ecologic fallacy related to use of averaged data. Cofactor data have been successfully added to the statistical models of occupational risk, e.g., tobacco smoking in analyses of the Finnish and the Norwegian data (38, 42).

In NOCCA, the following cofactor data are collected from the national sources:

1. Percentages of current/former/never *smokers* by birth cohort and calendar time. Quite good data are available from all Nordic countries, however, not from the 1950s and 1960s, that would be most important taken into account the long latency between the start of smoking and cancer outcome. Therefore, in analyses of non-lung cancer risk, the occupation-specific lung cancer incidence rates will be used as proxies of past smoking in the category.
2. The survey data on heavy *alcohol* drinking were considered to be qualitatively too heterogeneous and unreliable.

Therefore, alcohol consumption is estimated using a proxy variable, i.e., occupation-specific liver cirrhosis mortality, which can be estimated from the whole-population data. Therefore, we also link the NOCCA cohort with national cause of death registers. As a side-product, this additional linkage makes it technically possible and easy to study occupational variation in non-cancer outcomes such as risk of cardiovascular or accidental deaths.

3. Prevalence of *obese* person with body mass index >30 will be gathered from population survey data collected, e.g., in the context of cardiovascular risk factor studies.

4. Leisure time *physical activity*, defined as percentage of people who exercise at least three times a week would be useful but may be not be available in useful format in all Nordic countries. A decision has been made to drop this variable for the time being. Physical activity at work is included as one variable in the Nordic JEMs.

8.2.6. Optional Data on Confounders

In addition to the above variables to be similarly estimated for all Nordic citizens, there are additional useful factors that can be used on national settings. In Finland, one such group of variables is the use of medicaments registered by the Social Insurance Institution (Kela). For instance, every regimen of hormonal (replacement) therapy bought by Finnish women since 1994 has been registered. Long-term hormonal therapy increases the risks of cancer of breast, endometrium, and possibly other organs (15) and may, therefore, confound occupational risk estimates of these cancers.

The NOCCA data could also be linked with exact residential history data, but none of the factors related to living environment was considered so strong that adjustment for those factors would be necessary. However, municipalities of residence are recorded because cancer screening practices and coverage vary by municipality, and they evidently affect incidence rates and need to be taken into account in analyses on cervical and breast cancer.

There are plans to extend the Nordic occupational risk factor study to utilize possibilities of extensive Nordic biobanks. For instance, indicators of stress may be measured from historical samples, and cotinine offers another means of possibility to estimate the role of tobacco smoke.

8.3. Publication Plan

The overall results on cancer rates by main job titles were reporte in 2009 (51). A simplified example of the output format is given in Table 4; hundreds of more result tables can be found at http://astra.cancer.fi/NOCCA/tables-sites.html. The end result of the team of occupational hygienists is a publication on the prevalence of carcinogenic agents in the work environment in the Nordic countries (52). Another planned publication will focus

Table 4
Observed number (Obs) of cancers and standardized incidence ratio (SIR) for male waiters (*N*=8,974) in Denmark, Finland, Norway, and Sweden, 1971–1991, by cancer site (49). Reference rate (SIR=100): national male population

ICD-7	Site	Obs	SIR	95% CI
140	Lip	5	51	17–120
141	Tongue	21	716	443–1094
143–144	Mouth	25	547	354–808
145–148	Pharynx	40	656	469–894
150	Esophagus	45	505	368–675
151	Stomach	42	102	74–138
153	Colon	69	135	105–171
154	Rectum	49	125	92–165
155.0	Liver	41	498	357–676
155.1	Gallbladder	9	201	92–382
157	Pancreas	40	156	112–213
161	Larynx	47	366	269–487
162	Lung	246	185	163–210
177	Prostate	123	132	111–158
178	Testis	13	129	69–221
180.0	Kidney	33	135	93–190
181	Bladder	92	160	129–197
190	Skin melanoma	23	97	62–146
193	Brain	28	113	75–164
201	Hodgkin's lymphoma	4	72	20–185
200,202	Non-Hodgkin lymphoma	13	65	35–111
203	Multiple myeloma	9	87	40–164
204.3	Acute leukemia	7	93	37–191
204.0–2,4	Other leukemia	15	129	72–212
199	Unknown	22	143	90–217
140–204	All sites[a]	1092	159	149–168

[a]Excludes non-melanoma skin cancer

on socioeconomic differences in cancer incidence. The fourth publication will report variation in the health habits (see Subheading 8.2.6 above) by country and occupation.

Now when the basic data on both numbers of cancers and prevalence of occupational and non-occupational cancer-related factors have been published, it is time to go for in-depth dose–response analyses on selected combinations of occupational hazards and cancer-specific risks. There are tens of such topics on the current list of specific projects, each of them producing an article in a peer-reviewed journal, and the list will be extended all the time. The main limiting factor will be the capacity of educated epidemiologists.

8.4. Could Biobank Data Improve the NOCCA Study?

Access to large population-representative biobank materials would offer valuable additional information to the NOCCA study, both as cofactor data and as measures of direct occupational exposures.

It was demonstrated above how biobank materials have been used to create estimations of geographic variation of HPV prevalence (Fig. 3). If these samples were linked to occupation codes taken from population census, then we would get estimates of occupation-specific prevalence rates of HPV. These estimates could be used in interpretation of possible roles of other occupation-related factors in the occupational variation of cervical cancer (Table 5). There might be markers of other cofactors that might shed light on occupational variation of, e.g., markers of tobacco

Table 5
Selected occupations with high and low standardized incidence ratio (SIR), social class, adjusted for social class, ages 35–64 years, Finland, 1971–1985 (39)

Occupation	SIR	95% CI
Hotel/restaurant manageresses	4.57	1.48–10.7
Road transport	3.83	1.24–8.93
Woodworkers NOS	3.35	1.09–7.81
Waiters in restaurants	2.24	1.44–3.34
Greasers	2.20	0.89–4.54
Dressmakers	1.96	1.09–3.23
Plywood makers	1.89	1.06–3.10
Waiters in cafés, etc.	1.52	0.91–2.37
Private secretaries	1.48	0.68–2.81
Hygiene and beauty services	1.38	0.75–2.31
Agricultural workers	0.31	0.11–0.67

Table 6
Risk of pancreatic cancer in Finland by occupation, as defined in a specific case–referent study (45) and routine register-based study (39)

	Case–referent study			Register study		
Occupational branch	*N*	OR[a]	95% CI	*N*	SIR[b]	95% CI
Agriculture, forestry, fishing	169	0.8	0.7–1.0	555	0.9	0.8–0.9
Mining and quarrying	6	1.5	0.6–4.2	14	1.5	0.8–2.6
Transport and communication	54	1.0	0.7–1.5	204	1.2	1.0–1.3
Textiles and clothes	12	0.7	0.4–1.4	42	0.8	0.6–1.1
Sawmilling	10	1.3	0.6–2.9	17	1.0	0.6–1.6
Paper and board	17	1.4	0.8–2.5	15	1.6	0.9–2.6
Restaurants, cafés, snack bars	7	1.8	0.3–1.9	21	1.3	0.8–2.1
Hairdressing, manicure	4	1.8	0.5–6.4	14	2.1	1.2–2.3

[a]Odds ratio adjusted for age, gender, smoking, alcohol consumption, and diabetes
[b]Standardized incidence ratio, adjusted for social class

smoking (cotinine), alcohol consumption (CDT, GGT, MCV), markers of stress (cortisol), and light at night (melatonin).

Although industrial hygiene measurement databases provide an overview of exposure levels and can identify situations where exposure is unacceptably high, they often do not represent typical exposure levels in occupations. Therefore, it would be good to measure markers of direct occupational exposures such as organic solvents, gasoline/diesel exhaust, PAHs, or mycotoxins from population-representative biobank samples (Table 6).

Serologic measures of the actual agent or its metabolite represent markers of the internal dose of the chemical agent. Measures of macromolecular adducts reflect the integration over time of carcinogen exposure and interindividual carcinogen metabolism, DNA repair, and other factors. Their half-lives variy from a few weeks to several months for protein adducts, and from hours to years for DNA adducts (43). Biological agents may be measured by serologic markers or nucleic acid markers.

There are still only a limited number of validated biomarkers of health risk (44). Presently, chromosomal aberrations are the best validated predictors of risk. For most types of biomarkers, the most important consideration would not only be the stability of the substance with respect to time after exposure but also variation over time of the year or even time of the day. While persistent toxic substances may be found in body fluids for long periods of time, as a metabolite or protein or DNA adduct, non-persistent substances will disappear from blood quickly and possible adducts formed stay for shorter periods of time. When this is taken into consideration, the measurement of chemicals and their metabolites and of adducts to macromolecules in body fluids may be highly sensitive and specific to the exposure.

9. Final Remarks

The use of sensitive information in research is justifiable only when the studies serve widely acceptable aims and are designed and carried out to the highest possible standards of quality. One of the key issues of the entire register-based study line is expressed in a repeatedly asked question: Can register-based study produce reliable results? Answers to this question have been searched in settings where a setting of a specific study has been repeated using readily existing registry materials. Table 4 shows on its left panel occupation-specific estimates from a Finnish high-standard questionnaire study on risk factors of pancreatic cancers (45). The rates, adjusted for age, gender, smoking, alcohol consumption, and diabetes, are very similar to the simple register-based risk estimates from the very same time period (39). Only the confidence intervals are shorter in the latter one because there is no need to reduce the study size as it is in the questionnaire study. The use of existing administrative data in research is attractive, since the total study costs and the time spent on data collection can be reduced significantly.

9.1. Future Challenges

A governmental working party, which reviewed the current and planned future health and social welfare information system in Finland (46), concluded the importance of continuing the compilation of individual-based data, and supported more active utilization of the nationwide registers. The current health and social welfare information system with registers given in Table 1 will be kept unchanged. The working party proposed only one new register: the National Public Health Institute may initiate a nation-wide *Vaccination Register* to monitor immunization coverage and the possible harmful effects of vaccinations (47).

A threat to the current register practice and epidemiologic research is the tightening of data protection legislation. This may happen, for example, if a single leak occurs from one of the protected data sources such as the national health registers or from a research register. Decision making in such a scenario is political and its endpoint is thus hard to predict. In Finland, the data protection legislation rather promotes than prevents research. In general, use of data is based on informed consent. In case the biobank or other register is a public one and is derived from routine activities, the use may be granted by a general permission by health and social welfare authorities. The research-prone attitude of both the subject and the authority is one of the key elements to make the Nordic countries a world leader in epidemiologic research, and resulted in an improvement in health in the Nordic populations.

In discussions of the good science policy, especially in the biobank-related research where practices and traditions have no such long development history as in other types of register data, it has been criticized that the ethical committees and permission officials sometimes take more power than they morally should, by being slow and requiring unnecessary much bureaucracy work from the researchers to get to the right to start the real scientific work. It is said that such unnecessary work, slowness, and cost may make the study impossible, cause loss of the benefit of cutting-edge research the Nordic countries would have, and finally lead to loss of best research forces and external funding.

9.2. "Paradise of Epidemiology"

Nordic countries have unique possibilities to lead scientific development in many areas of cancer epidemiology. The PICs allow precise automatic linkages, complete population registration systems allow creation of non-selected research materials, and there are no losses to follow-up. The socio-demographic variables cover entire populations, and there are excellent specialized registers on cancer and other diseases. The modern statistical methods allow effective utilization of the scattered survey data on life habits and exposure, estimated both from questionnaire responses and historically collected and systematically stored biological samples of non-diseased representative population.

Since legislators understand the value of the register-based information and develop to a direction that promotes ethically justified use of the data infrastructure to benefit individuals and society, the term "paradise of epidemiology" used to describe the Nordic possibilities is justified. It also gives a global responsibility to do research that is not possible (or unnecessary slow and expensive) elsewhere.

Acknowledgments

Norwegian Journal of Epidemiology kindly gave permission to use parts of the excellent text by Mika Gissler and Jari Haukka – including their table presented as Table 1 in this chapter – published in its special issue "Epidemiological registries – access, possibilities and limitations" in 2004. Felipe Castro from Colombia offered unique materials to demonstrate possibilities to extend the Finnish list of register-based variables with data derived from biobanks. Toni Patama from Kuopio prepared the informative maps.

References

1. Parish S, Collins R, Peto R, Youngman L, Barton J, Jayne K, Clarke R, Appleby P, Lyon V, Cederholm-Williams S, Marshall J, Sleight P. (1995) Cigarette smoking, tar yields, and non-fatal myocardial infarction: 14,000 cases and 32,000 controls in the United Kingdom. The International Studies of Infarct Survival (ISIS) Collaborators. Br Med J. **311**, 471–477.
2. Pukkala E. (2007) Nordic biological specimen bank cohorts as basis for studies of cancer causes and control – quality control tools for study cohorts with more than 2 million sample donors and 100,000 prospective cancers. (chapter 3 in this book).
3. Gissler M, Haukka J. (2004) Finnish health and social welfare registers in epidemiological research. *Nor Epidemiol.* **14**, (1) 113–120.
4. Cappelen I, Daltveit AK (guest editors). (2004) Epidemiological registries – access, possibilities and limitations. *Norsk Epidemiology.* **14**, 3–128 (http://www.ub.ntnu.no:80/journals/norepid/2004-1.html).
5. Keskimäki I, Aro S. (1991) Accuracy of data on diagnoses, procedures and accidents in the Finnish Hospital Discharge Register. *Int J Health Sci.* **2**, 15–21.
6. Teppo L, Pukkala E, Lehtonen M. (1994) Data quality and quality control of a population-based cancer registry. *Acta Oncol.* **33**, 365–369.
7. Gissler M, Teperi J, Hemminki E, Meriläinen J. (1995) Data quality after restructuring a nationwide medical birth registry. *Scand J Soc Med.* **23**, 75–80.
8. Gissler M, Ulander V-M, Hemminki E, Rasimus A. (1996) Declining induced abortion rate in Finland: data-quality of the abortion register. *Int J Epidemiol.* **25**, 376–380.
9. Gissler M, Kauppila R, Meriläinen J, Toukomaa H, Hemminki E. (1997) Pregnancy-associated deaths in Finland in 1987–1994 – definition problems and benefits of record linkage. *Acta Obstet Gynaecol Scand.* **76**, 651–657.
10. Gissler M, Järvelin M-R, Hemminki E. (2000) Children's health in Northern Finland – a comparison of cohort and register based studies. *Eur J Epidemiol.* **16**, 59–66.
11. Pukkala E, Andersen A, Berglund G, Gislefoss R, Gudnason V, Hallmans G, Jellum E, Jousilahti P, Knekt P, Koskela P, Kyyrönen P, Lenner P, Luostarinen T, Löve A, Ögmundsdóttir H, Stattin P, Tenkanen L, Tryggvadóttir L, Virtamo J, Wadell G, Widell A, Lehtinen M, Dillner J. (2007) Nordic biological specimen banks as basis for studies of cancer causes and control – more than 2 million sample donors, 25 million person-years and 100,000 prospective cancers. *Acta Oncol.* **46**, 286–307.
12. Tolonen H, Helakorpi S, Talala K, Helasoja V, Martelin T, Prättälä R. (2006) 25-year trends and socio-demographic differences in response rates: Finnish adult health behavior survey. *Eur J Epidemiol.* **21**, 409–415.
13. Vesterinen E, Pukkala E, Timonen T, Aromaa A. (1993) Cancer incidence among 78,000 asthmatic patients. *Int J Epidemiol.* **22**, 976–982.
14. Pukkala E, Pönkä A. (2001) Increased incidence of cancer and asthma in houses built on a former dump area. *Environ Health Perspect.* **109**, 1121–1125.
15. Lyytinen H, Pukkala E, Ylikorkala O. (2006) Breast cancer risk in postmenopausal women using estrogen-only therapy. *Obstet Gynecol.* **108**, 1354–1360.
16. Kauppinen T. (2001) Finnish occupational exposure databases. *Appl Occup Environ Hyg.* **16**, 154–158.
17. Anttila A, Heikkilä P, Pukkala E. et al. (1995) Excess lung cancer among workers exposed to lead. *Scand J Work Environ Health.* **21**, 460–469.
18. Kauppinen T, Saalo A, Pukkala E, Virtanen S, Karjalainen A, Vuorela R. (2007) Evaluation of a national register on occupational exposure to carcinogens: effectiveness in the prevention of occupational cancer, and cancer risks among the exposed workers. *Ann Occup Hyg.* **51**, 463–470.
19. Frick MH, Elo O, Haapa K, Heinonen OP, Heinsalmi P, Helo P, Huttunen JK, Kaitaniemi P, Koskinen P, Manninen V. (1987) Helsinki Heart Study: primary-prevention trial with gemfibrozil in middle-aged men with dyslipidemia. Safety of treatment, changes in risk factors, and incidence of coronary heart disease. *N Engl J Med.* **317**, 1237–1245.
20. ATBC (The Alpha-Tocopherol B-CCPSG). (1994) The effect of vitamin E and beta carotene on the incidence of lung cancer and other cancers in male smokers. *N Engl J Med.* **330**, 1029–1035.
21. Baldwin JA, Acheson ED, Graham WJ. (ed.) (1987) *Textbook of medical record linkage.* Oxford Medical Publications. Oxford University Press, Oxford.
22. Castro F, Haimila K, Pasanen K, Kaasila M, Partanen J, Patama T, Partanen J, Surcel H-M, Pukkala E, Lehtinen M. (2007) Geographic distribution of cervical cancer associated HLA antigens and cervical cancer incidence in fertile-aged Finnish women. *Intl J STD AIDS.* **18**, 672–679.

23. Laukkanen P, Koskela P, Pukkala E, Dillner J, Läärä E, Knekt P, Lehtinen M. (2003) Time trends in incidence and prevalence of human papillomavirus type 6, 11 and 16 infections in Finland. *J Gen Virol.* **84**, 2105–2109.
24. Lehtinen M, Kaasila M, Pasanen K, Patama T, Palmroth J, Laukkanen P, Pukkala E, Koskela P. (2006) Seroprevalence ATLAS of HPV infections in Finland in the 1980's and 1990's. *Int J Cancer.* **120**, 2612–2619.
25. Luoto R, Raitanen J, Pukkala E, Anttila A. (2004) Effect of hysterectomy on incidence trends of endometrial and cervical cancer in Finland 1953–2010. *Br J Cancer.* **90**, 1756–1759.
26. Engholm, G., Ferlay, J., Christensen, N., Bray, F., Gjerstorff, M.L., Klint, Å., Køtlum, J.E., Ólafsdóttir, E., Pukkala, E., Storm, H.H. (2010) NORDCAN: cancer incidence, mortality, prevalence and prediction in the Nordic countries, Version 3.6. *Association of the Nordic Cancer Registries.* Danish Cancer Society: http://www.ancr.nu.
27. Mattsson B. (1984) Cancer registration in Sweden. Studies on completeness and validity of incidence and mortality registers. Dept of Oncolology and Cancer Epidemiology, Stockholm.
28. Finnish Cancer Registry. (2005) Cancer in Finland 2002 and 2003. Helsinki, Cancer Society of Finland Publication No. 66.
29. Auvinen A, Karjalainen S, Pukkala E. (1995) Social class and cancer patient survival in Finland. *Am J Epidemiol.* **142**, 1089–1102.
30. Travis LB, Andersson M, Gospodarowicz M, van Leeuwen FE, Bergfeldt K, Lynch CF, Curtis RE, Kohler BA, Wiklund T, Storm H, Holowaty E, Hall P, Pukkala E, Sleijfer DT, Clarke EA, Boice JD, Jr, Stovall M, Gilbert E. (2000) Treatment-associated leukemia following testicular cancer. *J Natl Cancer Inst.* **92**, 1165–1171.
31. Travis LB, Hill DA, Dores GM, Gospodarowicz M, van Leeuwen FE, Holowaty E, Glimelius B, Andersson M, Wiklund T, Lynch CF, van't Veer MB, Glimelius I, Storm H, Pukkala E, Stovall M, Curtis R, Boice JD, Jr, Gilbert E. (2003) Breast cancer following radiotherapy and chemotherapy among young women with Hodgkin disease. *JAMA.* **290**, 465–475.
32. Gilbert ES, Stovall M, Gospodarowicz M, van Leeuwen FE, Andersson M, Glimelius B, Joensuu T, Lynch CF, Curtis RE, Holowaty E, Storm H, Pukkala E, van't Veer MB, Fraumeini JF, Boice JD, Jr, Clarke EA, Travis LB. (2003) Lung cancer after treatment for Hodgkin´s disease: focus on radiation effects. *Radiat Res.* **159**, 161–173.
33. Hill DA, Gilbert E, Dores GM, Gospodarowicz M, van Leeuwen FE, Holowaty E, Glimelius B, Andersson M, Wiklund T, Lynch CF, van't Veer MB, Storm H, Pukkala E, Stovall M, Curtis RE, Allan JM, Boice JD, Travis LB. (2005) Breast cancer risk following radiotherapy for Hodgkin lymphoma: modification by other risk factors. *Blood.* **106**, 3358–3365.
34. Leonard DG, Travis LB, Addya K, Dores GM, Holowaty EJ, Bergfeldt K, Kohler BA, Lynch CF, Wiklund T, Stowall M, Hall P, Pukkala E, Slater DJ, Felix CA. (2002) p53 mutations in leukemia and myelodysplastic syndrome after ovarian cancer. *Clin Cancer Res.* **8**, 973–985.
35. Worrillow LJ, Travis LB, Smith AG, Rollinson S, Smith AJ, Wild CP, Holowaty EJ, Kohler BA, Wiklund T, Pukkala E, Roman E, Morgan GJ, Allan JM. (2003) An intron splice acceptor polymorphism in hMSH2 and risk of leukemia after treatment with chemotherapeutic alkylating agents. *Clin Cancer Res.* **9**, 3012–3020.
36. Rintala PE, Pukkala E, Paakkulainen HT, Vihko VJ. (2002) Self-experienced physical workload and risk of breast cancer. *Scand J Work Environ Health.* **28**, 158–162.
37. Kauppinen T, Toikkanen J, Pukkala E. (1998) From cross-tabulations to multipurpose exposure information systems: a new job-exposure matrix. *Am J Ind Med.* **33**, 409–417.
38. Pukkala E, Guo J, Kyyrönen P, Lindbohm M-L, Sallmén M, Kauppinen T. (2005) National job-exposure matrix in analyses of census-based estimates of occupational cancer risk. *Scand J Work Environ Health.* **31**, 97–107.
39. Pukkala E. (1995) Cancer risk by social class and occupation. A survey of 109,000 cancer cases among Finns of working age. *Contributions to Epidemiology and Biostatistics*, vol 7. Basel: Karger, pp. 1–288.
40. Kogevinas M, Pearce N, Susser M, Boffetta P. (eds.) (1997) *Social Inequalities and Cancer.* IARC Sci Pub No 138. Lyon, IARC.
41. Pukkala E, Weiderpass E. (1999) Time trends in socio-economic differences in incidence rates of cancers of the breast and female genital organs (Finland, 1971–1995). *Int J Cancer.* **81**, 56–61.
42. Haldorsen T, Andersen A, Boffetta P. (2004) Smoking-adjusted incidence of lung cancer by occupation among Norwegian men. *Cancer Causes Control.* **15**, 139–147.
43. Perera FP. (2000) Molecular epidemiology: on the path to prevention? *J Natl Cancer Inst.* **92**, 602–612.
44. Bonassi S, Au WW. (2002) Biomarkers in molecular epidemiology studies for health risk prediction. *Mutat Res.* **511**, (1) 73–86.

45. Partanen T, Kauppinen T, Degerth R, Moneta G, Mearelli I, Ojajärvi A, Hernberg S, Koskinen H, Pukkala E. (1994) Pancreatic cancer in industrial branches and occupations in Finland. *Am J Ind Med.* **25**, 851–866.
46. Gissler M, Muuri A, Hämäläinen H. (2004) How to make good even better? The Reform of Social and Health Care Information System in Finland. *Dialogi* **1B**, 20–22.
47. Lehtinen M, Herrero R, Mayaud P, Barnabas R, Dillner J, Paavonen J, Smith PG. (2006) Studies to assess long-term efficacy and effectiveness of HPV vaccination in developed and in developing countries. *Vaccine.* **24**, 233–241,
48. Valkonen S. (1999) *Annual statistics of biomonitoring services in 1998* (In Finnish). Finnish Institute of Occupational Health, Helsinki.
49. Andersen A, Barlow L, Engeland A, Kjaerheim K, Lynge E, Pukkala E. (1999) Work-related cancer in the Nordic countries. *Scand J Work Environ Health.* **25**, (2).
50. Pukkala E. (1992) Use of record linkage in small-area studies. In: Geographical & Environmental Epidemiology: Methods for Small-area Studies. Eds. P. Elliott, J. Cuzick, D. English, R. Stern. Oxford University Press, Oxford 1992, pp. 125–131.
51. Pukkala, E., Martinsen, J.I., Lynge, E., Gunnarsdottir, H.K., Sparén, P., Tryggvadottir, L., Weiderpass, E., Kjærheim, K. (2009) Occupation and cancer – follow-up of 15 million people in five Nordic Countries. *Acta Oncol.* **48**, 646–790.
52. Kauppinen, T., Heikkilä, P., Plato, N., Woldbaek, T., Lenvik, K., Hansen, J., Kristjansson, V., Pukkala, E. (2009) Construction of job-exposure matrices for the Nordic Occupational Cancer Study (NOCCA). *Acta Oncol.* **48**, 791–800.

Chapter 6

Study Designs for Biobank-Based Epidemiologic Research on Chronic Diseases

Esa Läärä

Abstract

A review is given on design options to be considered in epidemiologic studies on cancers or other chronic diseases in relation to risk factors, the measurement of which is based on stored specimens in large biobanks. The two major choices for valid and cost-efficient sampling of risk factor data from large biobank cohorts are provided by the *nested case–control* design, and the *case–cohort* design. The main features of both designs are outlined and their relative merits are compared. Special issues such as matching, stratification, and statistical analysis are also briefly discussed. It is concluded that the nested case–control design is better suited for studies involving biomarkers that can be influenced by analytic batch, long-term storage, and freeze-thaw cycles. The case–cohort design is useful, especially when several outcomes are of interest, given that the measurements on stored materials remain sufficiently stable during the study.

Key words: Nested case–control , Case–cohort, Matching, Stratification, Statistical analysis, Risk factors

1. Introduction

Epidemiologic studies of chronic diseases require large study populations and skillful planning on various aspects of study design, selection of the study subjects, measurements of the values of interesting risk factors and other variables, organization of the follow-up for identification of the study outcomes, and analysis of the results. Careful planning is even more demanding, when measurements are based on stored biological materials, such as tissue or blood specimens, considering the labor and costs associated with them.

In this paper, a review is presented on the choices of epidemiologic study designs to be considered in this kind of investigations. Our special focus is on the *nested case–control* (NCC)

Joakim Dillner (ed.), *Methods in Biobanking*, Methods in Molecular Biology, vol. 675,
DOI 10.1007/978-1-59745-423-0_6, © Springer Science+Business Media, LLC 2011

design and the *case–cohort* (CC) design. More detailed accounts on the various designs are given in many excellent textbooks, such as those of dos Santos Silva (1) and Rothman et al. (2). Important aspects of the two major designs from a more statistical perspective are concisely and quite untechnically treated, e.g., by Borgan and Samuelsen (3). Vineis et al. (4) provide an extensive discussion on the relative merits of the NCC and the CC designs with special reference to biobank-based studies, and they offer thoughtful guidelines for choosing between them.

As a concrete introduction to the theme, two representative examples of modern biobank-based epidemiologic research are briefly summarized.

Example 1. "Activation of maternal Epstein-Barr virus infection and risk of acute leukemia in the offspring" (5). The study population comprised a joint cohort of ca. 550,000 offspring, their mothers being identified from the Icelandic and the Finnish biobanks covering pregnant women. Serum samples were routinely taken from all these women in the first trimester of pregnancy, from 1975 to 1983 onwards in the two countries, respecively. Follow-up of the offspring began at birth and lasted until 1997. In the total of 7 million person-years of follow-up, 304 cases of acute lymphatic leukemia (ALL) and 39 cases of other leukemias (non-ALL) occurring in the offspring by 15 years of age were identified from the national cancer registries. Three or four control subjects for each case were sampled from the original cohorts by incidence density sampling. The control subjects were matched with the case on biobank/country, maternal age at serum sampling (±2 years), date of specimen collection (±2 months), as well as on gender, and date of birth (±2 months) of the offspring. The frozen sera from mothers of these cases and from 943 mothers of the control subjects were analyzed for antibodies to viral capsid antigen (VCA), early antigen, and EBV transactivator protein ZEBRA. One major result was that "EBV VCA IgM antibodies were associated with a statistically significant relative risk of childhood ALL (odds ratio = 1.9, 95% confidence interval: 1.2, 3.0)."

Example 2. "Risk alleles of *USF1*-gene predict cardiovascular disease" (6). The study population comprised two FINRISK cohorts in Finland, in total ca. 14,000 males and females, of initially 25–64 years of age. The cohorts were recruited in 1992 and 1997, respectively. The baseline measurements comprised a health examination and a structured questionnaire, and blood specimens were also taken at entry. A subcohort of 786 subjects was randomly sampled from the cohorts. The cohorts were followed-up from entry to 31 Dec 2001 and 31 Dec 2003, respectively. In the 112,000 years of total follow-up, 528 new cases of cardiovascular diseases (CVD) were identified in the cohorts, of which 72 were in the subcohort. The frozen blood specimens pertaining to the cases and the subchort members were genotyped. One of the main results was that "female carriers of a *USF1* risk haplotype had a twofold risk of a CVD event (hazard ratio (HR) 2.02; 95% confidence interval (CI) 1.16–3.53), after adjustment for conventional risk factors."

2. Validity and Efficiency of an Epidemiologic Study

An epidemiologic study is a measurement exercise (2). The object of measurement is some *parameter* of interest, such as the *hazard rate ratio* (HR or "relative risk") of a major coronary event between individuals with a high-risk and a low-risk haplotype, respectively. The result of this exercise is an *estimate* of the parameter, which is an empirical measure to be computed from the available data. Estimates of the HR include the *incidence rate ratio* (IR or *incidence density ratio*) obtained from a cohort study, or the *exposure odds ratio* (EOR) from a case–control study.

The estimation of a parameter is prone to error; we can express an estimate as a sum of three components:

$$\text{Estimate} = \text{true parameter value} + \text{bias} + \text{random error}.$$

Common sources of *bias* or *systematic error* include (a) confounding or non-comparability of the exposure groups, (b) measurement error and misclassification, (c) non-response, losses to follow-up, or otherwise incomplete data, and (d) sampling and selection of subjects to the study and to be measured. An educating presentation on various biases is given by Maclure and Schneeweiss (7). The main sources of *random error* are in turn (a) biological variation between and within individuals, (b) measurement variation, (c) sampling (whether random or non-random), and (d) division of exposure (whether properly randomized or non-randomized).

An epidemiologic study is said to be *valid*, when its design and methods would provide an *unbiased* estimate of the parameter (such as HR) of interest. Unbiased estimation means that the estimate (like IR or EOR) would equal the true parameter value (HR) if the study had no random error. For example, if the true HR on CVD events for high- vs. low-risk haplotype carriers was 2.5, this value would be exactly obtained by our estimate IR if we had unlimited amount of data and if our designs were valid. (NB. By exceptional luck, we could get an IR of 2.5 also with typical amount of data even with a biased design!)

The *precision* of an estimate means smallness of random error in estimation. Random error is measured by the variance or *standard error* (SE) of the estimate, or by the *confidence interval* (CI) of the parameter. The *efficiency* of a design means its ability to provide a precise estimate with given data. We say that design A is more efficient than design B if either (1) with the same amount of data, the estimate from A has a smaller random error than that from B, or (2) smaller amount of data is needed by design A to obtain the same precision as that obtained by B.

3. Cohort Studies

An outline of a typical cohort study or a full cohort design is as follows:

1. Subjects fulfilling the eligibility criteria are selected to the study cohort.
2. Risk factors of interest as well as relevant confounders and effect modifiers are measured in all cohort members.
3. New incident cases of outcome (e.g., cancer) are identified during the follow-up from the time of entry to until the time of exit from the follow-up.
4. Incidence rates = cases/person–time in the exposure groups, and the ratios (IRs) between them are computed.
5. Confounding and modification are controlled by stratification and Mantel–Haenszel methods, or nowadays more commonly by regression modeling: the Poisson regression, or the proportional hazards (Cox) model.

In both examples presented in the introduction, a full cohort design would imply that serologic assays for the EBV antibodies would have been performed on the sera of all the 550,000 mothers in Iceland and Finland, as well as genotyping for the USF1-gene would have been conducted for all the 14,000 members of the two FINRISK cohorts.

The principle of estimating the HRs of interest from a full cohort design is illustrated in the simplest possible setting: one single dichotomous risk factor. From the figures given in Table 1,

Table 1
Crude summary of follow-up results in a cohort study addressing the effect of a dichotomous risk factor ("exposed" vs. "unexposed") on the hazard of getting a given disease

	Exposed	Unexposed	Total
New cases	D_1	D_0	D
Person–time	Y_1	Y_0	Y
Incidence rate	$I_1 = D_1/Y_1$	$I_0 = D_0/Y_0$	$I = D/Y$

the target parameter, HR, is estimated by the ratio of the empirical incidence rates I_1 and I_0 in the two exposure groups.

$$\mathrm{IR} = \frac{I_1}{I_0} = \frac{D_1/Y_1}{D_0/Y_0} = \frac{D_1/D_0}{Y_1/Y_0}.$$

This crude estimation ignores the possible confounding caused by other risk factors of the outcome disease, but provides a convenient starting point to illustrate the precision and efficiency of different designs.

The precision in the estimation of the HR depends inversely on the numbers of cases. The estimated variance of the logarithm of the crude IR is, namely, expressed as

$$V = \frac{1}{D_1} + \frac{1}{D_0} = \frac{1}{\text{no. exposed cases}} + \frac{1}{\text{no. unexposed cases}}.$$

From this, we obtain the common approximate confidence limits for the hazard ratio:

$$\mathrm{IR} \times \exp(\pm 1.96 \times \sqrt{V}).$$

Note that the variance does not depend on the sizes of the exposure groups (or their person–times) as such, even if these were millions. However, for rare diseases with low rates, large cohorts are needed to obtain enough cases for adequate precision.

Collection and processing of data on exposure variables, confounders, and modifiers are very slow and expensive in large cohorts. It is relatively easy and cheap with data obtained by questionnaires or from readily available registers. However, it would be extremely costly and laborious for, e.g., measurements from biological specimens (like genotyping, antibody assays, etc.), dietary diaries, and occupational exposure histories in manual records. In our two example studies, the full cohort design would obviously be an imaginary possibility only.

Thus, a question arises whether we are able to obtain equally valid estimates of the interesting HRs with nearly as good precision as those obtained by some other, less costly strategies. The answer is "yes," and we shall justify this by first inspecting more closely the estimation of hazard ratios:

The crude IR in a cohort study can be expressed by

$$\begin{aligned} \mathrm{IR} = \frac{D_1 / D_0}{Y_1 / Y_0} &= \frac{\text{cases: exposed / unexposed}}{\text{person – times: exposed / unexposed}} \\ &= \frac{\text{exposure odds in cases}}{\text{exposure odds in person – times}} \\ &= \text{exposure odds ratio (EOR)} \end{aligned}$$

In practical terms, this estimator relates the exposure distribution observed in the cases vs. the exposure distribution prevailing in the whole cohort. A suggestion is thus given for the search of more efficient designs:

1. To obtain information on the *numerators* of the incidence rates in the two exposure groups, one should aim at collecting exposure data on all possible cases of the outcome disease.
2. As to the *denominators* of the rates, one may estimate with high precision the division of person–times Y_1/Y_0 into the exposure groups by appropriate sampling of referent or "control" subjects, on whom exposure data will be measured and collected, from the members of the whole cohort at risk. This idea leads us to the case–control designs.

4. Case–control Studies

The general principle in the so-called case–control or case–base, or case–referent designs is the following: The selection of study subjects from a given study population is stratified by the outcome (disease) under study.

The study population comprises subjects who *would be* included as cases *if they got* the outcome disease during the study. Hence, this population may also be called as the source population of the cases (2).

In cohort-based case–control studies, the study population is a well-defined closed population, the membership being fixed by entry to the cohort and lasting forever. These kinds of case–control studies are the focus of this article, and the so-called hospital-based and register-based case–control studies are left aside (1).

In all types of case–control studies, the data on interesting risk factors are collected separately from

1. The *case group*, comprising all (or a high proportion of) the D subjects in the study population (total N subjects) encountering the outcome disease during follow-up
2. The referent or *control group*, which is a *random sample* of C subjects from the whole population (C much smaller than N), such that the eligible controls must be *at risk,* i.e., alive, under follow-up and still free from the outcome at specified time points

Depending on how these time points are actually specified, different sampling schemes or designs for the selection of control subjects are obtained. The major sampling schemes or designs are the following:

(a) *Traditional* design ("case–noncase" sampling): Controls are chosen from these $N - D$ cohort members who are still at risk (healthy) *at the end* of the follow-up. We do not consider this design any further, which is typically used in studies of acute diseases (outbreaks). It also presupposes complete follow-up (no losses) of the cohort over a fixed-length risk period, which is rarely realized with chronic diseses.

(b) *Incidence density sampling* (or concurrent sampling) design: Controls are drawn at different times *t during the follow-up* from these N_t subjects at risk. An important special case is the *nested case–control* design (NCC), in which a set of controls is sampled in a *time-matched* manner from the *risk set at each time t of diagnosis* of a new case.

(c) *Case–cohort* design (CC): The control group – *subcohort* – is a random sample of the whole cohort (N) *at the beginning of the follow-up.*

It is worth mentioning that the term "nested case–control studies" has variable meanings. In biostatistical literature (3), it commonly refers to the most popular variant of density sampling, in which *time-matching* or *risk-set sampling* is employed: At each time t when a new case is found, a set of controls is sampled from the N_t members of the study population belonging to the *risk-set* at time t (see above). This is illustrated in Fig. 1. However, in some epidemiologic texts (1), the "nested case–control design" refers to any kind of control sampling when a study population is a well-defined cohort, covering thus also the traditional sampling as well as the case–cohort design. Here, the word "nested case–control design" is used in the first meaning, i.e., referring to the time-matched sampling of controls from risk sets (3).

Note that in this design, a control chosen at a time of some previous case can later on become a case, too.

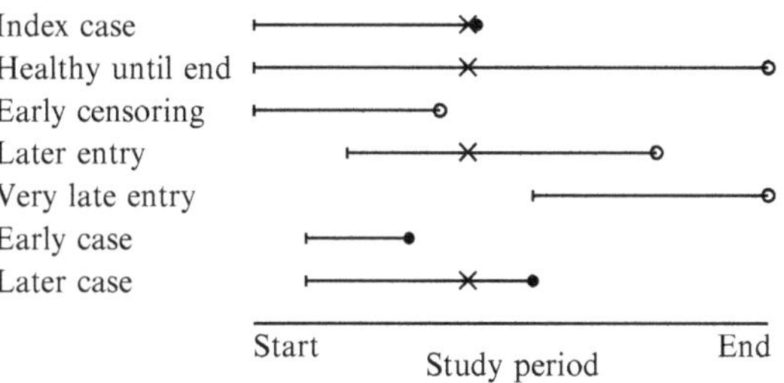

Fig. 1. Time-matched sampling from risk sets. Follow-up lines of seven subjects run vertically at different levels, and they may end either by the outcome event (*filled circle*) or censoring (*open circle*) due to deaths from other causes or emigration. The risk set from which controls are sampled at the time of diagnosis of the index case comprises subjects (marked by ×) who are alive, free from the outcome, and under follow-up at that time.

In order to guarantee a valid sampling frame for control selection from the relevant risk set at any time, it is very important to maintain accurate and complete follow-up also with respect to dates of deaths and emigrations occurring in the cohort, apart from the outcome events.

Example 1 in the introduction is a typical NCC study. Time-matched sampling of controls from the risk sets was employed, although not explicitly described in this paper that for each case, the chosen controls were alive, not censored, and free from leukemia at the date of diagnosis. Close time-matching was actually performed on the age scale, too, because the date of birth of each control was less than 2 months apart from that of the case. In addition to time and age, the selection of controls was matched on various other factors, too (more on this in Sect. 6).

Example 2 in the introduction is clearly a CC study. The subcohort, a random sample of 786 subjects from the whole cohort, selected at the outset, served as the control group for all subsequent cases. In this design, a subcohort member can become a case, too, as actually happened to 72 subjects.

The nested case–control variant of the density sampling design (b) is the most popular one in chronic disease epidemiology. The case–cohort design (c) is newer, but is gradually gaining in popularity. It is particularly recommended when several outcomes are of interest, and measurements of risk factors from any stored material are relatively stable.

5. Estimation, Precision, and Efficiency

Results from a case–control study are often summarized as in Table 2. From these four counts, the crude exposure odds ratio is computed:

$$\mathrm{EOR} = \frac{D_1 / D_0}{C_1 / C_0} = \frac{\text{cases: exposed / unexposed}}{\text{controls: exposed / unexposed}}.$$

A common but false doctrine, unfortunately still found in many elementary textbooks in epidemiology, is that the only parameter

Table 2
Crude summary of results in a case–control study with a dichotomous risk factor

	Exposed	Unexposed	Total
No. of cases	D_1	D_0	D
No. of controls	C_1	C_0	C

estimable from a case–control study is the odds ratio, meaning specifically the *risk odds ratio* (ROR)

$$\text{ROR} = \frac{\text{odds of disease in the exposed}}{\text{odds of disease in the unexposed}} = \frac{R_1/(1-R_1)}{R_0/(1-R_0)},$$

where R_1 and R_0 are the *risks* of disease over a fixed risk period in the two exposure groups. This holds indeed in the traditional "case–noncase" design. When the disease is "rare," the ROR is closely approximating the corresponding *risk ratio* RR = R_1/R_0 as well as the HR.

However, in case–control studies based on density sampling or case–cohort sampling, one can estimate directly the HR without any rare disease assumption. For the density sampling, the argument is simplified as follows (2): It can be shown that given certain assumptions, the exposure odds C_1/C_0 among the controls provide a statistically consistent estimate of the odds Y_1/Y_0 of person–times between the exposure groups in the whole cohort from which the cases and controls are sampled. Hence, EOR between cases and controls actually is a valid and efficient estimate of the unknown HR, which is the target of our interest.

In the case–cohort design, the principle is the same but the estimation of the hazard ratio is more complicated. Nevertheless, the argument above illustrates the true role of the controls: They are NOT representing the population of "non-cases," i.e., those who would remain healthy; instead, they are providing data on the distribution of exposures in the whole cohort.

As an aside, another common but misleading textbook wisdom says that absolute levels of incidence rates or risks cannot be estimated from a case–control study. This statement holds only for studies based on an ill-defined source population of cases, such as hospital-based case–control studies in USA. Suppose, however, that (1) a well-defined cohort is followed up for Y total person-years, (2) $D = D_1 + D_0$ cases plus $C = C_1 + C_0$ controls are drawn from it, and (3) their exposure assessed. In these circumstances, the person-years and the crude absolute incidence rates in the two exposure groups $k=0$, 1 would be estimated in a straightforward way:

$$\tilde{Y}_k = \frac{C_k}{C} \times Y, \quad \tilde{I}_k = \frac{D_k}{\tilde{Y}_k}.$$

These crude computations are, however, not useful in real-life studies with variable follow-up times over a wide age range. More refined methods for absolute risk estimation are available, though, as presented by Langholz and Borgan (8).

Consider next the precision and efficiency of the estimation of "relative risk" in case–control studies. In density sampling, or the NCC design, the estimated variance of the logarithm of the crude exposure odds ratio may be expressed as

$$V_{\text{NCC}} = \frac{1}{D_1} + \frac{1}{D_0} + \frac{1}{C_1} + \frac{1}{C_0}$$
$$= \text{cohort variance} + \text{sampling variance.}$$

The variance depends thus basically on the numbers of exposed and unexposed cases, whenever the numbers of controls C_1 and C_0 are clearly bigger than the numbers of cases. Hence, the variance is not much bigger than that in a full cohort study with the same number of cases. Usually, the gain to be obtained with more than four or five controls per case is marginal. This shows that the case–control design is very cost-efficient!

Some results from Example 1 are summarized in Table 3. Ignoring matching for the sake of illustration only, the crude estimate of the HR between the antibody positives and the antibody negatives is

$$\text{EOR} = \frac{30/274}{47/815} = 1.9$$

Even though one should not be content with reporting a crude estimate when really analyzing matched data, we note that this value happens to be numerically the same as the HR estimate (or "odds ratio," as the authors called it) reported in the original article, which was adjusted for matching factors and for some other covariates by conditional logistic regression model (see Sect. 7).

The estimated variance of log(EOR) is

$$V = \left(\frac{1}{30} + \frac{1}{274}\right) + \left(\frac{1}{47} + \frac{1}{815}\right) = 0.0370 + 0.0225 = 0.0595,$$

and the 95% confidence interval ranges from 1.2 to 3.1, these crude limits being again close to the reported ones. Thus, the variance

Table 3
Maternal IgM antibodies to the EBV VCA and the acute lymphatic leukemia (ALL) in the offspring. Numbers of antibody positive and negative cases and controls

	Maternal antibody status		
	Positive	Negative	Total
No. of cases of ALL	30	274	304
No. of controls	47	815	862

in the EOR estimation was increased only by 0.0225/0.037 = 61%, when antibody status was assessed in less than 900 controls, compared to the theoretically conceivable full cohort design, which would have required altogether 550,000 antibody assays.

6. Matching and Other Forms of Stratified Sampling

Matching is a procedure typically applied in nested case–control studies. It means stratified sampling of controls, such that for each individual case, the controls are chosen from, e.g., the same region, sex, and age group, etc., as the case.

The main reason for matching is that it creates similar distributions in controls and cases for the factors used as matching criteria, which leads to more balanced comparisons. Hence it tends to increase precision and efficiency in HR estimation, but only if the matching factors are (1) strong risk factors of the disease and (2) correlated with the exposure.

In addition, confounding due to observable but not quantifiable factors (like sibship, neighborhood, etc.) can be removed by close matching, but the bias is removed only if the data are properly analyzed. Especially in biobank studies matching the controls with each case on the storage time, freeze-thaw cycle and analytic batch improve comparability of measurements from frozen biological material (4).

As noted above, in Example 1, the control subjects were matched with the cases on time of diagnosis and age. Moreover, the controls were drawn from the same biobank/country and the same gender group, and the differences in maternal ages were less than 2 years compared to that in the cases. In addition, the dates of specimen collection were within ±2 months. Hence, matching on storage time was realized. It was not mentioned in the paper, whether the sera of each case and the matched controls were assayed in the same run, and whether they were matched on the freeze-thaw cycle, too.

Matching must always be accounted for in the statistical analysis of data either using simple Mantel–Haenszel estimators or by conditional logistic regression modeling (2).

A word of warning about *overmatching* should be said at this point. Matching a case with a control subject, namely, is a very different issue from matching an unexposed subject to an exposed one, e.g., in a randomized block experiment or in an observational cohort study – and is much trickier (2).

First, if one employs matching on an *intermediate* variable between exposure and outcome, a bias in effect estimation will be introduced. Second, matching on a *surrogate* or *correlate* of exposure, which is not a true risk factor of the outcome, would lead to loss of efficiency in estimation.

From the latter fact arises the idea of *counter-matching* (9): Choose a control which *is not similar* to the case with regard to the easily measured surrogate, which is strongly correlated with the exposure. This procedure tends to increase the statistical efficiency of the design, but necessitates a somewhat more complicated statistical analysis.

In CC studies, the efficiency may sometimes be improved by selecting the subcohort from the whole cohort at entry using stratified sampling, instead of simple random sampling (10). Useful stratification is based on a variable U, which is (a) surrogate of the main risk factor Z of interest, and (b) easy and cheap to measure, and available for the whole cohort. Stratification by U with few strata, the most informative of them getting the greatest sampling fractions, tends to increase efficiency in estimating the HRs associated with Z. Note, however, that this stratification may not be efficient for other risk factors.

7. Statistical Analysis of Case–control Data

In previous sections, we presented for illustrative purposes only very simple formulas used in crude estimation of the interesting hazard ratios. However, when analyzing case–control data arising from whatever design, more refined approaches are needed in order to propely allow for the specific sampling design used, including possible stratification or matching, as well as for confounding and effect-modification due to other relevant risk factors.

The most popular approach for statistical analysis is based on fitting the proportional hazards (PH) model, also known as the Cox model (3). In this model, the hazard (i.e., the theoretical incidence rate) of the outcome event at time (often age) t for a cohort member i possessing a risk factor profile $x_i = (x_{i1}, \ldots, x_{ip})$ is expressed as

$$\lambda_i\left(t, x_i; \beta\right) = \lambda_0(t)\exp\left(x_{i1}\beta_1 + \cdots + x_{ip}\beta_p\right).$$

In this model, $\lambda_0(t)$ is the baseline hazard depending on the basic time variable t. The parameters $\beta_1 \ldots, \beta_p$ are regression coefficients with the following interpretation. For each quantitative or binary explanatory variable (risk factor) X_j, the regression coefficient β_j is interpreted to be the logarithm of the hazard ratio (HR_j) associated with a unit change of the value of X_j. The hazard ratio itself is obtained as the antilogarithm: $\mathrm{HR}_j = \exp(\beta_j)$.

In the estimation of these parameters, the typical method for nested case–control studies is based on maximizing the partial likelihood function, which is equivalent to fitting the equivalent conditional logistic regression model (3). This can nowadays be easily done by appropriate procedures found in many statistical programs (like R, SAS, S-Plus, and Stata). In case–cohort studies, the estimation is based on an analogous weighted pseudo-likelihood. The computational tools for the partial likelihood mentioned above can be used here, too, but they must be supplemented by certain additional calculations in order to obtain valid standard errors and confidence intervals, which take into account the special features of this design. See Samuelsen et al. (10) for details of such computations using the R environment.

Estimation of "absolute" risks is also feasible by proper weighting, as shown by Langholz and Borgan (8).

Full-likelihood solutions have also been recently developed, but they tend to be computationally quite challenging (using methods such as, e.g., EM algorithm, and MCMC simulation for Bayesian data augmentation).

8. Concluding Remarks

The properties of NCC and CC designs are now briefly compared on a few selected dimensions, based on more detailed discussions found, e.g., in references (3, 4).

The statistical efficiency in the two designs is roughly similar with the same amount of cases and controls, apart from some exceptional circumstances. Statistical analysis and inference in NCC studies are fairly straightforward with widely available software fitting conditional logistic regression or PH models. In CC studies, the analysis is somewhat more complicated, although software for PH models can be used when augmented with additional tricks to get valid SE, etc.

In the NCC design, only the time scale used in the definition of risk sets can be the time variable t in the baseline hazard of the PH model. However, in the CC design, the analysis of outcome rates based on the PH model is possible to conduct on different time scales (e.g., age, time since first exposure, or time since entry), because the subcohort members are not time-matched to the cases.

Missing data on risk factors may induce bias and inefficiency in the estimation of interesting parameters. In a NCC study, whenever very close matching was employed, a whole matched case–control set would be lost if the case had data missing on the risk factor(s) of interest. In CC studies, missingness of a few data items is less serious.

Quality and comparability of biological measurements based on frozen biological material are a serious concern in biobank-based studies. The NCC design allows each case and its own controls to be matched for analytic batch, storage time, and freeze-thaw cycle. This has the virtue that differential misclassification (1, 2) of exposure may be removed. In CC studies, the measurements for the subcohort members are performed at different times – typically earlier – than for the cases. This may more easily lead to differential misclassification and bias with unpredictable direction.

The possibility of investigating many diseases using the same control group for each group of cases is complicated (11) in the NCC study, and even impossible with too refined matching. In CC design, the same control group can easily serve for several diseases, because when no matching (on time or any other factor) is employed, no subcohort member is "tied" with any case.

In conclusion, cost-efficient sampling designs based on "case–controlling" are available and widely used in large-scale epidemiologic studies based on biobank cohorts. The NCC design is better suited for studies involving biomarkers that can be influenced by analytic batch, long-term storage, and freeze-thaw cycles. The CC design is useful especially when several outcomes are of interest, given that the measurements on stored materials remain sufficiently stable during the study. Finally, proper application of these designs requires well-organized follow-up systems for accurate identification of cases, deaths, and migrations occurring in the study cohort, as well as adequate statistical expertise in both planning and analysis of specific studies.

References

1. dos Santos Silva, I. (1999). *Cancer Epidemiology: Principles and Methods.* International Agency for Research on Cancer, Lyon.
2. Rothman, KJ., Greenland, S., and Lash, TL. (2008). *Modern Epidemiology, 3rd ed.* Lippincott Williams and Wilkins, Philadelphia, PA.
3. Borgan, Ø and Samuelssen, S.-O. (2003). A review of cohort sampling designs for Cox's regression model: Potentials for epidemiology. *Norsk Epidemiologi* **13**, 239–248. http://www.medisin.ntnu.no/ism/nofe/norepid/2003(2)%2008-Borgan.pdf
4. Rundle, A.G., Vineis, P. and Ahsan, H. (2005). Design Options for Molecular Epidemiology Research within Cohort Studies. *Cancer Epidemiology, Biomarkers and Prevention* **14**, 1899–1907.
5. Tedeschi, R., Bloigu, A., Ögmundsdottir, H.M. *et al.* (2007). Activation of Maternal Epstein-Barr Virus Infection and Risk of Acute Leukemia in the Offspring. *American Journal of Epidemioloy* **165**, 134–137.
6. Komulainen, K., Alanne, M., Auro, K. *et al.* (2006). Risk Alleles of *USF1*-Gene Predict Cardiovascular Disease of Women in Two Prospective Studies. *PLoS Genetics* **2**, e69.
7. Maclure, M. and Schneeweiss, S. (2001). Causation of Bias: The Episcope. *Epidemiology* **12**, 114–122.
8. Langholz, B. and Borgan, Ø. (1997). Estimation of Absolute Risk from Nested Case–control Data. *Biometrics* **53**, 768–775.
9. Langholz B and Borgan Ø. (1995). Counter-Matching: A Stratified Nested Case–control Sampling Method. *Biometrika* **82**, 69–79.
10. Samuelsen, S.-O., Ånestad, H. and Skrondal, A. (2007). Stratified Case–cohort Analysis of General Cohort Sampling Designs. *Scandinavian Journal of Statistics* **34**, 103–119.
11. Saarela, O., Kulathinal, S., Arjas, E. and Läärä, E. (2008). Nested Case–control Data Utilized for Multiple Outcomes: A Likelihood Approach and Alternatives. *Statistics in Medicine* **27**, 5991–6008.

Chapter 7

The European Prospective Investigation into Cancer and Nutrition Biobank

Pierre Hainaut, Béatrice Vozar, Sabina Rinaldi, Elio Riboli, and Elodie Caboux

Abstract

The European Prospective Investigation into Cancer and Nutrition (EPIC) is a multi-center prospective cohort study designed to investigate the relationship between nutrition and cancer, with the potential for studying many etiologic or genetic factors as well as other disease end-points. The study includes 521,448 participants (367,993 women and 153,455 men, mostly aged 35–70 years) recruited in 23 centers located in ten European countries, who are followed up for cancer incidence and cause-specific mortality for several decades. At enrolment, which took place between 1992 and 2000 at each of the centers, information was collected through a non-dietary questionnaire on lifestyle variables and through a dietary questionnaire addressing usual diet. Anthropometric measurements were performed and blood samples taken, from which plasma, serum, red cells, and buffy coat fractions were separated and aliquoted. A central biobanking facility, located at the International Agency for Research on Cancer, Lyon, was developed for the long-term storage of the specimens in liquid nitrogen. The biobank operates as a service provider and sample distribution center for scientific consortia engaged in studies involving biomarker analyses. To date, EPIC represents the largest single resource worldwide for prospective investigations on the etiology of cancers that can integrate questionnaire data on lifestyle and diet, and can also provide access to measurements of biomarkers of diet and of endogenous metabolism (e.g., hormones and growth factors) and genetic polymorphisms. This chapter describes the building up of the EPIC central biobank and the mechanisms that have been developed to manage the access to specimens by a large number of different users.

Key words: Cancer, Nutrition, Prospective study, Biomarkers, Blood products, Biobank management, Consortium

1. EPIC Objectives and Structure

The European Prospective Investigation into Cancer and Nutrition (EPIC) was developed by the International Agency for Research on Cancer (IARC) as a long-term, multi-centric prospective study in Western Europe to investigate the relationships

Joakim Dillner (ed.), *Methods in Biobanking*, Methods in Molecular Biology, vol. 675,
DOI 10.1007/978-1-59745-423-0_7, © Springer Science+Business Media, LLC 2011

between nutrition and cancer. The principal objective of EPIC is to investigate, in a prospective manner, the etiology of cancers at various sites (as well as other forms of chronic disease) in relation to diet and lifestyle. The study takes advantage of the contrast in cancer rates and dietary habits between centers and countries and of its large overall size, which makes it possible to explore interactions between nutritional, genetic, hormonal, and lifestyle factors (1). The building up of the study has been supported by the Europe Against Cancer programme of the European Union.

The prospective cohort approach includes the collection of baseline questionnaire and interview data on dietary and non-dietary variables, as well as anthropometric measurements and blood samples for long-term storage from apparently healthy populations. The enrolment of subjects in all EPIC centers took place between 1992 and 2000. The cohort participants are followed over time for the occurrence of cancer and other diseases, as well as for overall mortality, to allow incidence and mortality comparisons by exposure variables. At regular intervals, follow-up questionnaires are used to update information on selected aspects of lifestyle that are known or strongly suspected to be related to cancer risk. The EPIC study has recruited 521,448 participants, in 23 centers located in ten European countries. The study started in 1992 with 17 research centers in seven core EPIC countries (France, Germany, Greece, Italy, The Netherlands, Spain, and the UK). Subsequently, these were joined by centers in three Scandinavian countries (Sweden, Denmark, and Norway) and one center in Italy (Naples) that were conducting broadly similar prospective studies. The study is coordinated by the EPIC steering committee, which includes representatives of each participating centers as well as key scientists involved in the management of integrated resources, such as the EPIC biobank or EPIC databases.

By the end of 2006, the follow-up for cancer incidence had already led to the identification of large numbers (about 19,000 cases) of subjects who developed cancer after cohort enrolment at one of the major sites (lung, colon–rectum, prostate, and breast). The EPIC is now in its "exploitation phase," during which data and stored biospecimens are being analyzed in the course studies developed by several large, international consortiums according to protocols reviewed and approved by the EPIC steering committee.

The overall objectives, structure, and data acquisition mechanisms of EPIC, as well as the scientific strategies for assessing dietary patterns, have been extensively described in by Riboli et al. (1) and Slimani et al. (2). In this chapter, we briefly summarize the characteristics of the EPIC study cohort, its source populations, and the baseline information collected from the participants. Further-more, we describe the mechanisms developed for the obtention, processing, and storage of biospecimens, as well as those being currently developed to ensure efficient, cost-effective access to biospecimens or derived products by a large group of users located in different countries.

2. Building the EPIC Biobank

2.1. Recruitment of Subjects

The EPIC study was constructed by the integration of different cohorts into a common framework defined essentially on a geographic basis. In the majority of study centers, subjects were invited from the general adult population residing in a given town or geographic area. There were, however, exceptions to this recruitment scheme. The French cohort was based on members of the health insurance for teachers (with the aim of facilitating follow-up for incidence of cancer and other diseases); components of the Italian and Spanish cohorts included members of local blood donor associations; the cohorts in Utrecht (The Netherlands) and Florence (Italy) included women invited for a local population-based breast cancer screening program. In Oxford (UK), half of the cohort was recruited among subjects who did not eat meat, including vegans (who consume no animal products), lacto-ovo vegetarians, and fish eaters (i.e., consumers of fish but not meat). In France, Norway, Utrecht (The Netherlands), and Naples (Italy), only women were recruited.

As a rule, participants were invited to participate either by mail or in person. Individuals who agreed to participate signed an informed consent and were mailed a questionnaire on diet and a questionnaire on lifestyle. Most participants completed these questionnaires at home and were then invited to a study center for an examination that included collection of the two completed questionnaires, blood donation, anthropometry, and measurement of blood pressure. There were, however, deviations from this general scheme in several centers according to the nature of cohort, as documented in Riboli et al. (1). For example, in France, the EPIC cohort was built from a study that started in 1990 and included lifestyle questions with self-reported anthropometry measurements. The participants enrolled in EPIC are those who answered the dietary questionnaire, a subset of whom (20,725 women close to a metropolitan area) later came to a field center, donated blood, and underwent blood pressure and anthropometry measurements. In Greece, most participants were actively recruited and completed an interviewer-administered questionnaire on diet and a questionnaire on lifestyle at the examination center. In Denmark and Malmö (Sweden), the participants filled in dietary questionnaires at home and lifestyle questionnaires at the study centers. In Umeå (Sweden), both questionnaires were completed at the study center. In Norway, participants completed an initial mailed questionnaire unrelated to EPIC, completed a subsequent mailed questionnaire for EPIC, and then had blood samples mailed to the study center in Tromsø for processing.

2.2. Biological Samples

Blood was obtained by venipucture and processed according to standard separation protocols. Biological samples included blood plasma, blood serum, white blood cells, and erythrocytes collected from 388,527 of the 521,448 EPIC study participants (Table 1). The procedure for storage of blood samples differed between the seven initial EPIC countries and the three Scandinavian countries that joined EPIC at a later stage.

In the seven initial EPIC countries and in Naples (Italy), blood samples were aliquoted into 28 plastic straws containing 0.5 ml each (12 plasma with sodium citrate, eight serum, four erythrocytes, and four buffy coat for DNA). Plastic straws (CryoBioSystems (CBS), Paris, France), made of chemically inert and biocompatible ionomeric resin, were designed for long-term storage. These straws carry a dual identification system. The primary identification of the straw is by its color, while its definitive identification is by its alpha-numerical code. Different colors of straws were used for each type of specimen (red, yellow, blue, and green). To ensure a high degree of standardization, the same materials (syringes, straws, etc.) were purchased centrally and distributed to the centres. The samples were then split into two mirror halves of 14 aliquots each. One set was stored locally, and one transported to IARC to be stored in liquid nitrogen (at −196°C) in a central biorepository. In Norway the biological samples were collected in 20 0.5-ml plastic straws; for 11,182 subjects, 12 of the 16 plasma and two of the four buffy coat samples were shipped to IARC for storage in the central repository. In Sweden and Denmark, blood samples were stored in tubes (not in plastic straws). For practical reasons they are stored only in local repositories since the central EPIC repository at IARC has been primarily developed for storing straws. In Sweden, the samples are kept in freezers at −80°C, and in Denmark in nitrogen vapour (−150°C).

2.3. Central EPIC Biobanking Infrastructure

The central EPIC biobank located at IARC consists in 33 liquid nitrogen (LN2) tanks equipped with straw storage systems and connected to an automated, central LN2 supply system. The specimens are kept under N2 liquid phase (−196°C). The biobank contains about 3.8 million straws with blood aliquots from 275,861 EPIC participants. The straws of each participant are stored together using the CBS™ visotube/goblet system (CryoBioSystems). The straws of one participant are stored inside a colored tube (visotube), which inturn is stored in, successively, a goblet, canister, and container. The canisters are arranged in color-coded concentric circles located in each LN2 containers.

Table 1
Biological specimens in the EPIC biobank from (1)

Center		n	Age range (years) (1st–99th percentile)	Female (%)	Achievement rate (%)[a]	Samples collected (number of 0.5-ml straws desired)[b]				Storage location	
						Plasma	Serum	White blood cells	Erythrocytes	IARC	Local
Core EPIC cohorts											
Greece	Nation-wide	28,500	29–76	58.2	99.8	12	8	4	4	Yes	Yes
Spain	Granada	6,892	35–66	77.0	87.5	12	8	4	4	Yes	Yes
	Murcia	8,146	35–65	68.7	95.7	12	8	4	4	Yes	Yes
	Navarra	7,799	36–64	51.5	96.5	12	8	4	4	Yes	Yes
	San Sebastian	8,325	36–65	50.6	98.9	12	8	4	4	Yes	Yes
	Asturias	8,417	35–65	64.0	98.5	12	8	4	4	Yes	Yes
Italy	Ragusa	6,396	35–65	52.3	99.9	12	8	4	4	Yes	Yes
	Florence	13,597	35–65	74.2	100.0	12	8	4	4	Yes	Yes
	Turin	10,604	35–64	43.0	100.0	12	8	4	4	Yes	Yes
	Varese	12,073	36–72	78.9	99.9	12	8	4	4	Yes	Yes
France		20,725	43–68	100.0	31.0[c]	12	8	4	4	Yes	Yes
Germany	Heidelberg	24,235	36–64	52.6	94.9	12	8	4	4	Yes	Yes
	Potsdam	26,444	35–66	59.8	95.9	12	8	4	4	Yes	Yes
Netherlands	Bilthoven	19,388	21–64	54.0	93.1[d]	12	8	4	4	Yes	Yes
	Utrecht	16,930	49–69	100.0	96.9	12	8	4	4	Yes	Yes
United Kingdom	Cambridge	24,035	41–76	54.3	93.8[e]	12	8	4	4	Yes	Yes
	Oxford	19,103	23–73	76.7	96.1[f]	12	8	4	4	Yes	Yes

(continued)

Table 1 (continued)

Center		n	Age range (years) (1st–99th percentile)	Female (%)	Achievement rate (%)[a]	Samples collected (number of 0.5-ml straws desired)[b]				Storage location	
						Plasma	Serum	White blood cells	Erythrocytes	IARC	Local
Associated EPIC cohorts											
Italy	Naples	5,055	34–68	100.0	99.9	12	8	4	4	No	Yes
Denmark	Aarhus	17,094	50–65	50.8	99.7	T^b	T^b	T^b	T^b	No	Yes
	Copenhagen	39,037	50–65	52.7	97.8	T^b	T^b	T^b	T^b	No	Yes
Sweden	Malmö	28,023	46–73	60.6	99.7	T^a	T^a	T^a	T^a	No	Yes
	Umeå	25,732	30–61	51.7	100.0	T^a	T^a	T^a	T^a	No	Yes
Norway	Tromsø	9,197	40–55	100.0	~60.0[g]	16	NC	4	NC	Yes	Yes

T^a – stored in 2-ml tubes at –80°C; T^b – stored in 1-ml tubes in nitrogen vapor at a temperature between –150°C and –160°C; *NC* not collected

[a]In all centers, except those in France, the UK, Bilthoven (Netherlands), and Norway, all EPIC participants were invited to donate blood

[b]In the core centers, biological samples are distributed equally between ARC and local storage, and are straws at –96°C

[c]In France, 66,858 EPIC participants living near a metropolitan area were asked to donate blood

[d]In Bilthoven, 13,451 EPIC participants recruited from Amsterdam or Doetinchem after 11 May 1993, and 7,364 EPIC participants recruited from Maastricht after 2 June 1993 were asked to donate blood

[e]In Cambridge (UK), 25,633 EPIC participants who attended a study examination were asked to donate blood

[f]In Oxford (UK), enrolment of the participants recruited by general practitioners from the local counties was based on a willingness to donate blood and the achievement rate is 96.1%; among the "health-conscious" sub-cohort, 24.4% donated blood

[g]In Norway, collection of biological samples is currently underway and will continue until samples have been collected from 12,000 participants

Each goblet contains 12 visotubes. Each straw is labeled with the participant's ID and color-coded to indicate its contents; in addition, the tube, goblet and canister are color-coded to aid in identifying the samples. Finally, a computer software program indicates the container, canister, goblet, and the location of the goblet and the canister within each container to track the stored biological samples of each participant. A Laboratory Information Management System (LIMS) has been developed to identify, track, and follow-up the different straws contained in each tube during analysis. The biobank is housed in three purpose-build, ventilated storage rooms. LN2 tanks are monitored for LN2 pressure and alarmed. Rooms are equipped with LN2 sensors to monitor potential LN2 health hazards.

2.4. Data Management and Databases

Information on individual dietary intake was assessed using different validated dietary assessment methods across participating countries. In order to adjust for possible systematic over- or underestimation in dietary intake measurements, a calibration approach was developed. This approach involved an additional dietary assessment common across study populations to re-express individual dietary intakes according to the same reference scale. A single 24-h diet recall was, therefore, collected, as the EPIC reference calibration method, from a stratified random sample of 36,900 subjects from the entire EPIC cohort, using a software program (EPIC–SOFT) specifically designed to standardize the dietary measurements across study populations (2). These studies showed that despite certain inherent methodological and logistic constraints, a study design such as this one works relatively well in practice. The overall results suggest that after adjustment for age, dietary intakes estimated from calibration samples can reasonably be interpreted as representative of the main cohorts in most of the EPIC centers.

An integrated database system was developed to hold and manage the different datasets pertaining to individual participants and to biospecimens. The EPIC core database comprises individual EPIC data, as well as the computer software (ORACLE) and the programs that store, track, and manage the biospecimens. Similar to the central biobank, the EPIC–ORACLE database was built as a platform integrating the information collected and gathered in participating centers. For practical reasons, 14 centers in the ten participating countries act as coordinating centers that interact with IARC for centralization of the EPIC data (in particular, all Spanish and Italian data are centralized in Barcelona and Milan, respectively). Information in the database was stored using the center-specific variable names and formats as well as variable names and formats standardized across EPIC. Center-specific data were loaded into the ORACLE system and transformed into the standard EPIC variables on which quality control checks were then run (1).

2.5. Follow-Up of Cohort Members and Database Updates

After enrolment, cohort members are contacted at regular intervals every 3–4 years to obtain information on various aspects of lifestyle that are known or strongly suspected of being related to cancer risk, and that may change over time. This includes tobacco smoking, alcohol drinking, physical activity, weight, menstruation, pregnancies, menopause, and other variables. In addition, a series of questions was added on whether the subjects had suffered from any major diseases. Follow-up aimed at identifying cancer cases is based on population cancer registries in seven of the participating countries (Denmark, Italy, The Netherlands, Norway, Spain, Sweden, and the UK); on a combination of methods including health insurance records, and cancer and pathology registries; and on active follow-up of study subjects and their next-of-kin in three countries (France, Germany, and Greece). A working group created in 1996 (End-Point Committee) prepared a detailed protocol for the collection and standardization of clinical and pathological data on each cancer site: *Guidelines for collection of end-point data in the EPIC study* (IARC, 1998). In parallel, data on total and cause-specific mortality are collected at the EPIC study centers through mortality registries or active follow-up and death-record collection.

3. Managing the EPIC Biobank

3.1. Decision Mechanisms in Allocating Biospecimens

The EPIC provides a framework for addressing a wide range of questions relevant to cancer. In addition, information on vital status and cause of death can be used to address endpoints other than cancer, in particular cardiovascular diseases, as well as survival after cancer diagnosis. When biological samples are involved, studies mostly use the nested case–control approach. Typically, "cases" are subjects who developed a particular pathology after they were recruited in the cohort (incident cases) and had not been diagnosed with cancer before or at the time of recruitment. "Controls" are usually chosen at random among all cohort members who were alive and without cancer at the time of diagnosis of the case subject. In contrast with traditional case–control studies, this prospective design avoids inverse causation bias that may occur when biomarkers are altered by the metabolic effects of a tumor or by antitumor treatments, psychological stress, or lifestyle changes after cancer diagnosis.

The management of EPIC is based on a collegial decision process by the EPIC Steering Committee, which includes the principal investigators of each participating center and key scientists involved in coordination tasks. A number of thematic EPIC working groups have been developed, focusing either on a particular pathology/tumor site or on generic issues, such as biomarker development and validation. Participation in working groups is on a voluntary basis. Overall, the working groups currently involve over 200 participants in Europe and beyond. In most instances, study protocols are developed by the working groups, and submitted to the EPIC steering committee that reviews, discusses, and formally endorses them. Within this framework, each EPIC center makes decisions regarding the inclusion of its data and biospecimens into a particular study. This scientific decision process is paralleled by an ethical review process, during which the Ethical Review Boards at each participating center are consulted in compliance with their national legal and regulatory requirements. Upon completion of these decision processes, a list of EPIC subject is extracted from the EPIC database and converted into a list of specimens to retrieve from the biobank.

3.2. Inter-center Variations

The very magnitude of the individual and total cohorts, the related lengthy period of subject recruitment, and the variety of local facilities have made it impossible to standardize all procedures strictly, as would be possible for smaller studies. However, considerable effort has been put into ensuring maximum comparability within and between cohorts, in particular where dietary information is concerned, by means of the large calibration subsample. Moreover, the use of EPIC samples in a large prospective investigation on gene–environment interactions raises a number of technical and practical issues. A review by Vineis and colleagues describes how these issues were approached within a case–control study nested into EPIC, the GenAir investigation. GenAir is aimed at measuring the effects of air pollution and environmental tobacco smoke on human health in EPIC with a nested design and with biological measures. Validation studies included comparisons between cotinine measurements, hemoglobin adducts, and questionnaire data to assess tobacco exposure; analysis of the determinants of DNA adduct concentration; and comparison among different genotyping method. Such validation studies have identified variations in the amount of biological materials that can be extracted from the straws originating from different centers (3). These variations are particularly marked in the case of plasma DNA. Small amount of free DNA can be retrieved in the plasma of most healthy subjects, and these amounts are often increased in subjects with chronic disease including cancer. There is evidence that most of this DNA originates from tissue damage. However, this parameter is very sensitive to conditions affecting sample processing since some DNA is also released from blood cell lysis during the preparation of plasma or serum. High plasma DNA concentrations were systematically observed in specimens from some centers, which may reflect differences in the treatment of the samples, in particular the time between blood collection, separation, and final cryopreservation. Such variations may have an important impact on assessing unstable molecular markers (4).

3.3. Biobanking Services

The logistical tasks related to specimen management, retrieval from the biobank, and distribution are handled by the Laboratory Services and Biobank Group (LSB) at IARC. Based on the lists of specimens and on their known position in the biorepository, the LSB technicians develop an ordered retrieval plan that minimizes the time of opening of each LN2 tank. Specimen retrieval is performed manually. It takes about 5 min to access one specific storage position and to retrieve either one or several straws of materials from the same subject. Standard operating procedures include double checking of 5% of retrieval positions and retrieved specimens to minimize the risk of individual error. On average, a trained technician can retrieve specimens for about 150 subjects over one normal working day. Specimen retrieval is a limiting factor in the pre-analytical processing of EPIC biospecimens, and its demand in terms of manpower entails important costs, in particular for studies in which several thousand specimens are included.

The LSB Group offers a range of biobanking services including automated DNA extraction (GENTRA autopure DNA extraction system), quantification, aliquoting in various tube or microplate formats, and specimen shipping. Shipment of biospecimens or derived products is carried out according to UN and IATA packaging regulations. Unless otherwise planned, the extraction of DNA from EPIC buffy coats samples is carried out at IARC, which ships aliquoted, quantified DNA to the laboratories that perform analytical processes. A recent, independent assessment of the performance of the EPIC retrieval/DNA extraction activities in the context of a multi-centric, genome-wide association studies has shown that EPIC biospecimens processed at IARC meet qualitative and quantitative criteria with a "pass rate" of 96% (unpublished data).

3.4. Storage of Derived Products and Leftovers

A major challenge in the long-term management of the EPIC biobank is the traceability and storage of derived products (in particular DNA) and of specimen leftovers. Given the limited amount of material available in the collection, each DNA extracted from an EPIC subject is stored in freezers as purified material in a microplate format. Thus, as more and more biospecimens are being processed for analysis, a parallel collection of stored DNA is developing, which will be easily accessible for further analyses. The suitability of DNA amplification by Whole Genome Amplification techniques is currently under evaluation. Storage of DNA in microplates allows for rapid, simple retrieval that is amenable to automation, thus reducing the workload and the cost of pre-analytical processing.

Although shipment protocols involve measures for the return of unused specimens and leftovers to the biobank, this has so far been difficult to implement in practice. Returned leftovers often come as diluted aliquots of uncertain quantity and quality, and their storage is time and space consuming. So far, no attempt has been made to re-qualify those left-overs for further use.

4. Conclusion: Lessons from 15 Years of EPIC Biobanking

Over 15 years after its inception, the EPIC study and its biobank have provided a basis for more than 100 published and ongoing studies involving biomarkers and addressing a very diverse array of questions related to cancer and other chronic diseases. These studies have demonstrated the quality of the baseline information and biological samples available within this framework. Over the next 10 years and beyond, it is expected that the cohort and biobank will continue to generate original study designs to investigate many aspects of cancer etiology and genetic susceptibility, as well as to discover, validate, and assess biomarkers of cancer risk. The storage of biological samples in multiple aliquots in liquid nitrogen represents the best available technology for maintaining long-term stability. Due to its particular structure, which is based on the assembly of a series of regional cohorts, the collection is uniquely poised to contribute specimens for many large-scale wide-genome association studies.

With hindsight, two main problems may be identified that should be considered as important lessons for the future development of large, multi-centric prospective collections. First, the biospecimens collected are representative of only one (pre-diagnostic) time point. The availability of biospecimens taken at diagnosis, paired with the pre-diagnosis specimen, would have added another dimension to EPIC, by enabling the development of studies assessing early biomarkers of disease in a strictly prospective design. Attempts at retrospective retrieval of archived paraffin blocks for EPIC cancer subjects have proven very difficult and time consuming. In future studies, it is recommended to plan strategies for the collection of diagnostic specimens from the very onset. Such an initiative would indeed have been difficult to implement at the time of initiation of EPIC, due to the lack of adequate specimen collection and biobanking infrastructures in many centers. Today, the widespread development and increased awareness of biobanking issues make it possible to interconnect prospective studies with tissue repositories and cancer registries. Such interconnections are crucial for the next generation of multicentric, large cohort studies.

Second, the cost and logistical implications of maintaining such a large cohort over a long period of time have not been fully taken into consideration at the onset of the study. Based on the costs of storage and specimen handling, it may be estimated that the total cost of maintaining and running the EPIC biobank over a period of 20 years will be in the range of 10 million Euros (about 30% of which represents the costs of LN2), excluding the initial investment for collecting the specimens and building the biobank infrastructure. Although this represents a relatively low cost-per-stored sample (in the range of 0.16 euro/sample straw/year), it should be considered that over 50% of the individual straws may actually never be used in studies addressing EPIC's primary objectives. Thus, the real costs of biobanking per analyzed subject will eventually reach 90–100 Euros. Mechanisms must be developed to apportion these costs appropriately between core biobank funding and research budgets of specific studies using the specimens.

In conclusion, the EPIC biobank provides a management and logistical model for prospective studies involving the banking of human biospecimens. Aside of the direct scientific impact of EPIC, the expertise gained in assembling, managing, and running such a large biobank will be one of the long-term benefits of this unique scientific and technical networking initiative.

Acknowledgments

This chapter is largely based on previous description of the EPIC study by Riboli et al. (2002) and Slimani et al. (2002). The authors are grateful to Paolo Vineis for comments and suggestions, and to Thomas Cler, Christophe Lallemand, and Elodie Colney for their technical support to the biobank. EPIC is a collective project involving many scientists throughout Europe. Their contribution to the development of biobanking infrastructures and processes at IARC is duly acknowledged.

References

1. Riboli, E., Hunt, K.J., Slimani, N., Ferrari, P., Norat, T., Fahey, M., Charrondiere, U.R., Hemon, B., Casagrande, C., Vignat, J., Overvad, K., Tjonneland, A., Clavel-Chapelon, F., Thiebaut, A., Wahrendorf, J., Boeing, H., Trichopoulos, D., Trichopoulou, A., Vineis, P., Palli, D., Bueno-De-Mesquita, H.B., Peeters, P.H., Lund, E., Engeset, D., Gonzalez, C.A., Barricarte, A., Berglund, G., Hallmans, G., Day, N.E., Key, T.J., Kaaks, R., and Saracci, R. (2002) European Prospective Investigation into Cancer and Nutrition (EPIC): study populations and data collection. *Public Health Nutr.* **5**, 1113–1124.
2. Slimani, N., Kaaks, R., Ferrari, P., Casagrande, C., Clavel-Chapelon, F., Lotze, G., Kroke, A., Trichopoulos, D., Trichopoulou, A., Lauria, C., Bellegotti, M., Ocke, M.C., Peeters, P.H., Engeset, D., Lund, E., Agudo, A., Larranaga, N., Mattisson, I., Andren, C., Johansson, I., Davey, G., Welch, A.A., Overvad, K., Tjonneland, A., Van Staveren, W.A., Saracci, R., and Riboli, E. (2002) European Prospective Investigation into Cancer and Nutrition (EPIC) calibration study: rationale, design and population characteristics. *Public Health Nutr.* **5**, 1125–1145.
3. Peluso, M., Hainaut, P., Airoldi, L., Autrup, H., Dunning, A., Garte, S., Gormally, E., Malaveille, C., Matullo, G., Munnia, A., Riboli, E., and Vineis, P. (2005) Methodology of laboratory

measurements in prospective studies on gene-environment interactions: the experience of GenAir. *Mutat Res.* **574**, 92–104.

4. Gormally, E., Hainaut, P., Caboux, E., Airoldi, L., Autrup, H., Malaveille, C., Dunning, A., Garte, S., Matullo, G., Overvad, K., Tjonneland, A., Clavel-Chapelon, F., Boffetta, P., Boeing, H., Trichopoulou, A., Palli, D., Krogh, V., Tumino, R., Panico, S., Bueno-De-Mesquita, H.B., Peeters, P.H., Lund, E., Gonzalez, C.A., Martinez, C., Dorronsoro, M., Barricarte, A., Tormo, M.J., Quiros, J.R., Berglund, G., Hallmans, G., Day, N.E., Key, T.J., Veglia, F., Peluso, M., Norat, T., Saracci, R., Kaaks, R., Riboli, E., and Vineis, P. (2004) Amount of DNA in plasma and cancer risk: a prospective study. *Int J Cancer* **111**, 746–749.

Chapter 8

The AIDS and Cancer Specimen Resource

Leona W. Ayers, Sylvia Silver, Jan M. Orenstein, Michael S. McGrath, and Debra L. Garcia

Abstract

The AIDS and Cancer Specimen Resource (ACSR) is a cooperative agreement among the United States National Cancer Institute (NCI) (Office of the Director, Office of HIV and AIDS Malignancy (OHAM)) and regional US consortia, University of California, San Francisco (West Coast), George Washington University (East Coast), and The Ohio State University (Mid-Region). The ACSR's main objective is to collect, preserve, and disperse HIV-related tissues and biologic fluids along with clinical data to qualified investigators with a focus on HIV/AIDS-related malignancies. The ACSR biorepository has more than 265,000 human HIV-positive and control samples available from 39 processing types, 16 specimen types, and 52 anatomical site types. These HIV-infected biological fluids and tissues are made available to funded approved investigators at no fee. Technical support such as HIV DNA identification in tissues and tissue microarray (TMA) blocks are available to assist approved investigators. Research needs may be filled through ACSR cooperative arrangements when not met by currently banked material. Those participating with the ACSR are expected to share their research findings with the scientific community. Some 117 abstract/poster and podium reports at national and international scientific meetings and 94 publications have been contributed to the scientific literature (as of 2010). Investigators can browse the ACSR Internet site at http://acsr.ucsf.edu for biospecimens to support their scientific initiatives, including basic, translational, biomarker discovery, and molecular epidemiology studies.

Key words: HIV, AIDS, Biospecimens, Biological specimens, Cancer tissue bank, TMA

1. The Vision/Goal

In the early 1990s, the USA and Europe were experiencing an escalation in the Acquired Immunodeficiency Syndrome (AIDS) epidemic along with an expanding AIDS-related cancer epidemic. In 1994, the National Cancer Institute (NCI) of the National Institutes of Health (NIH) in the USA in response established the AIDS Malignancy Bank (AMB) through a cooperative agreement to collect and disperse HIV-positive biological specimens

Joakim Dillner (ed.), *Methods in Biobanking*, Methods in Molecular Biology, vol. 675,
DOI 10.1007/978-1-59745-423-0_8, © Springer Science+Business Media, LLC 2011

with associated clinical data to approved researchers. This was to assure an adequate HIV-infected sample resource for translational research into HIV/AIDS-related malignancies (1). This AMB program continues today as the AIDS and Cancer Specimen Resource (ACSR) (2). ACSR supports the research community at large by providing HIV-infected and control biological specimens. Besides providing specimens the ACSR makes available selected technical resources required for the identification of mechanisms contributing to cancer risks in individuals who are HIV-infected, HIV/oncogenic virus coinfected, HIV-infected smokers and those treated with HAART. While the introduction of HAART in developed countries has been associated with a dramatic decrease in major HIV-related malignancies, such as Kaposi's sarcoma (KS) and primary central nervous system non-Hodgkin's lymphoma (NHL), developing countries continue to experience the escalation of HIV infection with the associated AIDS and cancer epidemics (3).

2. Ethical, Legal, and Social Issues

The ACSR central office and data coordinating center (CODCC) interacts directly with individual ACSR sites and maintains a centralized data base for tracking all ACSR-related regulatory, ethical, legal, and social issues. Each ACSR site utilizes their local Institutional Review Board (IRB), which is responsible for ACSR site-specific research protocol review, oversight of biospecimen, and associated data collection. Thus, each site program utilizes IRB approved research protocols for maintaining donor privacy and confidentiality of clinical data as required by the US federal standards, including the requirements of HIPAA (4). All clinical data are deidentified when they are incorporated into the national ACSR database and all samples and clinical data released to investigators through the ACSR executive committee and CODCC are anonymous.

3. Affected Populations

HIV is estimated to be present in 33.4 million (31.1-35.8 million) people worldwide at the end of 2008 (5). While AIDS is the most recognized consequence of infection, persons infected with HIV/AIDS have excess risks for the development of cancers, particularly in association with known oncogenic viruses: Epstein Barr virus (EBV) with non-Hodgkin's lymphoma, human herpes virus 8 (HHV8) with KS, Hepatitis B (HBV) and Hepatitis C (HCV) viruses with hepatocellular carcinomas, and human papilloma virus (HPV) with squamous cell carcinomas of the cervix, anal canal, conjunctiva, and possibly mouth. HIV-infected tobacco smokers have increased incidence of smoking-related cancers of lip, mouth, pharynx, and lung (6). With the use of highly active antiretroviral therapy (HAART) beginning in 1996, there was a dramatic reduction in the excess risk for non-Hodgkin's lymphomas and KS but not for other excess risk cancers, including Hodgkin's lymphoma and the smoking-related carcinomas (6). Perinatally HIV-infected children have higher cancer incidence compared to HIV-uninfected children even after the treatment with HAART (7). Extended exposure of such a large population of HIV-infected individuals of all ages to nucleoside analogs, such as azidothymidine (AZT), dideoxycytidine (DDC), and other powerful AIDS drugs, could induce DNA damage leading to other neoplasms such as renal carcinoma. There is currently a HIV/AIDS-related cancer epidemic in developing countries with high HIV infection prevalence (8, 9).

4. Recruitment of Biospecimens

Each ACSR site has a consortium of pathologists and clinicians that identify, collect, and contribute HIV-infected biospecimens and uninfected controls along with appropriate patient demographic and medical information to this biorepository.

4.1. Donations from Individuals

4.1.1. HIV+ Personal Donations

HIV-infected individuals with accessible KS skin malignancies and a desire to support AIDS research donate biopsy samples along with peripheral blood for peripheral blood mononuclear cells (PBMC) and plasma directly to this biorepository. Each donor signs the local IRB approved ACSR informed consent document and the accompanying approved HIPAA document. Other types of personal donations, such as urine and oral mucosal cells, may be made as needed.

4.1.2. HIV+ Surgical Remnants and Autopsy Tissues

These samples are donated fresh or frozen according to local IRB approved protocols that usually involve a third party, such as a *tissue procurement service* or an *honest broker*. Such samples are usually accompanied by limited patient demographic and medical data. These samples are deidentified if specific patient consent for tissue donation and the release of medical information was not obtained prior to tissue removal.

4.1.3. Archives

HIV/AIDS-associated malignant tissues (typically formalin-fixed, paraffin-embedded tissues) are available from the beginning of the HIV epidemic in the USA. One of the strengths of ACSR is that it has KS and NHL samples available from the early days of the epidemic in men who have sex with men (MSM). Samples both pre- and post-HAART are available for comparison to each other or for comparison to samples from countries where HAART has not been introduced or has only recently been introduced. All geographic regions, minorities and genders of the USA are represented in the biorepository offerings. Biospecimens providing such a rich representation of the US HIV epidemic are made possible by the ACSR consortium institutions supporting large, long-term pathology tissues archives.

4.2. Donations by HIV/AIDS Treatment and Epidemiology Groups

The ACSR is directly affiliated with the AIDS Malignancy Consortium (AMC), http://pub.emmes.com/study/amc/public/index.htm, a national consortium that conducts clinical trials on patients with AIDS malignancies. This affiliation allows the banking of well-characterized, longitudinal biospecimens from a variety of clinical trials for ultimate use by approved researchers. Using the AMC relationship as a model, further group affiliations were developed, such as relationships with the Women's Interagency HIV Study (WIHS), the AIDS Clinical Trials Group (ACTG), the San Francisco Gay Men's Health Study, the National Ano-genital Cancer Study, and a variety of smaller studies.

4.3. Collections Donated or Accessed Through a Referral Process

Besides providing specimens from its own collection, ACSR brokers specimens to its research applicants from large collections of specimens within established programs. Relationships exist with the National NeuroAIDS Bank (NNAB), the National NeuroAIDS Tissue Consortium (NNTC), the University of California Los Angeles Brain Bank program, various multicenter AIDS Cohort Study (MACS) Groups, the UCSF AIDS Specimen Bank (ASB), and the Hawaiian AIDS Natural History Cohort Study. The ACSR can accept transfer of banked HIV-infected biospecimen collections with their attendant databases, for inclusion in the overall ACSR program. Specimens have been transferred to the ACSR from the San Francisco Gay Men's HHV-8 Natural History Cohort, the Rwandan cohort connected with the Women's Interagency HIV Study, and the US Department of Defense Thailand Vaccine Trial Serum Specimen Bank.

4.4. Enhanced Collections

Special collections of biospecimens, such as samples from clinical studies with associated clinical data, tissue with documented HIV DNA integrations and paraffin-embedded tissues, assembled as TMAs are available. TMAs of common HIV/AIDS-related malignancies, such as NHL and KS are prepared and banked. ACSR quality control analysis of these TMAs assures that the included tissues from over the span of the HIV epidemic retain sufficient reactivity to be useful in IHC and ISH-based studies. Lymphoma tissues that were initially well fixed retain good levels of reactivity for IHC stains, such as CD45, CD3, CD20, CD10,

CD138, MIB1, MUM-1, and Bcl-6, over the historical span of the epidemic. The KS TMAs demonstrate expression of KSHV latency-associated nuclear antigen (LANA) in all tumors. Other markers, such as K1 protein and the marker of late lytic replication (KSHV Orf26), are variably expressed (10). Investigators with novel probes are encouraged to initially test the probes in tissues less than 3 years of age or in tissues known to be positive before requesting a TMA. Investigators with special interests may have customized TMAs constructed to meet their study needs (see http://acsr.ucsf.edu).

5. HIV/AIDS Researchers

5.1. Outreach Program

The CODCC coordinates the ACSR outreach program. The program targets national AIDS and cancer research centers as well as individual investigators interested in the study of malignancies in HIV and non-HIV associated diseases.

5.1.1. Advertising Venues

The CODCC has developed several different types of advertising materials aimed at informing research investigators about the ACSR and its available resources as well as how to apply for specimens. Written material is regularly updated and distributed to the directors and administrators of the NCI Cancer Centers and local and regional AIDS research institutes. Researchers who attend national AIDS and cancer-focused meetings, such as the NCI's International Conference on HIV/AIDS-related Malignancies and those sponsored by: Infectious Diseases Society of America (IDSA), Interscience Conference on Antimicrobial Agents and Chemotherapy (ICAAC), American Association of Cancer Research (AACR), American Society of Clinical Oncology (ASCO), American Society of Hematology (ASH), International AIDS meetings (IAS), and the Institute of Human Virology, are informed about ACSR services.

5.1.2. Quarterly Newsletter

The ACSR newsletter contains:

- Reports from local ACSR sites
- Spotlight on individual investigators or publications associated with ACSR procurements
- Highlights of special or underutilized collections
- Reports on new technologies that may be of interest to our investigators
- Reminders of upcoming events at which ACSR Principal Investigators are speakers or ACSR materials will be available.

5.1.3. Booth

An ACSR exhibit booth is displayed along with other NCI booths at national AIDS and cancer meetings. Such meetings include the American Association for Cancer Research (AACR), Federation of American Society of Experimental Biology (FASEB), the United States and Canadian Academy of Pathology (USCAP), or others as appropriate for distribution of HIV educational information and recruitment of interested researchers. ACSR staff and/or investigators manage this booth.

5.1.4. Abstracts and Posters

ACSR Principal Investigators present abstracts and posters at national AIDS and cancer meetings highlighting the specimens available through the ACSR. The ACSR has been successful in describing its mission and accomplishments to other scientists at major scientific meetings and research conferences. Presenters using ACSR resources in their reported investigations cite the *NCI's AIDS and Cancer Specimen Resource.*

5.1.5. Web site

The ACSR Web site is the source for updated information on the content of the ACSR biorepository. Other useful information on current services, collection and preservation methods, available ACSR materials, and application dates is available to researchers. The ACSR Web site (http://acsr.ucsf.edu) also links to other tissue procurement and banking groups.

5.1.6. Direct Contact

Investigators needing more information on the availability of biospecimens for their research or to discuss research questions are highly encouraged to address their inquiries directly to the ACSR CODCC for complete and up-to-date information. Contact information is available through the ACSR Web site (http://acsr.ucsf.edu).

5.1.7. Identifying Researcher Needs

Researchers are directed to contact the ACSR CODCC to determine whether needed samples and associated data are available. The CODCC's Principal Investigator then contacts the researcher to discuss the proposed study plan and how best to utilize the ACSR specimens to meet their research needs. Upon completion of this consultation and confirmation of specimen availability, the researcher is encouraged to submit an application or Letter of Intent (LOI) to acquire the specimens. The LOI is reviewed by an independent review panel and biostatistician for scientific merit. If specimens are not available, the researcher is referred to another resource. Such requests often influence specimen recruitment. Some of the general potential uses of ACSR specimens are outlined in Table 1 (11).

5.1.8. Letters of Intent

LOI from investigators are accepted throughout the year. Investigators can use a simpler *short form LOI* to request relatively small numbers of specimens. Forms can be downloaded from the

ACSR Web site, http://acsr.ucsf.edu. The application process requires minimal paper work and is outlined in detail on the ACSR Web site. When LOI applications are approved, specimens are packaged and shipped (on dry ice, when appropriate) according to the US Federal regulations. The ACSR works with other biological material resources to refer investigators if specimens are not available within the ACSR cooperative group.

5.2. Scientific Publications

Investigators using ACSR resources cite the *NCI's AIDS and Cancer Specimen Resource* in the publications describing their results.

6. Quality

The quality of stored biological samples can generally be anticipated based on specimen type, type of fixative or processing, length of storage, storage method, and the type of testing anticipated by the investigator (12, 13). Frozen and formalin-fixed, paraffin-embedded tissues have been quality tested for DNA and RNA content, immunoreactivity (IHC) and performance with in situ hybridization (ISH) probes as well as HIV copy numbers. Many of the banked samples are too small to allow pretesting that requires destructive sampling. Any individual tissue may fail to perform because of initial delays in processing, poor fixation and processing, or faulty storage. Formalin-fixed, paraffin-embedded tissues from different international sites may vary widely in the preservation of DNA for genomic studies. DNA may range from the modest retrieval of short fragments to only amino acids. Newer tissue fixatives may greatly enhance the recovery of DNA as well as RNA and proteins (14). Investigators are encouraged to assure the performance of the requested biological sample type before undertaking their experiments. Investigators who receive ACSR specimens assist in the quality control process by grading the quality of specimens they receive and returning a written evaluation document to the ACSR. Faulty samples are replaced if appropriate samples are available.

7. Financing

The ACSR biorepository is fully supported by the USA NCI/NIH so that biospecimens can be provided to approved, funded investigators working in non-profit research settings at no fee. Investigators working in a commercial setting may also obtain

Table 1
Research use of selected ACSR banked specimens by type

Specimen type	Research study types	
Frozen lymphoma, KS, and tumor tissues	DNA, RNA, and protein array; viral discovery/strain variation	
Autopsies (multisite) frozen and fixed	DNA and protein within individuals; involved vs. uninvolved tissues	
	Tissue array analyses	Comparison of antigen expression between many patient tumors
		Diseased tissue-specific cytokine, virus, antigen expression
Plastic-embedded tissue suitable for transmission electron microscopy	Evaluation of virus identity, morphogenesis, and cytopathology in various disease states	
Non-Hodgkin's lymphoma (AIDS and non-AIDS) epidemiology study (serum and fixed tissue)	Serologic on cases vs. controls for cytokine, viral antigen, serum proteins. Coupled with epidemiologic data in collaboration with Dr. Elizabeth Holly test disease associations, risk factors, transmission risk factors, resistance factors, lymphoma tissue, DNA/protein correlation with serum factors/disease associations	
Ano-genital specimens from men and women HIV+/HIV−	Role of HIV strains in early stages of ano-genital carcinogenesis	
AIDS Malignancy Consortium clinical trial specimens	Longitudinal, trial-associated specimens for the analyses of disease-specific markers in collaboration with AMC	
Serum specimens from cross-sectional survey in Thailand	Low KS prevalence untreated HIV and serum cohort infected vs. uninfected age and sex-matched specimen comparisons HIV, non-US viral isolates. Repeat blood draws from HIV+ individuals available for the rate of variation	

samples for research if approved by the REDP but a fee-for-sample and service may be applied.

7.1. Grant Support

The NCI's OHAM supports HIV and AIDS malignancy research and coordinates all AIDS and AIDS oncology efforts across NCI, including the development of extramural initiatives and AIDS cofunding agreements. The NCI OHAM works closely with the Centers for AIDS Research (CFAR) at the NIH, providing administrative and research support for AIDS research projects. The NCI OHAM is part of the Office of the Director at the NCI of the NIH, Washington, District of Columbia, USA. NCI, U01-CA066529, U01-CA096230, U01-CA066535 and U01-CA066531.

8. Scientific Access and Productivity

Since 1996, nationally and internationally funded researchers have used ACSR HIV-infected and noninfected control biospecimens and clinical data to contribute significant discoveries to the scientific literature. Included are: markers of selective B cell activation during lymphomagenesis (14), effect of HIV integration site on cancer development (15), role of macrophages (16), chemokines, cytokines, and the growth factors in cancer (17), KSHV-induced transcriptional reprogramming in KS cell types (18), correlation of interleukins, CD4+ lymphopenia, viral load, and disease progression (19), persistent infections associated with cancer, especially EBV (20) and human papillomaviruses (21), and diagnostic assay development and validation (22).

Investigations into the pathobiology of HIV infection have used ACSR-supplied human tissue to provide translation from important discoveries made in cell culture or animal studies. HIV insertion sites within human tissues, including somatic cells (23), macrophages and malignancies (24) have been defined using ACSR-provided infected human tissue. The neuropathogenesis of AIDS dementia was explored (25) using HIV-infected brain tissue and technical resources within the ACSR. Likewise, human associated vasculopathy (26) and cardiomyopathy were elucidated by translation to HIV/AIDS human tissues relevant findings from a *murine AIDS* model (27).

9. Summary

Many highly productive investigations have come from the use of biospecimens and clinical data resources. Funded investigators or investigators in the areas of HIV/AIDS-related malignancies are recruited to use the resources of the ACSR in their research. The ACSR will work with such researchers to identify specimens if not currently available in the ACSR biorepository. The ACSR Web site, http://acsr.ucsf.edu, allows immediate access to banked specimen resources with potential for use in their research.

Requests for Reprints

Debra Garcia, ACSR CODCC, 1001 Potrero Avenue, Building 3, Room 207, San Francisco, CA 94110.

References

1. National Cancer Institute. (1994) *Tissue and Biological Fluids Banks of HIV-Related Malignancies NIH Guide RFA: CA-94-003 P.T. 34.* 23. Available from http://grants.nih.gov/grants/guide/rfa-files/RFA-CA-94-003.html.
2. National Cancer Institute. (2001) *Tissue and Biological Fluids Banks of HIV-Related Malignancies NIH Guide RFA: CA-02-001.* Available from http://grants.nih.gov/grants/guide/rfa-files/RFA-CA-02-001.html.
3. Gondos, A., Brenner, H., Wabinga, H., and Parkin, D. M. (2005) Cancer survival in Kampala, Uganda. *Br. J. Cancer.* **92**, 1808–1812.
4. Department of Health and Human Services. (2002) *45 CFR (Code of Federal Regulations), 164.514 (6)(2)(i). Standards for Privacy of Individually Identifiable Health Information (final).* US Government Printing Office, Washington, DC.
5. Joint United Nations Programme on HIV/AIDS (UNAIDS). (2008) *Report on the Global AIDS Epidemic: Executive Summary.* UNAIDS Information Centre, Geneva, Switzerland.
6. Clifford, G. M., Polesel, J., Rickenback, M., Dal Maso, L., Keiser, O., Kofler, A., Rapiti, E., Levi, F., Jundt, G., Fisch, T., Bordoni, A., De Weck, D., Franceschi, S., and Swiss HIV Cohort. (2005) Cancer risk in the Swiss HIV cohort study: associations with immunodeficiency, smoking, and highly active antiretroviral therapy. *J. Natl. Cancer Inst.* **97**, 425–432.
7. Kest, H., Brogly, S., McSherry, G., Dashefsky, B., Oleske, J., and Sege, G. R. 3rd. (2005) Malignancy in perinatally human immunodeficiency virus-infected children in the United States. *Pediatr. Infect. Dis. J.* **24**, 237–242.
8. Engels, E. A., Pfeiffer, R. M., Goedert, J. J., Virgo, P., McNeel, T. S., Scoppa, S. M., and Biggar, R. J. for the HIV/AIDS Cancer Match Study. (2006) Trends in cancer risk among people with AIDS in the United States 1980–2002. *AIDS.* **20**, 1645–1654.
9. Mbulaiteye, S. M., Katabira, E. T., Wabinga, H., Parkin, D. M., Virgo, P., Ochai, R., Workneh, M., Coutinho, A., and Engels, E. A. (2006) Spectrum of cancers among HIV-infected persons in Africa: the Uganda AIDS-Cancer Registry match study. *Int. J. Cancer.* **118**, 985–990.
10. Wang, L., Dittmer, D. P., Tomlinson, C. C., Fakhari, F. D., and Damania, B. (2006) Immortalization of primary endothelial cells by the K1 protein of Kaposi's sarcoma-associated herpesvirus. *Cancer Res.* **66**, 3658–3666.
11. Ayers, L. W., Silver, S., McGrath, M. S., Orenstein, J. M., and The AIDS Cancer and Specimen Resource (ACSR). (2007) The AIDS and Cancer Specimen Resource: role in HIV/AIDS Scientific Discovery. *Infect. Agents Cancer.* **2**, 7.
12. Srinivasan, M., Sedmak, D., and Jewell, S. (2002) Effect of fixatives and tissue processing on the content and integrity of nucleic acids. *Am. J. Path.* **161**, 1961–1971.
13. Jewell, S., Srinivasan, M., McCart, L., Williams, N., Grizzle, W. H., LiVolsi, V., MacLennan, G., and Sedmak, D. D. (2002) Analysis of the molecular quality of human tissue. An experience from the Cooperative Human Tissue Network. *Am. J. Clin. Path.* **118**, 733–740.
14. Ng, V., Hurt, M., Herndier, B. H., Fry, K. E., and McGrath, M. S. (1997) Vh gene used by HIV-1 associated lymphoproliferations. *AIDS Res. Hum. Retroviruses.* **13**, 135–149.
15. Lykidis, D., Van Noorden, S., Armstrong, A., Spencer-Dene, B., Li, J., Zhuang, Z., and Stamp, G. W. (2007) Novel zinc-based fixative for high quality DNA, RNA and protein analysis. *Nucleic Acids Res.* **35**(12), e85.
16. McGrath, M. S., Shiramizu, B., and Herndier, B. G. (2000) Clonal HIV in the pathogenesis of AIDS-related lymphoma: sequential pathogenesis, in *Infectious causes of cancer: targets for intervention.* (Goedert, J., ed.) Humana Press, Totowa, NJ, pp. 231–242.
17. Zenger, E., Abbey, N. W., Weinstein, M. D., Gofman, I., Millward, C., Gascon, R., Elbaggari, A., Herndier, B. G., and McGrath, M. S. (2002) Injection of human primary effusion lymphoma cells or associated mac-

rophages into SCID mice causes murine lymphomas. *Cancer Res.* **62**, 5536–5542.

18. Aoki, Y., Yarchoan, R., Braun, J., Iwamoto, A., and Tosato, G. (2000) Viral and cellular cytokines in AIDS-related malignant lymphomatous effusions. *Blood.* **96**, 1599–1601.
19. Wang, H. S., Totter, M. W. B., Lagos, D., Bourboulia, D., Henderson, S., Makinen, T., Elliman, S., Flanagan, A. M., Alitalo, K., and Boshoff, C. (2004) Kaposi sarcoma herpesvirus induced cellular reprogramming contributes to the lymphatic endothelial gene expression in Kaposi sarcoma. *Nat. Genet.* **36**, 687–693.
20. Napolitano, L. A., Grant, R. M., Schmidt, D., DeRosa, S., Herzenberg, L., Deeks, S., and Loftsu, R., and McCune, J. M. (2001) Circulating interleukin-7 levels are correlated with CD4+ lymphopenia and vial load in HIV-1 infected individuals: implications for disease progression. *Nature Med.* **7**, 73–79.
21. Przybylski, G. K., Goldman, J., Ng, V. L., McGrath, M. S., Herndier, B. G., Schenkein, D. P., Monroe, J. G., and Silberstein, L. E. (1996) Evidence for early B-cell activation preceding the development of Epstein-Barr virus-negative acquired immunodeficiency syndrome-related lymphoma. *Blood.* **88**, 4620–4629.
22. Palefsky, J. M., Holly, E. A., Hogeboom, C. J., Ralston, M. L., DaCosta, M. M., Botts, R., Berry, J. M., Jay, N., and Darragh, T. M. (1998) Virologic, immunologic, and clinical parameters in the incidence and progression of anal squamous intraepithelial lesions in HIV-positive and HIV-negative homosexual men. *J. Acquir. Immune Defic. Syndr. Hum. Retrovirol.* **17**, 314–319.
23. Martin, J., Amad, Z., Cossen, C., Lam, P. K., Kedes, D., Page-Shafer, K., Osmond, D., and Forghani, B. (2000) Use of epidemiologically well-defined subjects and existing immunofluorescence assays to calibrate a new enzyme immunoassay for human herpesvirus 8 antibodies. *J. Clin. Microbiol.* **38**, 696–701.
24. Mack, K. D., Jin, X., Yu, S., Wei, R., Kapp, L., Green, C., Herndier, B., Abbey, N. W., Elbaggari, A., Liu, Y., and McGrath, M. S. (2003) HIV insertions within and proximal to host cell genes are a common finding in tissues containing high levels of HIV DNA and macrophage-associated p24 antigen expression. *J. Acquir. Immune. Defic. Syndr.* **33**, 308–320.
25. Killebrew, D. A., Troelstrup, D., and Shiramizu, B. (2004) Preferential HIV-1 integration sites in macrophages and HIV-associated malignancies. *Cell Mol. Biol. (Noisy-le-grand)* **50**, OL581–OL589.
26. Salemi, M., Lamers, S. L., Yu, S., de Oliveira, T., Fitch, W. M., and McGrath, M. S. (2005) HIV-1 phylodynamic analysis in distinct brain compartments provides a model for the neuropathogenesis of AIDS. *J. Virol.* **79**, 11343–11352
27. Baliga, R. S., Chaves, A. A., Jing, L., Ayers, L. W., and Bauer, J. A. (2005) AIDS-related vasculopathy: evidence for oxidative and inflammatory pathways in murine and human AIDS. *Am. J. Physiol. Heart Circ. Physiol.* **289**, H1373–H1380.
28. Chaves, A. A., Mihm, M. J., Schanbacher, B. L., Basuray, A., Liu, C., Ayers, L. W., and Bauer, J. A. (2003) Cardiomyopathy in a murine model of AIDS: evidence of reactive nitrogen species and corroboration in human HIV/AIDS cardiac tissues. *Cardiovasc. Res.* **60**, 108–118.

Chapter 9

Specific Advantages of Twin Registries and Biobanks

Jaakko Kaprio

Abstract

This chapter briefly reviews the role of twin studies and study designs based on using twins in different settings. In the Nordic countries, twin registers and cohorts have existed already for many decades. These are a unique resource for scientific studies; a major strength being their unselected and representative nature. In the past years, biological samples are also being collected within the studies conducted on the Nordic twins.

Key words: Register, Twin research, Genetics, Environmental factors, Epigenetic

1. Introduction

There is a long history of twin research to investigate the contribution of genetic and environmental factors to traits and diseases (1). Many different kinds of scientific questions can be asked using genetically informative data sets, such as twin pairs and families of twins (2, 3). In this chapter, I do not review all the possible study designs, but provide a brief review of the main principles of twin studies. Also, I discuss the specific advantages of twin studies, and provide a brief overview of the current status of twin studies in the Nordic countries.

Many cohort and case-control studies on individuals that examine the relationship between specific genes and specific diseases ignore the developmental aspects of the disease and the contribution of risk factors. We often lack knowledge of the dynamics of gene action and of specific environmental conditions that modify gene expression. Longitudinal twin and twin-family studies with multiple measurements can permit a more detailed assessment of the developmental aspects of risk factors and diseases, and how the relative roles of genes and environment unfold

Joakim Dillner (ed.), *Methods in Biobanking*, Methods in Molecular Biology, vol. 675,
DOI 10.1007/978-1-59745-423-0_9, © Springer Science+Business Media, LLC 2011

over time. The long-standing Nordic twin studies offer a unique resource to carry out such studies.

2. Basic Twin Design for Estimating Genetic Effects

A first step in the exploration of the genetic architecture of a trait or disease is to establish that genetic factors are of importance. Twin studies are the prime, but not only, method used to establish whether familial, in particular genetic factors are of relevance for the trait, and to what degree genetic variation accounts for the total variance of a trait (4). The total variance in a behavior, trait, or liability to disease can be divided into (a) additive genetic, (b) nonadditive genetic, (c) common environmental, and (d) unique environmental variance. Additive genetic effects occur when the effects of each gene are adding up to affect the phenotype, whereas nonadditive (dominance) genetic effects denote interactions between the alleles at a genetic locus. These interactions produce deviations between the expected, additive genotypic value and actual genotypic value in the heterozygote. The additive and nonadditive effects add up over all the genes contributing to the phenotype. Interactions between genes (also known as epistatic effects) are generally seen as nonadditive genetic effects.

Environmental variance can be divided into shared (also called sometimes common) and unique components. Shared or common environmental effects denote all those aspects of the environment which cause family members, also cotwins to be similar. These shared effects can be derived from familial influences, such as parental socioeconomic status or common family home characteristics. They can also be peer effects that both twins share as they attend the same school or from being in the same occupation. In contrast, unique environmental factors affect only the member of the family in question. Unique environment refers to environmental experiences and exposures that do not contribute to familial resemblance. The estimate of unique environmental variance also contains error variance because random measurement error decreases correlations between family members.

The twin method is based on differences between the two types of twins: monozygotic (MZ) twins, who are genetically identical and dizygotic (DZ) twins, who share on average 50% of their segregating genes, like any other siblings. The comparison of trait similarity between the cotwins of the two types, measured using the correlations between the cotwins, provides first pass information on the genetic and environmental contribution to the phenotypic variation of that behavior. A MZ twin correlation double the DZ twin correlation indicates additive genetic effects, whereas genetic dominance reduces the DZ twin correlation to below half of the MZ twin correlation. DZ correlations more than one-half of the MZ correlation provide the evidence for shared environmental effects. While comparing MZ and DZ correlations are useful initial guides to the partitioning variance, the evaluation of different genetic models is best done by formal statistical models. Using Mx (5, 6) a tailor-made program for genetically informative data or other structural equation modeling programs (Lisrel, Mplus), alternative models can be compared in which different components of variance are specified, and goodness of fit statistics assess how well the various models fit the data. Scripts for different designs and models are available at the GenomEUtwin Mx-script library (http://www.psy.vu.nl/mxbib/), at the Mx home page (http://www.vcu.edu/mx/) and elsewhere.

When the data permit, the twin model can be extended to analyze much more detailed questions about the variance and covariance structures in the data (2). If information on specific genes is available, the contribution of the given gene can be distinguished from that of the remaining (unmeasured) polygenic effects using a twin model. Other genetic designs do not generally permit this.

3. Nordic Twin Registries and Cohorts

I do not describe the twin registries and cohorts in details, as these are well described in two theme issues (October 2002 and December 2006) of the journal *Twin Research and Human Genetics*. Based on those articles, I summarize the key features of these Nordic twin collections; the text below is modified from the published abstracts. In all the countries, there is great potential in each country for linking the twin data with other health registries and with information in national population-based biobanks. Also, all of the twin cohorts have collected actively biological samples. Internationally, platforms such as those developed within GenomEUtwin, for data standards and data sharing and access, and ethical frameworks are greatly facilitating international collaborations.

Twin cohorts in the Nordic countries are thus truly population-based, and do not represent selected parts of the populations which occur, for example, if employees or university students are used as the target sample. Twins are representative of the population also because twins are born into all social groups, and the morbidity and mortality after infancy is equivalent to the mortality experience of the general population. It has also been extensively shown that they are representative for practically all somatic, behavioral, and psychopathology characteristics. In addition, twins are above average active and enthusiastic study participants, and thus contribute with high participation rates at a time when voluntary participation in research appears to be declining strongly in many settings and countries. Furthermore, they are willing to recruit family members into research, and thus offer a good way to establish family data sets.

3.1. Denmark

The Danish Twin Registry is the oldest national twin register in the world, initiated in 1954 by the ascertainment of twins born from 1870 to 1910. During a number of studies, birth cohorts have been added to the register, and currently (in 2005), the Registry comprises 135 birth cohorts of twins from 1870 to 2004, with a total of more than 75,000 twin pairs included. In all cohorts, the ascertainment has been population-based and independent of the traits studied, although different procedures of ascertainment have been employed. In the oldest cohorts, only twin pairs with both twins surviving at age 6 have been included, while from 1931 all ascertained twins are included. The completeness of the ascertainment after the adjustment for infant mortality is high, with approximately 90% ascertained up to 1968, and complete ascertainment of all live born twin pairs since 1968.

The Danish Twin Registry is used as a source for large studies on genetic influence on aging and age-related health problems, normal variation in clinical parameters associated with the metabolic syndrome and cardiovascular diseases, and clinical studies of specific diseases. The combination of survey data with data obtained by linkage to national health-related registers enables follow-up studies both of the general twin population and of twins from clinical studies. Two papers summarize features of the register and give examples of recent developments and phenotypes studied. (7, 8)

3.2. Finland

In studies on the Finnish Twin Cohorts, genetic and environmental determinants of common, complex diseases, and their behavioral risk factors have been investigated in Finland. In 1974, the older twins were identified, with a total of 13,888 like-sexed pairs of known zygosity. They have participated since 1975 in mail surveys, clinical examinations for subsamples, have been used to recruit families and have been followed up for morbidity using national medical registers. Opposite twin pairs were added later. Two longitudinal studies of adolescent twins and their families, known as the FinnTwin12 and FinnTwin16 studies have focused on the determinants of health-related behaviors and disease in adolescents and young adults. Each has some 3,000 twin pairs, their parents, and sibs. Data collection and analyses are described elsewhere in detail (9–11).

3.3. Norway

In Norway, there have been historically a number of separate studies. One of the most significant recent developments is that an agreement is now in place to centralize the Norwegian twin data into a national Norwegian Twin Registry. This new registry will include twin cohorts born from 1905 onward.

The Norwegian Twin Registers include several sets of population-based subregisters, and covers twin pairs born between 1895 and today. Except for the missing birth years 1960–1967, the register is almost complete. Most of the register contains information about both same-sexed and opposite-sexed twin pairs, except for twin pairs born between 1946 and 1960, where only same-sexed twins are registered. In a substantial part of the register, information about zygosity is obtained, mainly by a mailed questionnaire and in some cases supported by DNA testing. These are the birth years 1915–1960 and the birth years 1967–1979. Zygosity information is further obtained in the different twin studies derived from the twin register. In 1990, the whole register was made available in a computerized form. Several twin studies have been derived from the different parts of the register (12).

One subregister has been the responsibility of the Norwegian Institute of Public Health in Oslo, which has an ongoing program of twin research using population-based cohorts of twins since 1992. The current database includes information on twins identified through the Medical Birth Registry of Norway and born from 1967 to 1979, altogether 15,370 twins. This is a longitudinal study with a cohort sequential design, whereby new cohorts are recruited into the study at 5- to 6-year intervals. It consists of a number of questionnaire and clinical interview projects exploring a broad array of mental and physical health outcomes. In the most recent years, a large effort has concentrated on completing a mental health interview study of Axis I psychiatric and substance use disorders and Axis II personality disorders (13).

3.4. Sweden

The Swedish Twin Registry was first established in the late 1950s. Today, it includes more than 170,000 twins – in principle, all twins born in Sweden since 1886. The first studies examined the importance of smoking and alcohol consumption on cancer and cardiovascular diseases, but since then it has been expanded and updated on several occasions. The focus has similarly broadened to most common complex diseases. In Pedersen et al. (14), the content of the database is described, ongoing projects based on the registry are summarized, and some principal findings on aging, cancer, and cardiovascular disease that have come from the registry are reviewed (14). In recent years, there has been extensive blood collection and genotyping to study the genetic bases of complex diseases, and in-depth studies of selected diseases, such as Parkinson's disease and chronic fatigue syndrome. Lichtenstein et al. (15)

describe current ongoing and recently completed projects based on the registry. All twins born between 1959 and 1985, and young twin pairs when they turn 9 and 12 years of age are being screened (15).

4. Special Aspects of Twin Studies

4.1. Assumptions of the Twin Design

The equal environment assumption (EEA) is a central tenet in twin studies. Briefly, the EEA posits that environmental influences that affect the trait of interest are not shared to a greater extent among MZ than DZ twins. Violation of this assumption could mean that increased similarity among MZ pairs leads to inflated estimates of genetic influences in twin studies. The tenability of the EEA is most likely phenotype-specific and should be examined for each phenotype and age group of interest. For most behavioral traits, the assumption is rarely violated, and twin studies remain a core approach even in psychiatric research (16).

Another assumption of the classic twin analysis is that there is random mating of the parents of twins with respect to the trait being studied. Under that assumption, the expected value of genetic resemblance of DZ twins is 0.5, i.e., they share 50% of their segregating genes in common. If there is assortative mating whereby phenotypic similarities affects partner selection, then the spousal correlation is greater than 0, and the parents may thus resemble each other with respect to the genes of the trait being studied. This is known as *phenotypic assortment*. On the other hand, parents can resemble each other because they share the same social background, which causes them to resemble each other to a greater existent without being genetically more alike than expected. This is known as *social homogamy*, for example, both effects are present for BMI (17). If nonrandom mating occurs, it needs to be taken into account. The Nordic twin registries have also collected information from parents in many substudies, and this source of potential bias can thus be taken into account.

4.2. Epigenetics, Environmental Effects, and Discordance in MZ Twin Pairs

One of the few approaches by which the effects or causes of a disease or trait can be studied in the absence of confounding due to genetic effects is to study MZ twin pairs discordant for that trait or disease. The diseased and the healthy cotwins share the same genes and differ only by environmental exposures; the environmental exposure discordance may already arise from very early events in utero (1). Moreover, the effective genotype of MZ twins may begin to diverge over time as epigenetic effects modify gene expression in the twins, even though their genomic DNA remains unchanged, except for possible somatic mutations (1).

As an example of this kind of cotwin control design, we were able to identify and study 15 healthy MZ pairs with 10–25 kg differences in weight from the Finntwin16 studies (9) of nearly 3,000 twin pairs. A control group of nine normal-weight or obesity concordant MZ pairs was also studied. These studies show that acquired obesity is associated with increased liver fat content, insulin resistance, various vascular abnormalities, and several changes in adipose tissue metabolism and lipid profiles using lipidomics (18–23). The Nordic twin studies have carried out many studies using the MZ cotwin-control design, expanding on the pioneering work by Rune Cederlöf and his colleagues at the Swedish Twin Registry in the 1960s. They examined the relationship of smoking with medical outcomes using twin pairs discordant for smoking.

One may ask why this handful of MZ twin pairs was so strikingly discordant for obesity. Their growth was normal in childhood and adolescence until after puberty when intrapair weight differences began to appear (24). This might suggest that the proximate cause of their obesity relates to changes after midadolescence, possibly related to differences in their individual experiences. There are differences in their physical activity at ages 16–17 years preceding their weight change (25), suggesting that physical activity may be a proximal causal factor for future obesity. It is also possible that the manifestation of physical inactivity in adolescence and obesity in adulthood is preceded by much earlier events. One mechanism may be through epigenetic modification of gene expression in these MZ pairs. Fraga et al. (26) suggested that epigenetic changes increase with increasing age in trait discordant MZ pairs, but the epigenetic effects relevant to obesity could possibly develop in childhood or even prenatally. One could speculate that specific environmental factors such as dietary components or sustained physical inactivity could induce the development of obesity through modulation of expression of genes regulating satiety. Once excess weight development sets in, physical inactivity increases and a vicious circle is ready leading to obesity development.

4.3. Variability Genes

Genetic factors play an important role in the responsiveness to changing environmental conditions, and gene–environmental interactions are probably very important in most conditions. Some 20 years ago, Professor Kåre Berg from the Norwegian Twin Registry put forward the variability gene concept and indicated empirical evidence in its favor based on studies of intrapair differences in blood lipids of monozygotic (MZ) pairs differing in genotype (27). Thus, persons with certain genotypes would be more sensitive to environmental determinants than others. Recent examples with psychiatric outcomes and specific neurotransmitter receptor gene variants come from the studies of Caspi et al. (28, 29).

In the GenomEUtwin project, Nordic twins will be included in a large analysis of variability genes in over 2,000 MZ pairs that is in progress in 2007 using genome-wide SNP-mapping.

5. Conclusion

The Nordic twin registers and cohorts have many strengths. One in particular is that they are unselected for disease status, when many molecular genetic family studies are ascertained on the basis of a disease proband, and consist often of multiplex families. These cohorts also have extensive information on environmental factors, and are an established resource with decades of follow-up. Epidemiologically they can be analyzed as individuals, but also have family relationships inbuilt. Thus, genetically informative analyses can be conducted. The registries are building up an extensive DNA collection, which will permit even more sophisticated analyses in the future.

References

1. Martin NG, Boomsma DI, Machin G. (1997) A twin-pronged attack on complex traits. *Nat Genet* **17**, 387–392.
2. Boomsma D, Busjahn A, Peltonen L. (2002) Classical twin studies and beyond. *Nat Rev Genet* **3**(11), 872–882.
3. Posthuma D, Beem AL, De Geus EJ, van Baal GC, von Hjelmborg JB, Iachine I et al. (2003) Theory and practice in quantitative genetics. *Twin Res* **6**(5), 361–376.
4. Thomas DC. (2004) Statistical methods in genetic epidemiology. Oxford: Oxford University Press.
5. Neale MC, Cardon LR. (1992) Methodology for genetic studies of twins and families. Dordrecht: Kluwer Academic.
6. Mx: Statistical Modelling. Box 710 MCV, Richmond, Virginia 23298: Department of Psychiatry (1994).
7. Skytthe A, Kyvik K, Holm NV, Vaupel JW, Christensen K. (2002) The Danish Twin Registry: 127 birth cohorts of twins. *Twin Res* **5**(5), 352–357.
8. Skytthe A, Kyvik K, Bathum L, Holm N, Vaupel JW, Christensen K. (2006) The Danish Twin Registry in the new millennium. *Twin Res Hum Genet* **9**(6), 763–771.
9. Kaprio J, Pulkkinen L, Rose RJ. (2002) Genetic and environmental factors in health-related behaviors: studies on Finnish twins and twin families. *Twin Res* **5**(5), 366–371.
10. Kaprio J, Koskenvuo M. (2002) Genetic and environmental factors in complex diseases: the older Finnish Twin Cohort. *Twin Res* **5**(5), 358–365.
11. Kaprio J. (2006) Twin studies in Finland 2006. *Twin Res Hum Genet* **9**(6), 772–777.
12. Bergem AL. (2002) Norwegian Twin Registers and Norwegian twin studies – an overview. *Twin Res* **5**(5), 407–414.
13. Harris JR, Magnus P, Tambs K. (2006) The Norwegian Institute of Public Health twin program of research: an update. *Twin Res Hum Genet* **9**(6), 858–864.
14. Pedersen NL, Lichtenstein P, Svedberg P. (2002) The Swedish Twin Registry in the third millennium. *Twin Res* **5**(5), 427–432.
15. Lichtenstein P, Sullivan PF, Cnattingius S, Gatz M, Johansson S, Carlstrom E et al. (2006) The Swedish Twin Registry in the third millennium: an update. *Twin Res Hum Genet* **9**(6), 875–882.
16. Kendler KS. (2001) Twin studies of psychiatric illness: an update. *Arch Gen Psychiatry* **58**(11), 1005–1014.
17. Silventoinen K, Kaprio J, Lahelma E, Viken RJ, Rose RJ. (2003) Assortative mating by body height and BMI: Finnish twins and their spouses. *Am J Hum Biol* **15**(5), 620–627.
18. Gertow K, Pietilainen KH, Yki-Jarvinen H, Kaprio J, Rissanen A, Eriksson P et al. (2004)

Expression of fatty-acid-handling proteins in human adipose tissue in relation to obesity and insulin resistance. *Diabetologia* **47**(6), 1118–1125.

19. Kannisto K, Pietilainen KH, Ehrenborg E, Rissanen A, Kaprio J, Hamsten A et al. (2004) Overexpression of 11beta-hydroxysteroid dehydrogenase-1 in adipose tissue is associated with acquired obesity and features of insulin resistance: studies in young adult monozygotic twins. *J Clin Endocrinol Metab* **89**(9), 4414–4421.
20. Pietilainen KH, Rissanen A, Kaprio J, Makimattila S, Hakkinen AM, Westerbacka J et al. (2005) Acquired obesity is associated with increased liver fat, intra-abdominal fat, and insulin resistance in young adult monozygotic twins. *Am J Physiol Endocrinol Metab* **288**(4), E768–E774.
21. Pietilainen KH, Sysi-Aho M, Rissanen A, Seppanen-Laakso T, Yki-Jarvinen H, Kaprio J et al. (2007) Acquired obesity is associated with changes in the serum lipidomic profile independent of genetic effects – a monozygotic twin study. *PLoS ONE* **2**, e218.
22. Pietilainen KH, Bergholm R, Rissanen A, Kaprio J, Hakkinen AM, Sattar N et al. (2006) Effects of acquired obesity on endothelial function in monozygotic twins. *Obesity (Silver Spring)* **14**(5), 826–837.
23. Pietilainen KH, Kannisto K, Korsheninnikova E, Rissanen A, Kaprio J, Ehrenborg E et al. (2006) Acquired obesity increases CD68 and tumor necrosis factor-alpha and decreases adiponectin gene expression in adipose tissue: a study in monozygotic twins. *J Clin Endocrinol Metab* **91**(7), 2776–2781.
24. Pietilainen KH, Rissanen A, Laamanen M, Lindholm AK, Markkula H, Yki-Jarvinen H et al. (2004) Growth patterns in young adult monozygotic twin pairs discordant and concordant for obesity. *Twin Res* **7**(5), 421–429.
25. Pietiläinen KH, Kaprio J, Borg P, Plasqui G, Yki-Järvinen H, Kujala UM, Rose RJ, Westerterp KR, Rissanen A. (2008) Physical inactivity and obesity: a vicious circle. *Obesity* **16**, 409–414.
26. Fraga MF, Ballestar E, Paz MF, Ropero S, Setien F, Ballestar ML et al. (2005) Epigenetic differences arise during the lifetime of monozygotic twins. *Proc Natl Acad Sci U S A* **102**(30), 10604–10609.
27. Berg K. (1992) Introductory remarks – risk factor levels and variability. *Ann Med* **24**, 343–347.
28. Caspi A, Sugden K, Moffitt TE, Taylor A, Craig IW, Harrington H et al. (2003) Influence of life stress on depression: moderation by a polymorphism in the 5-HTT gene. *Science* **301**(5631), 386–389.
29. Caspi A, McClay J, Moffitt TE, Mill J, Martin J, Craig IW et al. (2002) Role of genotype in the cycle of violence in maltreated children. *Science* **297**(5582), 851–854.

Chapter 10

The Swedish Multi-generation Register

Anders Ekbom

Abstract

The Swedish Multi-generation Register consists of data of more than nine million individuals, with information available on mothers in 97% and on fathers in 95% of index persons. Index persons are confined to those born from 1932 onwards and those alive on January 1, 1961. This register is a unique resource but is still underutilized.

Key words: Genealogy, Family history, Register

1. Introduction

There has been a growing realization that multi-generational information can provide valuable information on the etiology of different diseases. This is the case regarding both single and complex gene diseases. The Icelandic genealogy database serves as an example of how this can be done.

The Swedish Multi-generation Register is a part of the register system for the total population at Statistics Sweden and, during the last years, has been documented with regard to both content and quality, and currently contains data of more than nine million index persons.

2. Process of Creation

2.1. Background

In 1947, a national registration number was introduced in Sweden. In the same year, personal records were established for all persons who were registered in a parish registry in Sweden in 1947. If a person was 15 years or younger in 1947, information

Joakim Dillner (ed.), *Methods in Biobanking*, Methods in Molecular Biology, vol. 675,
DOI 10.1007/978-1-59745-423-0_10, © Springer Science+Business Media, LLC 2011

of their parents were recorded in their personal record. This means that parental information mainly exists for persons born in 1932 or later. However, the work to establish personal records for all persons who were registered in 1947 was carried out during 1947 and 1948. This means that parental information can be missing for index persons born in 1932 or 1933, because these persons were over 15 years of age when their personal records were established. Between the years 1947 and the first half year of 1991, personal records were established for all persons born in Sweden or for those who had immigrated to Sweden.

In 1961, the first census was taken which was computerized and later used to serve as a basis for the Multi-generation Register, thus the index persons must have been registered at some time since 1961 in order to be included.

In 1968, Statistics Sweden established a register called the Total Population Register for the nationally registered population in Sweden. In this computerized national register, information on biological or adopted parents was not included.

In 1991, restructuring of a national register was carried out. The restructuring meant that the national registration at local levels was moved from a parish office to the local tax office. Personal records of persons registered on June 30, 1991, were computerized. However, personal records of persons deceased between 1947 and June 30, 1991, were not computerized. For persons who had emigrated during this time period, between 1947 and June 30, 1991, information were computerized if they were born in 1920 or later. In this new computerized system, data on parents and children were included.

In 1998, a total system was reorganized and a total population registry with better quality was put into operation.

In 2000, in order to make up for the shortfalls in the data on family relation which have not been included in the original base retrieval, a supplementary data retrieval was done concerning data on the biological parents of persons who were either deceased or had emigrated. In addition, a child's relationship to a person other than the parent or guardian was also included. This includes information regarding adoption with links between the child and his or her adoptive parents.

2.2. Supplementary Work After 2000

The lack of parental information for those deceased between 1961 and 1991 led to concerns that missing data of those deceased at young age due to different diseases, at least in some instances, could be differential with regard to exposures or outcomes under study. Therefore, a study was carried out in 2001 that assessed the problem and its magnitude, and identified strategies to complement the register. Through funds provided by the Wallenberg consortium and the Swedish Cancer Foundation, personal records of persons born between 1942 and 1967 and of those who died during this period were computerized. This work was completed during 2004

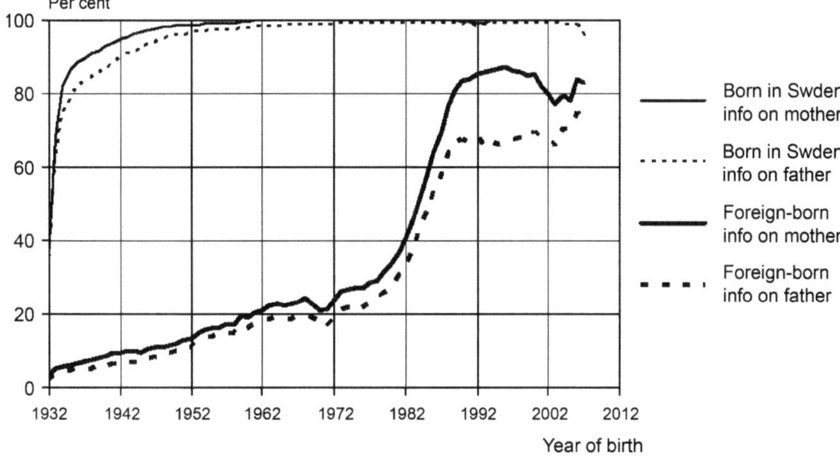

Fig. 1. Share of index persons with data on parents. Source: Statistics Sweden, Background Facts, Population and Welfare Statistics 2007:3, *Multi-generation register 2006. A description of contents and quality* (2).

and a new completed version of the Multi-generation Register was ready for use in 2005 (1) and updated 2006 (Fig. 1) (2). For index persons whose paternal information was still missing, Statistics Sweden can provide additional data, for instance, from census taken in 1970, 1975, 1980, and 1990 and the Medical Birth Register.

3. Contents of Multi-generation Register

National registration number of index person
Sex of index person
Country of birth of index person
Registration number of biological father
Date of birth of biological father
Country of birth of biological father
Registration number of biological mother
Date of birth of biological mother
Country of birth of biological mother
Child's position in the mother's family
Number of children – mother
Number of children – father
Registration number of adoptive father
Date of birth of adoptive father
Country of birth of adoptive father
Registration number of adoptive mother
Date of birth of adoptive mother
Country of birth of adoptive mother
Date of immigration
Date of adoption

Paternity is established in the national register in the following way: If a mother is married or recently widowed at the time of a child's birth, the husband is seen to be the father of the child. In other cases, paternity is normally established by acknowledgment or, in certain cases, by a court and is then reported later, not at the time of birth.

The child's position in the family shows the child's ordinal position among biological children in the register of biological mother and father. For index persons with parents born in Sweden in 1915 or later, information on their position in the family is of a good standard. If the parents were born before 1910, the Multi-generation Register should not be used to identify siblings. For index persons with foreign-born parents, their position in the family is of a worse standard and siblings who were registered from 1961 onwards in Sweden and who, at the time of immigration, were younger than 18 years of age are included.

4. Coverage

As of December 2005, the Multi-generation Register contains information on 9,371,000 individuals. According to this register, a total of 7.7 million people were born in Sweden, of which information on mother is found in 97% and on father is found in 95% of the index persons. The equivalent frequencies of those born outside Sweden are 27% and 22%. Lower frequencies are present for mothers and fathers of those born during the 1930s and this is illustrated in Fig. 1. This is to be expected as some parents would have died before January 1, 1947, when the national registration number was introduced. This means that information on biological mothers is available in 98% of individuals born in 1947 and 100% of those born in 1961 and onwards. The equivalent number for fathers is 94% for those born in 1947 and 98% for those born in 1961 and onwards. This means that an excellent coverage and additional complement efforts have not been deemed cost-efficient.

4.1. Ethical Issues

The principal rule for Statistics Sweden is that identifiable individual data in statistical activities should be kept confidential and should not be given out. However, exceptions to this rule are possible, as when information is needed for research purposes. However, information is given out only after approval from an Ethical Board but in addition a ruling within Statistics Sweden also must be made in order to ascertain that persons concerned or connected to the information should not be in any way harmed or injured.

A good working relationship has, however, emerged between the research community and Statistics Sweden with a mutual respect of the needs of the different stake-holders.

4.2. Example of Successful Research Utilizing the Swedish Multi-generation Register

More than a couple of hundreds of scientific publications, especially in medical research, have already been published utilizing the Swedish Multi-generation alteration register. These include assessments of the attributable risk of family cancer (3, 4), cancer rates in second-generation immigrants (5), cancer risks in gene carriers, for hereditary hematochromatosis (6), risks of suicide among different groups of second-generation immigrants (7), and to what extent a family history of colorectal cancer is an independent risk factor for individuals already at an increased risk for inflammatory bowel diseases (8).

5. Prospective for the Future

Although the Swedish Multi-generation Register has already been utilized to a great extent, there is still a need to educate researchers within Sweden and abroad about its existence and possibilities. Moreover, there is a potential for enlargement, and a need to find ways to link information in the Swedish Multi-generation Register to the existing biobanks. Three different venues are, therefore, presently pursued:

1. To ensure good documentation of the Swedish Multi-generation Register and a continuous update on a yearly basis. This has been made possible within the budget of Statistics Sweden.
2. To provide ways to link index cases and their relatives with an interesting pedigree with specimens already included in existing biobanks. In collaboration with Statistics Sweden, we have identified pathways taking into account ethical and logistic concerns, and a few pilot projects are presently underway.
3. To enlarge the Multi-generation Register in order to include individuals born in 1947 but deceased before January 1, 1960, and individuals born from 1920 onwards. A pilot study has been performed, which has resulted in a time-plan and an evaluation of the costs. Funding, in collaboration with Riksarkivet, is presently sought for at the Swedish Research Council.

6. Conclusions

Swedish Multi-generation Register is a unique resource for both cutting-edge medical research and geographic, economic, sociologic, and historic endeavors. It is still underutilized to a large extent but hopefully it will in the future be used with an increasing frequency by different research groups, both nationally and internationally.

References

1. Multi-Generation Register 2005. A description of contents and quality. Population and Welfare Statistics (2006:6). *Statistics Sweden.*
2. Statistics Sweden, Background Facts, Population and Welfare Statistics (2007:3), *Multi-generation register 2006. A description of contents and quality.*
3. K. Hemminki, K. Czene (2002) Attributable risks of familial cancer from the Family-Cancer Database. *Cancer Epidemiol Biomarkers Prev.*;**11**(12):1638–44.
4. K. Czene, P. Lichtenstein, K. Hemminki (2002) Environmental and heritable causes of cancer among 9.6 million individuals in the Swedish Family-Cancer Database. *Int J Cancer.*;**99**(2):260–6.
5. S.M. Montgomery, F. Granath, A. Ehlin, P. Sparén, A. Ekbom (2005) Germ cell testicular in offspring of Finnish immigrants to Sweden. *Cancer Epidemiol Biomarkers Prev.*;**14**(1):280–2.
6. M. Elmberg, R. Hultcrantz, A. Ekbom, L. Brandt, S. Olsson, R. Olsson, S. Lindgren, L. Lööf, P. Stål, S. Wallerstedt, S. Almer, H. Sandberg-Gertzén, J. Askling (2003) Cancer risk in patients with hereditary hematochromatosis and in their first-degree relatives. *Gastroenterology.*;**125**(6):1733–41.
7. A. Hjern, P. Allebeck (2002) Suicide in first- and second-generation immigrants in Sweden: a comparative study. *Soc Psychiatr Epidemiol.*;**37**(9):423–9.
8. J. Askling, P.W. Dickman, P. Karlén, O. Broström, A. Lapidus, R. Löfberg, A. Ekbom (2001) Family history as a risk factor for colorectal cancer in inflammatory bowel disease. *Gastroenterology.*;**120**(6):1356–62.

Chapter 11

Multigenerational Information: The Example of the Icelandic Genealogy Database

Hrafn Tulinius

Abstract

The first part of the chapter describes the Icelandic Genealogical Database, how it was created, what it contains, and how it operates. In the second part, an overview of research accomplished with material from the database is given.

Key words: Genetics, Familiality, Iceland, Population, Nation, Cancer, Cardiovascular diseases, Rheumatic diseases

1. Introduction

The genealogy database created by the Genetical Committee of the University of Iceland (GCU) was constructed by linking the census of 1910 to the Icelandic National Registry, which was founded in 1953. The data were made complete for all Icelanders born after 1845 by adding information on the period 1840–1910 from parish records and censuses.

The database has been used for numerous scientific investigations and is one of very few genealogy databases covering a defined population (a whole nation), which can reliably trace and link persons as far back as to 1845.

2. Process of Creation

2.1. Historical Background

The Icelandic nation has always been interested in genealogy. The country was settled, between 874 and 930, mainly from Norway,

Joakim Dillner (ed.), *Methods in Biobanking*, Methods in Molecular Biology, vol. 675,
DOI 10.1007/978-1-59745-423-0_11, © Springer Science+Business Media, LLC 2011

and also from other Nordic countries and from the Norse settlements on the British Isles.

The Icelanders founded a society governed by law in 930 with the establishment of the parliament, the Alþingi.

"The Book of Settlements" (Landnámabók), which is among the Icelandic sagas, describes in detail the area settled by each settler, the boundaries between the settlers land and that of the neighbours, as well as names of some of those who arrived with the settler. The oldest manuscripts for this book, remaining today, are from around 1300, but the writing of the book had started around 1100.

The law on heredity described how land and other positions should be distributed among the heirs. The information in Landnámabók explains the relationship between family members and sometimes includes more information on members of individual families. For this reason, sound knowledge in genealogy has been necessary. This tradition continued through the centuries, which is among the reasons for the continuing genealogical interest of Icelanders.

Life in Iceland was hard. This was due to changes in weather conditions, volcanic eruptions, and epidemics of infectious diseases. In 1702, the King of Denmark, who was also the King of Iceland, ordered an investigation of the economy and other conditions of the nation. This included making a census of the Icelandic population and descriptions of every farm in the country. This was accomplished and the census of 1703 is the first census ever to cover a whole nation, including names, ages, and status, i.e., father, mother, agricultural worker, etc., of every member of the household. Then, three-time censuses were taken of a few counties only, but in 1801 a general census was taken. From 1835, censuses were taken every 5 years until 1860 and from then, every 10 years until 1960.

Parish records were of two types. One was kept by the minister of the church registering all clerical work performed in his parish, including births, baptisms, and deaths, and some information about people moving in or out of the parish. The oldest such book dates back to 1664. In 1812, another type of parish books was added. This was to be kept by a person residing where the church was located. It included much of the same information as that found in the book kept by the minister for all churches he served. Even though some church records have perished during the ages, for example, in fires, due to this double bookkeeping, the information for the nineteenth century is reasonably complete.

2.2. The Process of Constructing the Database

The law on the National Registry (Þjóðskrá) was enacted in 1952, and the National Registry has been in operation since 1 January 1953.

After the Second World War, the US government was interested in fostering research on genetics, especially related to the effects of radiation. The United States Atomic Energy Commission brought together several world-renowned population geneticists to advise them on how to help accomplish such research. Professor Luca Cavalli-Sforza, a well-known Italian geneticist, was one of them. He was acquainted with the genealogical records of Iceland and of the interest of the people in genealogy, and suggested that work on making this information useful for research be financially supported.

The Genetical Committee of the University of Iceland was formed in 1965. The work on the genealogical database was initiated by Dr. Sturla Friðriksson and Prof. Niels Dungal. Other persons who participated in the preparations were Dr. Sigurður Sigurðsson, Director of health; Klemens Tryggvason, Chief of the Bureau of Statistics; Prof. Magnus Magnusson; and Áki Pétursson. The work was partly financed during 1965–1983 by the Atomic Energy Commission of USA.

The work started by computerising the census of 1910. The next step was to link the census of 1910 to the National Registry of 1953. For that information, the records of births, deaths, and marriages from 1910 to 1952 were used, as well as information about immigration and emigration from other records.

Information on the period from 1840 to 1910 was obtained from the parish records and censuses. This work was accomplished by a group of experienced genealogists who worked in the National Archives. All births, deaths, and marriages were recorded as well as available information on migration. A "mothers record" was created for each woman who had given birth during the period. This record contained her name and identification as well as names and identification of each child she had borne, and name and identification of the father of each child.

The database has direct access to the National Registry, which is updated daily.

At present, the database is financed by the Landspitali University Hospital and by sales of information.

3. Which Type of Information is Included

In the database, the following information is included for each individual:

- Unique personal identification number
- Name
- Gender

- Date of birth (yymmdd)
- Place of birth
- Twinning
- Legitimacy
- Date of death
- Identification number of father
- Date of birth of father (yymmdd)
- Name of father
- Identification number of mother
- Date of birth of mother (yymmdd)
- Name of mother
- Note
- Other identification

The National Registry has additional information which can be accessed.

4. Retrieval Mechanisms

The retrieval mechanisms used depend on the intended application. The most usual research requires all first, second, and third degree relatives of a defined group of people of a certain trait, such as a disease, for estimation of familiality.

When information is being requested from the genealogical database, the researchers are required to furnish the Genealogical Committee of the University with a copy of the permissions from the Bioethics Committee and the Data Protection Authority for that particular research for which the requested data will be used.

5. Ethical Considerations

Since the creation of the GCU in 1965, the legislation on ethics in research has radically changed. Since then, laws have been passed on ethics in research.

The first law on protecting the indivduals, as regards the documentation of personal data, was enacted in 1981. Now the most important acts are Act no. 74/1997 on the Rights of Patients; the Act no. 77/2000 on the Protection of Privacy as regards the Processing of Personal Data as amended in 2001, 2002, and 2003; and Act no. 110/2000 on Biobanks.

6. Examples of Succesful Research Performed

In 1967, GCU started a chromosome laboratory, the first in the country. The technician was sent to Birmingham, England, for training. This laboratory was later taken over by the Department of Pathology of the University of Iceland. Further laboratory work, partly in collaboration with the Icelandic Blood Bank, started around 1970.

A publication by Jensson, Palsdottir, Thorsteinsson, and Arnason (1) "The saga of cystatin C gene mutation causing amyloid angiopathy and brain haemorrhage – clinical genetics in Iceland" appeared in 1989. In addition to describing the research on amyloid angiopathy, it gave an overview of studies on clinical genetics performed by the committee until 1989. There are 15 references on hereditary amyloid angiopathy, work accomplished in collaboration with researchers in Sweden, UK, and USA.

As found from the studies on autosomal dominant diseases, there are four haematological disorders: hereditary elliptocytosis (2), hereditary spherocytosis (3), Pelger anomaly (4), and von Willebrand disease (5), then there are atrophia areata (6), osteogenesis imperfecta (7), Huntington's chorea (8), hypertrophic cardiomyopathy (9), and shizophrenia (10, 11). Autosomal recessive conditions are complement (C2) deficiency (12), isovaleric acidemia (13), 2,8-dihydroxyadenine crystalluria in Iceland (14), holocarboxylase synthetase deficiency (15), recessive osteopetrosis (16), autosomal recessive ichthyosis (17), homocystinuria (18), serum cholinesterase variants (19, 20), and 21-hydroxylase deficiency (21). The following X-linked diseases have also been studied: X-linked cleft palate (22–24), X-linked anencephaly and spina bifida (25), placental sulfatase deficiency (26), hereditary juvenile retinoschisis in an Icelandic family (27), hemophilia A (28), hemophilia B (28), X-linked thyroxine-binding globulin (29), and hypogammaglobulinemia (30).

Among the publications of the Genetical Committee are studies on diseases and genetic markers, such as familial leukaemia (31), familial macroglobulinaemia (32, 33), rheumatic diseases (34–36), insulin-dependent diabetes mellitus (37, 38), multiple sclerosis (39), thyroid disease, IgA deficiency (40), and leukaemia (41).

After 1989, the most productive collaboration has been on malignant diseases. This work started in 1972 with collaboration between The Icelandic Cancer Registry, the Department of Pathology of the University of Iceland, the GCU, and the International Agency for Research on Cancer by collecting families of breast cancer patients. An early publication (42) on The Breast Cancer Family Collection of the Icelandic Cancer Registry appeared in 1982. It has since been the basis of several publications (42–61). This material was used in the process of defining the BRCA2 gene, which is a founder mutation in the Icelandic population. It has also been used to show that the carriers of this gene are at increased risk for cancer of the prostate and that their prognosis is worse than that of non-carriers (70).

In the collaboration between the GCU and the Icelandic Cancer Registry, familial risk of papillary cancer of the thyroid (62) has also been studied, and in collaboration also with the Mayo foundation, familiality of gliomas of the brain (63) has been investigated.

The Iceland Genomics Corporation has conducted studies using the database of the Genetical Committee on familial risk of malignant diseases such as cutaneous melanoma and (64), colon and rectum (65), ovaries (66), prostate (67), and gastric cancer (68).

The Icelandic cardiovascular risk factor study, also known as the Reykjavik study, is a prospective cohort study started in 1967. In collaboration with the Genealogical Committee of the University of Iceland, it investigates the conventional risk factors in the descendants of victims of coronary disease (69).

References

1. Jensson, O., Palsdottir, A., Thorsteinsson, L. and Arnason, A. (1989) The saga of cystatin C gene mutation causing amyloid angiopathy and brain hemorrhage – clinical genetics in Iceland. *Clin. Genet.* **47**, 368–377.
2. Jensson, O., Jonasson, T. and Olafsson, O. (1967) Hereditary elliptocytosis in Iceland. *Br. J. Haematol.* **13**, 844–854.
3. Jensson, O., Jonasson, J. L. and Magnusson, S. (1977) Studies on herediatary spherocytosis in Iceland. *Acta Med. Scand.* **201**, 187–195.
4. Jensson, O., Arnason, K., Johannesson G. M. and Ulfarsson J. (1977) Studies on the Pelger anomaly in Iceland. *Acta Med. Scand.* **201**, 183–185.
5. Jensson, O. and Wallett, L. H. (1970) Von Willebrand's disease in an Icelandic family. *Acta Med. Scand.* **187**, 229–234.
6. Magnusson, L. (1981) Atrophia areata: a variant of peripapillary chorioretinal degerneration. *Acta Ophthalmol. (Copenh)* **59**, 659–664.
7. Sykes, B., Ogilvie, D., Wordsworth, P., Wallis, G., Mathes, C., Beighton, P., Nichols, A., Pope, M., Thomson, E., Tsipouras, P., Schwartz, R., Jensson, O., Arnason, A., Borresen, A. L., Heiberg, A., Frey, D. and Steinmann, B. (1990) Consistent linkage of dominantly inherited osteogenesis imperfecta to the two collagen 1 loci: COLIA1 and 47 COLIA2. *Am. J. Hum. Genet.* **47**, 592–594.
8. Gudmundsson, G., Jensson, O., Arnason, A. and Stefansson, K. (1989) Huntington's chorea in Iceland: epidemiological study. *Clin. Genet.* **35**, 225.
9. Bjarnason, I., Jonsson, S. and Hardarson, Th. (1982). Mode of inheritance of Hypertrophic cardiomyopathy in Iceland. Echographic study. *Br. Heart J.* **47**, 122–129.
10. Karlsson, J. L. (1988). Partly dominant transmission of Schizophrenia in Iceland. *Br. J. Psychiatry.* **152**, 324–329.
11. Sherrington, T., Brynjolfsson, J., Petursson, H., Potter, M., Duleston, K., Barraclough, B., Wasmuth, J., Dobbs, M. and Gurling, H. (1988). Localization of a susceptibility locus for schizophrenia on chromosome 5. *Nature* **336**, 164–167.
12. Arnason, A., Steinsson, K., Fossdal, R., Gunnarsdottir, H., Thorsteinsson, L., Palsdottir, A., Valdimarsson, H., Erlendsson, K. and Jensson, O. (1989). Deficiency of the second component (C2) in Iceland – a study of two cases. *Clin. Genet.* **35**, 215.
13. Gütller, F., Ludvigsson, P., Gregersen, N. and Gerdes, A. M. (1989). The first published cases of isovaleric acidaemia in Scandinavia are from Iceland. *Clin. Genet.* **35**, 218.
14. Laxdal, Th. and Jonasson, T. A. (1989). 2.8-Dihydroxyadenine crystalluria in Iceland. *Clin. Genet.* **35**, 219.
15. Gütller, F., Laxdal, Th. and Sweetman, L. (1989). A new variant of biotin responsive

holocarboxylase synthetase deficiency found in Iceland. *Clin. Genet.* **35**, 220.
16. Jensson, O. and Arnason, A. (1989). Recessive osteopetrosis in Iceland. *Clin Genet.* **35**, 223.
17. Jensson, O., Baldursson, B. T. and Arnason, A. (1989) Autosomal recessive ichtyosis, lamellar type, in Iceland. *Clin. Genet.* **35**, 224.
18. Jonsson, J. J., Jonasson, J. G., Sigurdsson, Th., Stefansdottir, A. and Halldorsson, S. (1989). Homocystinuria (cystathionine beta-synthase deficiency) in Iceland. *Clin. Genet.* **35**, 224.
19. Arnason, A., Jensson, O. and Gudmundsson, S. (1975) Serum esterases of Icelanders I. A "silent" pseudocholinesterase gene in an Icelalndic family. *Clin. Genet.* **7**, 405–412.
20. Arnason, A., Fossdal, R., Thorsteinsson, A. and Jensson, O. (1989) Serum cholinesterase at the E_1 locus in Icelanders. *Clin. Genet.* **35**, 226.
21. Arnason, A., Gudmundsson, Th., Fossdal, R. and Jensson, O. (1989) Chromosome 6 markers and congenital adrenal hyperplasia (CAH) due to 21-hydroxylase deficiency in Iceland *Clin. Genet.* **5**, 227.
22. Moore, G. E., Ivens, A., Chambers, J., Farrall, M., Williamson, R., Page, D. C., Björnsson, A., Arnason, A. and Jensson, O. (1987) Linkage of an X-chromosome cleft palate gene. *Nature* **326**, 91–92.
23. Ivens, A., Moore, G. E, Chambers J., Arnason A., Jensson, O., Björnsson, A. and Williamson, R. (1987) X-linked cleft palate: the gene is localized between polymorphic DNA markers DXYS12 and DXS17. *Hum. Genet.* **78**, 356–358.
24. Moore, G., Ivens, A., Chambers, J., Björnsson, A., Arnason, A., Jensson, O. and Williamson, R. (1988) The application of molecular genetics to detection of craniofacial abnormality. *Development.* **103**, 233–239.
25. Jensson, O., Arnason, A., Gunnarsdottir, I., Petursdottir, I., Fossdal, R. and Hreidarsson, S. (1988) A family showing apparent X linked inheritance of both anencephaly and spina bifida. *J. Med. Genet.* **25**, 227–229.
26. Hardardottir, H., Geirsson, R. T. and Hreidarsson, S. (1989) Placental sulfatase deficiency. *Clin. Genet.* **35**, 224.
27. Thordarson, K. and Gislason, I. (1989). Herreditary juvenile retionschisis in an Icelandic family. *Clin. Genet.* **35**, 225.
28. Jensson, O. (1979) Haemophilia A. *Nordic Council Arct. Res. Rep.* **26**, 29–33.
29. Sigurdsson, G. A., Arnason, A., Gudmundsson Th. V., Kjeld, M., Franzson, L. and Sigurdsson, G. (1984) Familial elevation of serum thyroxine binding globulin in an Icelandic Family. *Acta Endocirnol.* **107**, 352–356.
30. Thorsteinsson, L., Ögmundsdottir, H., Sigfusson, A., Arnason, A., Eyjolfsson, G. and Jensson, O. (1989) An Icelandic family with hypogammaglobulinemia, genetical and immunological study. *Clin. Genet.* **35**, 226.
31. Juginder, K., Catovsky, D., Valdimarsson, H., Jensson, O. and Spiers, A. S. D. (1972) Familial acute myeloid leukaemia with acquired Pelger-Huët anomaly and aneuploidy of C group.*Br. Med. J.* **4**, 327–331.
32. Björnsson, O. G., Arnason, A., Gudmundsson, S., Jensson, O, Olafsson, S. and Valdimarsson, H. (1978) Macroglobulinaemia in an Icelandic Family. *Acta Med. Scand.* **203**, 283–288.
33. Jensson, O., Björnsson, O. G., Arnason, A., Birgisdottir, B. and Pepys, M. B. (1982) Serum amyloid p-component and C-reactive protein in serumof healthy Icelanders and members of an Icelandic family with macroglogulinaemia. *Acta Med. Scand.* **211**, 341–345.
34. Teitsson, I., Thorsteinsson, J., Arnason, A. and Valdimarsson, H. (1985) Rheumatic diseases in an Icelandic family. *Scand. J. Rheumatol.* **14**, 109–118.
35. Palsdottir, A., Arnason, A., Fossdal, R. and Jensson, O. (1987) Gene organization of haplotypes expressing two different C4A allotypes. *Hum. Genet.* **76**, 220–224.
36. Arnason, A., Thorsteinsson, J. and Sigurbergsson, K. (1978) Ankylosing spondylitis, HLA-B27 and Bf. *Lancet.* **1**, 339–340.
37. Arnason, A., Helgason, T. and Jensson, O. (1979) HLA in insulin-dependent diabetes in Iceland. *Icelandic Med. J.* **4**, 189–190.
38. Danielsen, R., Helgason, T., Arnason, A. and Jonasson, F. (1982) HLA and retinopathy in type 1 (insulin-dependent) diabetic patients in Iceland. *Diabetologia* **22**, 297–298.
39. Arnason, A., Jensson, O., Skaftadottir, I., Birgisdottir, B., Gudmundsson, G. and Johanneson, G. (1980) HLA types, Gc protein and other genetic markers in multiple sclerosis and two other neurological diseases in Iceland. *Acta Neurol. Scand.* **62**, 39–40.
40. Ulfarsson, J., Gudmundsson, S., Birgisdottir, B., Kjeld, M. J. and Jensson, O. (1982) Selective serum IgA deficiency in Icelanders: frequency, family studies and IgA levels. *Acta Med. Scand.* **211**, 481–487.
41. Arnason, A., Jonmundsson, G. K., Skaftadottir, I., Thorsteinsson, L., Johannesson, G. M., Eyjolfsson, G., Fossdal, R., Petursdottir, I. and Jensson, O. (1985) Genetical markers and leukaemia in Iceland – an association

between homozygosity of HLA B7, HLA B40 and leukaemia. *Scand. J. Immunol.* **22**, 441.

42. Tulinius, H., Day, N., Bjarnason, Ó., Geirsson, G., Jóhannesson, G., Conzales, M., Sigvaldason, H., Bjarnadóttir, G. and Grímsdóttir, K. (1982) Familial breast cancer in Iceland. *Int. J. Cancer* **29**, 365–371.
43. Tulinius, H. Familial Risk of Breast Cancer in Iceland. In: Müller, Weber (eds.). Familial Cancer. 1st Int. Res. Conf., Basel 1985, p. 25–26 (Karger, Basel 1985).
44. Tulinius, H. Familial Cancer Registration in Iceland. In: Müller, Weber (eds.). Familial Cancer. 1st Int. Res. Conf., Basel 1985, p. 263–267 (Karger, Basel 1985).
45. Tryggvadóttir, L., Tulinius, H. and Robertson. J. M. (1988) Familial and sporadic breast cancer cases in Iceland: a comparison with respect to ABO blood groups and the risk of bilateral cancer of the breast. *Int. J. Cancer* **42**, 499–501.
46. Tulinius, H., Egilsson, V., Olafsdottir, G. H. and Sigvaldason, H. (1992) Risque de cancer de la prostate, de l'ovaire et de l'endométre en cas d'antécédent familial de cancer du sein. *le Journal International de Médecine* **151**, 25–28.
47. Tulinius, H., Olafsdottir, G. H., Sigvaldason, H., Tryggvadottir, L. and Bjarnadottir, K. (1994) Neoplastic diseases in families of breast cancer patients. *J. Med. Genet.* **31**, 618–621.
48. Thorlacius, S., Tryggvadottir, L., Olafsdottir, G. H., Jonasson, J. G., Ogmundsdottir, H. M., Tulinius, H. and Eyfjord, J. E. (1995) Linkage to BRCA2 region in hereditary male breast cancer. *Lancet* **346**, 544–545.
49. Tulinius, H. (1995) Impact of genetics on cancer. In: Parr, E., Næss, A. and Rossland, T: Kreftepidemiologi. Oslo 1995, 56–59.
50. Thorlacius, S., Olafsdóttir, G. H., Tryggvadóttir, L., Neuhausen, S., Jónasson, J. G., Tavigian, S. V., Tulinius, H., Ögmundsdóttir, H. M. and Eyfjörð, J. E. (1996) A single BRCA2 mutation in male and female breast cancer families from Iceland with varied cancer phenotypes. *Nat. Genet.* **13**, 117–119.
51. Thorlacius, S., Sigurdsson, S., Bjarnadottir, H., Olafsdottir, G., Jonasson, J. G., Tryggvadottir, L., Tulinius, H. and Eyfjord, J. E. (1997) Study of a single BRCA2 mutation with high carrier frequency in a small population. *Am. J. Hum. Genet.* **60**, 1079–1084.
52. Thorlacius, S., Struewing, J. P., Hartge, P., Ólafsdóttir, G. H., Sigvaldason, H., Tryggvadóttir, L., Wacholder, S., Tulinius, H. and Eyfjörð, J. E. (1998) Population-based study of risk of breast cancer in carriers of BRCA2 mutation. *Lancet* **352**, 1337–1339.
53. Tulinius, H., Sigvaldason, H., Ólafsdóttir, G., Tryggvadóttir, L. and Bjarnadóttir, K. (1999) Breast cancer incidence and familality in Iceland during 75 Years from 1921 to 1995. *J. Med. Genet.* **36**, 103–107.
54. Baffoe-Bonnie A., Beaty, T. H., Bailey-Wilson, J. E., Kiemeney, L., Sigvaldason, H., Ólafsdóttir, G., Tryggvadóttir, L. and Tulinius, H. (2000) Genetic epidemiology of breast cancer: segregation analysis of 389 icelandic pedigrees. *Genet. Epidemiol.* **18**, 81–94.
55. Tulinius, H., Olafsdottir, G. H., Sigvaldason, H., Arason, A., Barkardottir, R. B., Egilsson, V., Ogmundsdottir, H. M., Tryggvadottir, L., Gudlaugsdottir, S. and Eyfjord, J. E. (2002) The effect of a single BRCA2 mutation on cancer in Iceland. *J. Med. Genet.* **39**, 457–452.
56. Baffoe-Bonnie, A. B., Kiemeney, L. A. L. M., Beaty, T. H., Bailey-Wilson, J. E., Schnell, A. H., Sigvaldason, H., Ólafsdóttir, G., Tryggvadóttir, L. and Tulinius, H. (2002) Segregation analysis of 389 Icelandic pedigrees with breast and prostate cancer. *Genet. Epidemiol.* **23**, 349–363.
57. Antoniou, A., Pharoah, P. D., Narod, S., Risch, H. A. et al. (2003) Averge risks of breast and ovarian cancer associated with BRCA1 or BRCA2 mutations detected in case series unselected for family history: a combined analysing of 22 studies. *Am. J. Hum. Genet.* **72**, 5, 1117–1130.
58. Tryggvadottir, L., Olafsdottir, E. J., Gudlaugsdottir, S., Thorlacius, S., Jonasson, J. G., Tulinius, H. and Eyfjord, J. E. (2003) BRCA2 mutation carriers, reproductive factors and breast cancer risk. *Breast Cancer Res.* **5**, 121–128. http://breast-cancer-research.com/content/5/5/R121
59. Antoniou, A. C., Pharoah, P. D., Narod, S., Risch, H. A. et al. (2005) Breast and ovarian cancer risks to carriers of the BRCA1 1382insC and 185delAG and VRCA2 6174delT mutations: a combined analysis of 22 population based studies. *J. Med. Genet.* **42**, 602–603.
60. Ögmundsdóttir, H., Haraldsdóttir, V., Jóhannesson, G. M., Ólafsdóttir, G., Bjarnadóttir, K., Sigvaldason, H. and Tulinius, H. (2005) Familiality of benign and malignant paraproteinemias. A population-based cancer-registry study of multiple myeloma families. *Haematologica* **90**, 66–71.
61. Tryggvadottir, L., Sigvaldason, H., Olafsdottir, G. H., Jonasson, J. G., Jonsson, T., Tulinius, H. and Eyfjord, J. E. (2006) Population-based study of changing breast cancer risk in

Icelandic BRCA2 mutation carriers, 1920–2000. *J. Natl. Cancer Inst.* **98**, 116–122.

62. Hrafnkelsson, J., Tulinius, H., Jonasson, J. G., Ólafsdóttir, G. and Sigvaldason, H. (1989) Papillary thyroid carcinoma in Iceland. A study of the occurrence in families and the coexistence of other primary tumours. *Acta Oncologica* **28**, 785–758.
63. O'Neill, B. P., Blondal, H., Yang, P., Olafsdottir, G. H., Sigvaldason, H., Jenkins, R. B., Kimmel, D. W., Scheithauer, B. W., Rocca, W. A., Bjornsson, J. and Tulinius, H. (2002) Risk of cancer among relatives of patients with glioma. *Cancer Epidemiol. Biomarkers Prev.* **11**, 921–924.
64. Eldon, B. J., Thorlacius, S., Jonsson, T., Jonasson, J. G., Kjartansson, J., Bodvarsson, S., Steingrimsson, E. and Rafnar, T. (2006) A population-based study on the familial aggregation of cutaneous malignant melanoma in Iceland. *Eur. J. Cancer.* **42**, 922–926.
65. Stefansson, T., Moller, P. H., Sigurdsson, F., Steingrimsson, E. and Eldon B. J. (2006) Familial risk of colon and rectal cancer in Iceland: evidence for different etiologic factors? *Int. J. Cancer.* **119**, 304–308.
66. Rafnar, T., Benediktsdottir, K. R., Eldon, B. J., Gestsson, T., Saemundsson, H., Olafsson, K., Salvarsdottir, A., Steingrimsson, E. and Thorlacius, S. (2004) BRCA2, but not BRCA1, mutations account for familial ovarian cancer in Iceland: a population-based study. *Eur. J. Cancer.* **40**, 2788–2793.
67. Eldon, B. J., Jonsson, E., Tomasson, J., Tryggvadottir, L. and Tulinius, H. (2003) Familial risk of prostate cancer in Iceland. *BJU Int.* **92**, 915–919.
68. Imsland, A. K., Eldon, B. J., Arinbjarnarson, S., Egilsson, V., Tulinius, H., Tryggvadottir, L., Arngrimsson, R. and Magnusson, J. (2002) Genetic epidemiologic aspects of gastric cancer in Iceland. *J. Am. Coll. Surg.* **195**, 181–186; discussion: 186–187.
69. Andresdottir, M. B., Sigurdsson, G., Sigvaldason, H. and Gudnason, V. (2002) Fifteen percent of myocardial infarctions and coronary revascularizations explained by family history unrelated to conventional risk factors. *Eur. Heart J.* **23**, 1655–1663.
70. Tryggvadóttir, L., Vidarsdóttir, L., Thorgeirsson, J. G., Olafsdóttir, E. J., Olafsdóttir, G. H., Rafnar, T., Thorlacius, S., Jonsson, E., Eyfjord, J. E. and Tulinius, H. J. (2007) Prostate cancer progression and survival in BRCA2 mutation carriers. *Natl Cancer Inst.* Jun 20; **99**(12), 929–935.

Chapter 12

Creation of a New Prospective Research Biobank: The Example of HUNT3

Kristian Hveem

Abstract

When establishing a biobank, there are a number of issues to consider with significant impact on the outcome of your efforts and the quality of your work. Some of the most relevant are: (1) study design, (2) the size and composition of your cohort, (3) the nature and desired quality of the biological material, and (4) how to handle, store, and retrieve your samples to maintain the best quality for future analyses.

In this chapter, we in particular describe the organisation of biobanks derived from population-based prospective health surveys illustrated by the Norwegian HUNT study as a basis for more general considerations.

Key words: HUNT3, Establishing biobank, Sample collection, Study design, Staffing

1. Introduction

THE NORD-TRØNDELAG HEALTH STUDY (The HUNT study) was originally established to evaluate the detection of hypertension in an adult, total population as a cross-sectional design in 1984–86 (1). In addition, screening of diabetics was included (2). In 1995–97, a second health survey was conducted (HUNT 2), as a prospective, longitudinal follow up of HUNT 1 (3). From being basically a cardiovascular screening study, the HUNT study now developed to be one of the largest, longitudinal, and comprehensive health surveys for a total population ever conducted (Table 1). In contrast to HUNT 1, HUNT 2 also included blood sampling stored as serum, blood clots, and EDTA full blood. In October 2006, a third health survey (HUNT 3) was launched, screening the same total population as the two previous ones.

Joakim Dillner (ed.), *Methods in Biobanking*, Methods in Molecular Biology, vol. 675,
DOI 10.1007/978-1-59745-423-0_12, © Springer Science+Business Media, LLC 2011

Table 1
A schematic presentation of the Nord-Trøndelag Health Survey (The HUNT studies) from 1984–2008

	Age	*n*	Participation rate	Questionnaires	Clinical exam	Blood-samples
HUNT 1 (84–86)	20+	75,000	88%	+	+	–
HUNT 2 (95–97)	20+	65,000	72%	+	+	+
Young-HUNT	13–19	9,000		+	+	
HUNT 1+2	30+	46,000	61% from HUNT 1 re-attended in HUNT 2	×2	×2	+
HUNT 3 (06–08)	13+	60,000	60%	+	+	+

Contrary to both HUNT 1 and HUNT 2, the HUNT 3 study also has a strong focus on the establishment of a modern, "state-of-the-art" biobank, especially designed, built, and organised to meet all the potential requests from the international research society based on genetic and post-genetic analyses.

2. Planning – Selection of Target Participants and Study Design

In a prospective longitudinal design, there are at least two aspects to consider regarding the target participants, (1) whether to include the total or just a selected part of the population and (2) what is the most appropriate focus with respect to disease categories and clinical outcomes (endpoints).

Two essential elements contribute to the development of all diseases, genetic variation, and environmental exposures. If our targeted disease is a common flu and not the risk of cancer, the target population may well be recruited from all ages with a relatively limited observation period. However, we will need to include both genders and preferably an ethnic diverse population. In cases of rare diseases, we certainly need a larger cohort and a substantially longer observation period. Efforts are ongoing to establish trans-national large cohorts that can be genetically well classified, offer longitudinal profiles as early-detection markers, and precise identification of phenotypes as well as monitoring a diversity of exposure (4). Thus, a more complete pattern of human disease susceptibility may be set up. New technology opens up for a multiplicity of methods to collect health information, including linkage to electronic medical records, updated information through cell phones and SMS and small data chips with real-time registration of biological functions. These are all a great challenge to avoid violation of the personal integrity of each participant requiring high ethical standards in the conduct and governance of such studies.

2.1. Study Design

In a nested case–control study, cases of a definite disease category that occurs within the already defined and described cohort are identified and, for each case, matched with an adequate number of controls. The controls would normally be selected from the same cohort and among those who have not developed the disease at focus by the time of selection. For many research protocols, the nested case–control design potentially offers impressive reductions in costs and efforts of data collection and analysis compared with the full cohort approach, with relatively minor loss in statistical efficiency. The nested case–control design is particularly advantageous for the studies of biologic predecessors of disease and useful in the studies of large cohorts, since the time and cost involved in collecting exposure and covariate information for all members of a large study cohort may be substantial. By drawing a sample of controls for each case, the number of study subjects for whom exposure information needs to be obtained is reduced (5).

An epidemiological study with a nested case–control design, addressing a cohort based on the total population, allows for a greater variety of diseases to be studied provided the cohort is large enough to ensure sufficient statistical power.

3. Invitation to Participate

A fundamental premise for a successful attendance in population studies is high confidence and legitimacy in the study population. Also, the planning and organisation of large health surveys should be thoroughly secured and promoted broadly in the community/county. If possible, the establishment of an enthusiastic atmosphere and a broad scientific, public, and political support will contribute to a positive response. When screening a total population, it is helpful inviting family members, neighbours, and working colleagues within the same time period, creating a social environment in favour of participating.

Most commonly, the participants are invited through a personal letter, sent by ordinary mail, and often accompanied by a questionnaire. New technologies and Web-based solutions open up for a different approach, but so far not as a first contact (6) since e-mail addresses are not fully available in a comprehensive register.

Younger participants may prefer to answer questionnaires through a Web-based solution, recently shown to increase the total response rate (7, 8).

4. Informed Consent – Broad and Future Consent?

In most societies, there is a general agreement of the necessity of informed consent in medical research on human beings (9). Informed consent can be interpreted as a way of respecting individual autonomy, defending the individual against power abuse and bad research and ensuring trust between researchers and research subjects. The central question in the future will not be whether informed consent matters or not, but rather *how much* it should matter. Informed consent can be *active* or *passive*, and it can be *general* or *specific*. The golden rule in medical ethics is to demand active, specific informed consent. This gives the best protection of the research subjects. Active and specific informed consent for every new research project in a biobank may however represent an ethical "overload". On the other hand, since it is impossible for the donor to make an informed choice about the risks and benefits of unspecified future research protocols, it is not unlikely to question whether we should call such permission informed consent at all (10). The future task will be to ensure that biobanks and the expected outcome of biobank research – the common good – are founded on respect for individual rights and dignity but not unreasonably restricted by individualistic eccentricity and an overstated ethical focus in our society.

Active informed consent from participants in a biobank should be viewed as the start of a continuous dialogue between the research society, the donor society, and the regional ethics committee. If this dialogue is well managed, new use of the samples should not necessarily have the need of renewed, active, informed consent (11) provided there is a transparent, open-minded research governance to ensure a public insight in ongoing research protocols and results.

5. Selection of Samples to Obtain

There are numerous considerations to anticipate when selecting the bio-samples to be collected. In setting up a study protocol, the purpose of the study, the study budget, the availability of the samples, the character of the informed consent, the quality of the storage facility, the desired physiological coverage, and the future analytic strategies will all have significant impact (12). In addition, the *quality assurance* (QA) of the sample collection and further handling of the bio-material will be crucial to suit the wide range of possible assay technologies and to ensure that the biological information obtained is representative for the participants and not primarily reflecting a potential poor quality of the sample handling.

When obtaining blood samples, most often the desired fractions will be serum, plasma, white cells, red cells, and peripheral blood lymphocytes enabling analyses on bio-molecules as DNA, RNA, proteins, and other analytes. Also from fresh urine samples, bio-molecules, and analytes, including pharmaceuticals will be available, analysis as proteomics, and metabolomics can readily be performed, and the collecting procedure is at low costs. Other bio-specimens to be considered are faeces, saliva, hair, and nails, but the additional biological information may not be of such significance that it justifies the considerable extra efforts involved as well as the possibility of jeopardising the partaking in broad population-based studies. When studying an adolescent population, collection of saliva or buccal cells may be the preferred DNA source in contrast to venous puncture or capillary sampling. The latter may create some reluctance to participate in the younger age groups.

Having decided which bio-samples to be collected, there is an additional challenge involved in selecting among the numerous sample collection tubes available, with a variety of preservatives and anti-coagulants that may affect the results and quality of the bio-marker assays (13).

The temperatures at which the samples are handled before cryopreservation are similarly important. When handling both SST and EDTA tubes at 4°C before freezing, the UK biobank sample handling study has demonstrated that numerous assays can still be performed within 36 h (14). In general, it is also important that these processes are traceable, offering necessary information about the temperature of samples in various time periods prior to cryopreservation.

6. Standardising Sample Handling and Storage Protocols

The lack of high-quality clinically related bio-specimens is seen as a major bottleneck in medical research and a barrier in drug development. Sample collection, processing, and storage have a major impact on sample quality and utility for future analyses. There are very few standardised and quality-controlled protocols for pre-analytical procedures, which makes it difficult to compare and share samples from different studies, particularly as the sample sizes needed are likely to be very large. There is a need for international efforts to agree on standardised, harmonised, and exchangeable protocols, proper infrastructure, and sample formats to ensure that these valuable resources can be utilised to their fullest extent. To reach this goal, pan-European quality assurance schemes and guidelines for pre-analytical procedures for sample collection, handling, transport, processing, and storage are being worked out.

7. Sample Storage and Retrieval Systems

Safe and efficient storage and retrieval systems are an absolute pre-requisite for biobanks storing millions of samples in multiple tubes and vial formats. Whereas flexible automated solutions have been developed for some storage formats, (e.g. for DNA archives), the technology is still immature for ultra-low temperatures which may be necessary for more long-term storage of fluids and tissue samples. The lack of automation becomes a significant practical challenge when large sample sets are to be retrieved and an obstacle to the effective delivery of high-quality samples to the scientific community. Greater emphasis also needs to be put on the development of modern, scalable effective storage and retrieval systems since available systems are very expensive with an immature technology for ultra-low temperatures.

8. The Need for Large Sample Size

Recent findings arising from genome-wide association (GWA) scans have provided insights into the aetiology and pathogenesis of complex diseases (15) From this, it has become clear that susceptibility variants come in several different formats. The more common ones often have modest relative risk, but may be widespread and thus achieve a significant population-attributable fraction, while others are rare but with substantial relative risk to their carriers. There is a growing realisation among scientists that, for most complex traits, the power of genetic approaches to detect the first category, which is important from a preventive point of view, relies on the availability of very large sample sets, extending far beyond the reach of any single initiative or nation (16). Therefore, there is an urgent need for a European or even worldwide initiative for collaborative research. Integrating individual biobanking resources into a coherent research infrastructure will significantly facilitate the access to large sample sets and augment the research value of any biobank within the network. We are seeing the beginning of such a movement across international boundaries both within the European Community and on a global level.

9. Analyses to be Performed at Recruitment and Communication of Results

Ideally, a wide range of bio-medical markers may well be analysed prior to cryopreservation, both for reducing the costs involved in organising and retrieving samples as well as minimising the number of thaw/freeze cycles. The need for funding of the actual analyses and the magnitude of complex technology and methodology involved may, however, counteract this approach. It is likewise a challenge to foresee the analyses that will be more relevant.

Information of glucose and lipid levels and thyroid function may be of more immediate interest to the participants. It will possibly also improve the participation rate in a survey if these results are reported directly back to the contributor. It does, however, require an increased level of awareness from the investigators when taking on this responsibility. This is well illustrated when participants with previously unknown diabetes present themselves with clinically unacceptable high blood glucose levels. A roster among physicians involved is then necessary for immediate notification. It also calls for a continuous and good communication with the GPs in the follow up of these newly recruited patients.

New genotyping platforms and strategies (15), allowing genome-wide association studies (GWAS) to be performed at continuously lower costs, may open up for a genome-wide genotyping of all the participants. Since these analyses also necessitate DNA extraction, the total costs may still effectively hinder this approach.

As soon as the results from contracted analyses have been reported back to the donors, the cohorts will normally be de-identified and closed. Results from subsequent analyses will not be reported on an individual bases, but only communicated through scientific reports/publications. Succeeding studies identifying bio-markers with a consequence for morbidity and even mortality on an individual level will raise an ethically challenging dilemma – should this be reported directly to those at risk, or should this information also be revealed to the public through a general and anonymous approach.

10. Samples to Store – and How – Quality Assurance

Bio-samples may be stored at different temperatures, dependent on the material itself, its "robustness" and anticipated length of storage. Ideally, most samples should be stored below the re-crystallisation temperature of pure water at −130°C. Maintaining an ultra-low temperature, far lower than the re-crystallisation point, maintains vitrification of the samples without crystallisation. Thus, an archive (long term) storage on liquid Nitrogen (LIN), varying from −152°C to −196°C (vapour face or liquid face), seems both logical and economically favourable. Ultra-freezers at −152°C will suit the same purpose, but at higher costs. In cryo-preservation and immortalization of peripheral blood circulating cells, ultra-low temperatures are required, but also preferred as long-term storage temperature for other bio-samples as serum, plasma, and buffy coat. A working archive (short time) for basically the same set of samples is often established at higher temperatures as −80°C. Genomic DNA may be stored at +4°C, but preferably at −20°C. This is especially important in storage of already amplified DNA.

A number of studies have investigated the effect of freezing and the number of thaw/freeze cycle involved on cell recovery and viability (17, 18), but few have documented data on the effect of storage periods exceeding 2–3 years. There are inconsistent results to what extent the storage material/containers will have an effect on cell quality, but there are many documented reports on improved quality if the samples are stored at temperatures below −152°C.

11. Necessary Support Functions (Staffing)

Human resources needed in a modern, state-of-the-art biobank are highly dependent on a number of factors as complexity of the collection, sample size, heterogeneity of samples, desired throughput, analytic capacity/ambitions, and IT solutions. Dedicated laboratory personnel with specialised competence as well as respect and appreciation for the long-time efforts often involved in the recruitment of samples is invaluable. A balanced combination of a scientifically and technically skilled staff is also beneficial. Data management is critical, and in-house IT-personnel/LIMS management is essential. A diversity of scientists with training and background in molecular biology, epidemiology, bio-statistics, clinical medicine, and bio-medicine is a significant advantage.

Genomics and genetic epidemiology evolving from a modern and comprehensive population-based biobank is a rapidly growing research field of enormous interest that also has the potential of bridging the gap between basic research, clinical medicine, and public health opening up for true translational research (19).

References

1. Holmen J, Forsen L, Hjort PF, Midthjell K, Waaler HT and Bjorndal A. (1991) Detecting hypertension: screening versus case finding in Norway. *BMJ* **302**(6770), 219–22.
2. Midthjell K, Bjorndal A, Holmen J, Kruger O and Bjartveit K. (1995) Prevalence of known and previously unknown diabetes mellitus and impaired glucose tolerance in an adult Norwegian population. Indications of an increasing diabetes prevalence. The Nord-Trondelag Diabetes Study. *Scand J Prim Health Care* **13**(3), 229–35.
3. Holmen J et al. (2003) The Nord-Trøndelag Health Study 1995–97 (HUNT 2). *Nor Epidemiol* **33**, 19–32.
4. Potter JD. (2004) Toward the last cohort. *Cancer Epidemiol Biomarkers Prev* **13**(6), 895–7.
5. Richardson DB. (2004) An incidence density sampling program for nested case-control analyses. *Occup Environ Med* **61**(12), e59.
6. Ekman A and Litton JE. (2007) New times, new needs; e-epidemiology. *Eur J Epidemiol* **22**(5), 285–92.
7. Ekman A, Klint A, Dickman PW, Adami HO and Litton JE. (2007) Optimizing the design of web-based questionnaires – experience from a population-based study among 50,000 women. *Eur J Epidemiol* **22**(5), 293–300.
8. Ekman A, Dickman PW, Klint A, Weiderpass E and Litton JE. (2006) Feasibility of using web-based questionnaires in large population-based epidemiological studies. *Eur J Epidemiol* **21**(2), 103–11.
9. Tranoy KE. (1993) Research ethics: a European perspective. *Bull Med Ethics* **92**, 28–33.
10. Winickoff DE and Winickoff RN. (2003) The charitable trust as a model for genomic biobanks. *N Engl J Med* **349**(12), 1180–4.
11. Lindmo T, Hveem K et al. (2004) Health Surveys and Biobanking. A Foresight Analysis towards 2020 Report.
12. Landi MT and Caporaso N. (1997) Sample collection, processing and storage. *IARC Sci Publ* **142**, 223–36.
13. Ollier W, Sprosen T and Peakman T. (2005) UK Biobank: from concept to reality. *Pharmacogenomics* **6**, 639–46.
14. UK Biobank: protocol for a large-scale prospective epidemiological resource. (2006).
15. Chanock SJ, Manolio T, Boehnke M, Boerwinkle E, Hunter DJ, Thomas G et al. (2007) Replicating genotype-phenotype associations. *Nature* **447**(7145), 655–60.
16. Hattersley AT and McCarthy MI. (2005) What makes a good genetic association study? *Lancet* **366**(9493), 1315–23.
17. Valeri CR, Srey R, Lane JP and Ragno G. (2003) Effect of WBC reduction and storage temperature on PLTs frozen with 6 percent DMSO for as long as 3 years. *Transfusion* **43**(8), 1162–7.
18. Berz D, McCormack EM, Winer ES, Colvin GA and Quesenberry PJ. (2007) Cryopreservation of hematopoietic stem cells. *Am J Hematol* **82**(6), 463–72.
19. Khoury MJ, Gwinn M, Burke W, Bowen S and Zimmern R. (2007) Will genomics widen or help heal the schism between medicine and public health? *Am J Prev Med* **33**(4), 310–7.

Chapter 13

Best Practices for Establishing a Biobank

Göran Hallmans and Jimmie B. Vaught

Abstract

A biobank may be defined as the long-term storage of biological samples for research or clinical purposes. In addition to storage facilities, a biobank may comprise a complete organization with biological samples, data, personnel, policies, and procedures for handling specimens and performing other services, such as the management of the database and the planning of scientific studies. This combination of facilities, policies, and processes may also be called a biological resource center (BRC) (www.iarc.fr). Research using specimens from biobanks is regulated by European Union (EU) recommendations (Recommendations on Research on Human Biological Materials. The draft recommendation on research on human biological materials was approved by CDBI at its plenary meeting on 20 October 2005) and by voluntary best practices from the U.S. National Cancer Institute (NCI) (http://biospecimens.cancer.gov) and other organizations. Best practices for the management of research biobanks vary according to the institution and differing international regulations and standards. However, there are many areas of agreement that have resulted in best practices that should be followed in order to establish a biobank for the custodianship of high-quality specimens and data.

Key words: Biobank, Biological resource center, Specimen, Liquid nitrogen, Biological sample, Cryovials, Bar code

1. Introduction

Biobanks have been in existence for many years, both as large formal commercial organizations with well-established policies and procedures, and as small facilities operated with very few standards or rules. In recent years, the situation has changed with the proliferation of biobanks for research and clinical applications (1). As the necessity for access to high-quality specimens has increased, so has the necessity for standards to guide the proper collection, processing, storage, and dissemination of the specimens. In Europe, a major set of recommendations was established by

Joakim Dillner (ed.), *Methods in Biobanking*, Methods in Molecular Biology, vol. 675,
DOI 10.1007/978-1-59745-423-0_13, © Springer Science+Business Media, LLC 2011

the European Union (EU). In the U.S.A., the National Cancer Institute (NCI) has taken the lead in establishing best practices for its biobanks (also called biorepositories or biospecimen resources in the U.S.A.). Internationally, several other major sets of best biobanking practices have been published or are under development (see Subheading 3.10).

The EU recommendation on research on human biological materials was approved by the Committee of Ministers on 15 March 2006. In article 18 – "*Independent Examination,*" it states that a population biobank should be "subject to an independent examination of its compliance with the requirements of this recommendation." The independent examination should be interpreted as an appropriate auditing system. In article 19 – "*Oversight of population biobanks*" it is recommended that "each population biobank be subject to independent oversight" to protect the interests and rights of individuals and organizations involved in, or influenced by, the activities of the biobank. This statement supports the need for a system of quality control combined with a system for auditing. It states that population biobanks should establish the policies and procedures to determine whether a proposed research project is using the material in appropriate way, in particular when the material is rare or scarce. In article 20 – "*Access to population biobanks*" the importance of access by researchers to biological materials stored in population biobanks is emphasized, and it is recommended to member states to take appropriate measures to facilitate the access for scientists to various biobanks. These two statements strengthen the need for expert evaluation of individual projects within a biobank.

The NCI Best Practices for Biospecimen Resources development process was initiated through a multiyear undertaking that began in 2002, including a 2004 presentation to the National Cancer Advisory Board (NCAB) of a study that showed substantial heterogeneity in management practices across NCI-supported biospecimen resources. In 2005, the NCI took several actions approved by the NCAB to respond to these findings, including the establishment of the Biorepository Coordinating Committee (BCC) in an advisory role to the NCI Office of Biorepositories and Biospecimen Research (OBBR), and the development of the NCI Best Practices for Biospecimen Resources in the interest of ensuring sufficient biospecimens of documented quality to support NCI-sponsored research. The NCI Best Practices contain technical recommendations for specimen handling, collecting clinical data, quality assurance, biosafety and informatics, as well as ethical, legal, and policy practices concerning informed consent, access to specimens and data, intellectual property, custodianship, and privacy protection. Although the NCI Best Practices are voluntary, the NCI expects widespread adoption as its biobanks realize that better standards will result in higher quality specimens.

2. Materials

Biobanking work, including associated laboratory handling of specimens, should be performed in a methodical way, where all procedures are documented in standard operating procedures (SOPs), and with the aim of a continuous evaluation of methods, materials, and equipment. In principle, the recommendations that are valid for accredited laboratories should be followed, and it is an advantage if the biobank and its laboratory are accredited, or operated as a part of an accredited laboratory. Details concerning laboratory operations in a biobank are given in the description of Good Biobanking Practices (GBP) and the NCI's OBBR Web site (2).

2.1. Automation

Automation may be employed in several ways in the biobank. The traceability of samples may be facilitated by using bar codes. Bar coding greatly reduces errors in all specimen handling processes. Secondly, the whole process of DNA extraction and handling DNA may be automated using various robotic platforms. For laboratories with less financial resources, manual extraction and handling of DNA and other samples is needed. By planning the process using computerized programs with software provided for data handling and laboratory routines, the laboratory work may be streamlined to prevent errors. Biobanks may also benefit from automated aliquoting, especially if the specimens are processed in a standard manner in standard-sized containers or cryovials. Automated freezers are also available, but usually require a large investment of funds and are most amenable to processing large numbers of specimens of identical type using the same size and type of storage vessels.

2.2. Specimen Handling

Biobanks should record important information on all individual samples as they arrive at the facility, for example, occurrence of hemolysis of blood, missing or thawed tubes. All events are noted for each individual sample, e.g., retrieval to analyse, thawing, participation in a project, etc. If DNA is among the biobank's specimens, a separate DNA database may be a part of the system consisting of information on variables, such as DNA concentration and amount and quality parameters, such as A260/280 ratio and the evidence of degradation. The biobank database should be continuously updated with respect to incoming samples and sample withdrawals and also with supplementary data as applicable (see Subheading 3.3).

Specimens should be collected and processed using standard protocols that are suitable for the intended use and under conditions that preserve the quality and stability of the specimen. Specific best practices for specimen collection and processing may be found in the NCI and ISBER best practices documents (3). However, to provide one example, the use of various anticoagulants (EDTA, heparin, or citrate) may affect analytical results (4). Regarding proteomic analyses, plasma is preferred since it is estimated that approximately 40% of peptides are serum specific and derived from the clotting process (5, 6).

2.3. Storage Materials and Equipment

Proper storage requires the use of cryovials and labeling systems that will withstand the intended storage conditions. This requires that when using new storage vessels, labels and bar codes, or other printing systems, the materials are tested to assure that they are stable if, for example, the vessels will be stored in liquid nitrogen freezers for extended periods (2, 3).

2.3.1. Storage Containers

Samples should be deposited in freezers or other appropriate storage containers according to a specific storage system developed by the biobank. Depending on the specimen type, intended use, and estimated length of storage, specimens may be stored at room temperature, at 4°C, at –20°C, at –80°C, or at –150°C or below in liquid nitrogen tanks (3). For large scale biobanking, a computerized inventory system is strongly recommended. Preferably, all samples should be bar-coded and the inventory system should provide their exact locations in the biobank's storage containers.

2.3.2. Security for Storage Facilities

For storage in larger biobanks, the samples should be divided and kept in at least two different storage containers for security reasons. The storage containers should be located if possible in different buildings or clearly separated in the same building. In addition to general building security to protect against fire, unauthorized entrance and other usual hazards, all freezers should be protected with an alarm system against significant increases in temperature. Independent alarms should also be connected to a central alarm system, which must be manned for response at all hours of the day. In the event of freezer failure, an appropriate number (10% is recommended) of empty, operating backup freezers should be available. Any interruption of electrical power must be compensated for within minutes by an independent system, such as a generator with locally controlled production of electrical power. An additional backup system using liquid nitrogen is recommended, if no such electrical system is available. In systems using liquid nitrogen, the availability of an adequate supply is necessary, in the event of an interruption in delivery.

2.4. Specimen Shipment

Shipment of samples is a critical aspect of the biobanking operation, requiring strict adherence to the relevant rules and regulations. Specimen shipping is regulated as described in the International Society for Biological and Environmental Repositories Best Practices (3) and by the International Air transport Association (7).

Appropriate shipping temperature will depend on the analyses to be performed, and can range from ambient temperature to liquid nitrogen temperatures in "dry LN2 shippers" (3, 4).

The shipping and receiving biobanks must agree on the shipping schedule and must communicate promptly when delivery is delayed. It must be strictly agreed upon who has the responsibility of tracking the sample shipment if it is delayed. In a critical situation, where specimens are at risk of thawing, the representatives of the two organizations must solve the problem in cooperation with the transport company. Tracking systems and tracking numbers that identify shipments are provided by the transport companies to assist in preventing such problems.

3. Methods

3.1. Technical Practices for Specimen Handling

The following is a brief list of technical best practices for specimen collection, processing, storage, and retrieval. See other subheadings of the chapter, and ISBER and NCI best practices for detailed guidance.

- Collect specimens under conditions appropriate for the study.
- Involve a pathologist for expertise in collecting and processing surgical and autopsy specimens.
- Develop SOPs for all policies and procedures.
- Develop a comprehensive quality management system (QMS) (Subheading 3.2).
- Utilize a computerized specimen inventory and tracking system.
- Collect associated data for each specimen as appropriate.
- Minimize collection and processing time as appropriate for the specimen type.
- Provide a training system for biobank personnel.
- Store specimens in a stabilized state (e.g., formalin fixed or frozen).
- Automatically record storage conditions.
- Provide alarm systems and backup power.
- Establish rules for timely biospecimen disposal.
- Conduct periodic review of storage equipment performance.
- Choose biospecimen containers with analytical goals in mind.
- Adhere to biosafety, packaging, and shipping regulations.
- Use appropriate Material Transfer Agreements (MTAs) (see Subheading 3.8).

- Be familiar with specimen shipping regulations; use proper shipping temperatures; and train personnel in shipping regulations (see Subheading 2.4).

3.2. Quality Assurance/Quality Control

QMS is critical to the success of a biobank. QMS generally comprise quality assurance (QA) and quality control (QC). QA is the set of written standards that establishes the SOPs for managing the biobank. QC is the set of procedures used to measure adherence to SOPs and other standards established by the QA system.

For example, a quality system was developed within the Medical Biobank of Umeå University Hospital during the 1990s. In 2001, the quality assurance system GBP was formally introduced for QA, and today the system covers all activities of the biobank. The methods of GBP are applied in sample handling, database handling, laboratory work, research project planning and implementation, deviation reporting, information security, informed consent, and other formal aspects of research project administration. It also describes the structure of decision making, regarding overall strategic planning, as well as the evaluation of the scientific strength of different research initiatives.

Other widely used QMS include Good Manufacturing Practices (GMP) and ISO9001 (8). ISO9001, in particular, has been widely adopted as a quality system that is recognized for international biobanking collaborations. Both GMP and ISO require strict adherence to SOPs and other well-documented practices concerning materials acquisition and handling and equipment maintenance. See ISBER and NCI best practices for additional details.

3.2.1. Auditing of the QA System

Internal preauditing by the biobank staff: Together with the PI, all the managers should have regular group meetings, where quality problems are discussed at the practical level.

3.2.1.1. Auditing Systems

There are different systems for auditing available. Some of the preexciting systems like Good Laboratory Practice (GLP) can partly be used for this purpose. The following options have been identified:

1. Involving a specific auditor for each individual project is usually too expensive for academic biobanks. This system may be an option for commercially oriented projects if needed.
2. An internal auditing system can be established within individual projects and a report should be made after the completion of each project. A report of mistakes/deviations should be produced. Reports should be distributed to the participating scientists, laboratories, expert groups, and the steering committee.

3. Large-scale auditing systems may be carried out by recognized national certification bodies. This procedure applies to projects attached to large laboratories, where this procedure is a part of the internal routine.
4. Biobanking regulations in various countries offer a third alternative for auditing, where the responsible person is appointed by the regulatory authorities.

For the purpose of avoiding mistakes in sampling of blood, especially mixing samples, various systems have been developed. One system developed within the Medical Biobank at Umeå university hospital in Sweden is based on color-coded centrifuge tubes that correspond to cryovials with corresponding colors placed in a specific order. The system permits sampling to be performed by several hundred collection personnel with a minimum of mistakes.

3.2.2. Deviation Reports

The managers of different biobank sections are responsible for writing deviation reports whenever a significant error has occurred. Minor deviations are normally not reported. A copy of the deviation report is sent to the QA manager. The consequences of a deviation report must always be evaluated and measures must be taken. If necessary, the relevant SOPs must be changed. Some examples of deviations that should be reported are:

- Mistakes in processing samples;
- Specimen delivery and receipt errors;
- Errors that result in significant cost increases;
- Errors that result in the loss of samples;
- Established routines that are not followed or are no longer satisfactory.
- Errors in the security system

Deviation reports are always evaluated by the biobank auditor if such a person/system is available. In some countries, for example, Sweden, the Biobank Act provides a system which can be used for this purpose.

3.3. Collecting and Managing Data

Appropriate annotation of biospecimens is crucial to the overall usefulness of the biospecimen resource as a tool for scientific research (9). Biobanks store biospecimens collected using multiple methodologies and procedures. Researchers rely on banked specimens for a wide variety of purposes, including target discovery and validation, prevention research, research on early detection, genetic studies, and epidemiologic analyses. The data recorded by investigators and biospecimen resources depend on the types of biospecimens collected and the studies' objectives. The following types of data collection are recommended, as appropriate for the study design: specimen location and quality;

clinical data; demographic information; lifestyle factors; pathology data; treatment information; and any other relevant information necessary to accomplish the study goals. However, data collection must be in accordance with informed consent and relevant privacy rules and regulations (2, 3).

3.4. Biosafety

Laboratories and biobanks that handle biospecimens expose their employees to risks involving infectious agents and chemicals, as well as the general dangers of a laboratory. A predictable, yet small, percentage of specimens will pose a risk to biobank personnel who process them. All biospecimens should be treated as biohazards (10). In addition to taking biosafety precautions, biospecimen resources should adhere to key principles of general laboratory safety. In the U.S. laboratory, safety is regulated by the Centers for Disease Control and Prevention and the Occupational Health and Safety Administration (11).

3.5. Informed Consent

The policies and procedures for informed consent vary widely among countries (3). And the level of consent, i.e., narrow or broad consent to use a study participant's specimen for research, may vary depending on the study goals and local rules and regulations. Informed consent rules and regulations may also vary over time according to changes in legislation, demands from ethics committees, and the ethical or institutional review board (ERB or IRB depending on the local designation). For some research projects, the informed consent is outdated and does not cover current research activities. The IRB may, in such situations, demand renewed approval from the donors. The essential elements of informed consent are outlined in Chapter 2 of this volume.

Informed consent protects both the biobank and the study participant. As such a signed informed consent is necessary whenever possible. However, note that in the U.S.A., there are certain circumstances, where informed consent may be waived by an IRB (2). The protection of the donor should follow the rules of the Helsinki declaration (12), which outlines important principles to protect safety and integrity, and provide moral relevance for the donors.

It is evident that most study participants donate biological samples for the purpose of supporting science in an altruistic manner, to support the community in its public health efforts. The donation is usually given for academic research and not for personal profit of the scientists involved in the project (13). However, the study participant should be informed of the possibility of commercial use of specimens or specimen derivatives resulting from their donation (2, 14).

3.6. Custodianship and Access to Biospecimens and Data

3.6.1. Custodianship

Issues related to the ownership of specimens and data are not specifically regulated in the legislation of most countries. In the U.S.A., several highly publicized legal cases have resulted from disagreements over the ownership of donated specimens (15, 16). The NCI best practices recommend the term *custodianship* rather than ownership (2). A custodianship policy should be developed in each biobank that outlines policies and procedures to ensure the long-term physical integrity of the biospecimens while maintaining the privacy and confidentiality of research participants. The custodianship policy should include details concerning the disposal or sharing of specimens and data when a study ends, or when funding shortages force the reduction in the size of the biobank or its closure.

Other policies have been developed in various countries. In general terms, the "owners" are those organizations that funded the biobank, as stated by the British MRC (17, 18) while the scientists who have built up the biobank are the custodians or the PI of the biobank (13) and as such legally responsible for many of the activities within the biobank. This relationship has clearly been expressed by the EPIC Steering Committee (19): "...the principal investigators in each country have legal responsibilities for the proper custody and use of both biological samples and data from their cohort, wherever held..." (Statement of the EPIC Steering Committee 15 October 2003).

3.6.2. Access

One of the primary missions of a biobank is to disseminate specimens and data to the research community in a responsible manner. As stated in the NCI best practices (2), to best serve the needs of the research community, biobanks should establish the guidelines for sample distribution and clinical data sharing consistent with ethical principles, governing statutes and regulations, and, if applicable, informed consent language. These guidelines should be:

- *Clear* to ensure their comprehension and adoption;
- *Flexible* so that biobanks are responsive to changing scientific needs;
- *Amendable* to facilitate their adaptability over time; and
- *General* enough, so they may be applied to different kinds of biobanks.

In addition, access guidelines should delineate when biospecimens and clinical data are narrowly or broadly accessible and what justifications for access to specimens are expected.

3.7. Biobank Administration and Control of Access

Administrative personnel and procedures will vary from one biobank to another and from one country to another. The following is a general description that may be applied to a research biobank that provides for the review of applications to access its materials.

The overall decision-making process in larger biobanks is usually led by a steering committee, while in smaller biobanks the scientists themselves are the responsible persons. For routine activities and writing of SOPs, the principal investigator (biobanks initiated by scientists) or the director (some biobanks initiated by organizations) of the biobank is responsible. Priority to use the specimens is established by the responsible scientists in smaller biobanks and by the steering committee or specially appointed expert advisory groups in larger biobanks. The QA system also establishes practices to be followed for the management of the biobank, to give stability to the organization and to optimize the scientific output. The biobank management system should be transparent and well defined, and the systems should be audited internally or preferably by independent experts. It is mandatory for all individual employees and members of steering committees and expert groups to report deviations to the standard practices of the biobank. The deviations should be evaluated by the auditor, the PI/director, and the steering committee, and they should be reported in a standardized fashion.

In larger biobanks, the PI, research coordinator, the database manager, and laboratory manager may comprise a project management group responsible for contact and discussion with researchers or research groups. An important issue in the initial discussion is to coordinate different research projects in order to maximize the effective use of specimens, to avoid duplication of research goals, to facilitate for downstream handling, and to synchronize activities in the laboratory.

An application for accessing specimens from the biobank for a research project is documented, and the steering committee and the project management group are informed. The appropriate expert review group is selected to evaluate the scientific value of the project, i.e., if there is more than one expert group within the biobank. In some biobanks, the projects are evaluated using the members of the project team, without the involvement of independent evaluators.

The expert group members have different and selectively identified qualifications to cover the needed area of medical research. A decision on access to specimens may be based on the following criteria:

- Overall scientific value of the project as well as the individual sample analysis;
- Has the project plan been reviewed by a statistician? Is there a sufficient power to meet the scientific hypothesis of the project?
- What sample volumes are needed for the analysis? Are there other laboratory methods available which may reduce the sample volumes needed?

- What are the options to coordinate the different analyses of the project between different laboratories?
- Has the suggested laboratory (or laboratories) demonstrated high standards of quality for the specific laboratory method(s)?
- Have the applicant scientists demonstrated sufficient competence for the individual research project or do they need additional expertise?
- If there are conflicts of interest with previously approved projects special rules should be applied. In certain instances, rules must be defined to avoid the publication of very similar studies using the same biobank materials.
- If there is a conflict of interest related to an individual expert, that person must disclose the conflict and not participate in the decision for that specific project.

The experts must sign a confidentiality agreement when they enter the expert group concerning their evaluations, and their individual opinions are not public unless they issue a written statement. Sometimes, the expert groups may suggest changes in the project plan. If the expert group approves the project, the decision must be reconfirmed by the steering committee. However, the steering committee cannot approve an application if it has been rejected by the expert group.

Before the biobank launches a project, the following criteria must be fulfilled:

- Approval from the expert group and/or the steering committee;
- Approval by the PI, according to a formal agreement signed by the PI;
- Approval by the Ethical Review Board and the Data Inspection Board whenever appropriate;
- Permissions from those who are responsible for the involved disease registries or cohorts and from other parties as appropriate;
- New informed consents from the donors and other information provided to the donors if needed and when appropriate;
- Completion of a Material Transfer Agreement (MTA) or other appropriate documentation of the transfer.
- In general, the biobank rules must be in accordance with the specific regulations that control its operation, such as the National Biobank Acts, Personal Data Acts, EU laws in Europe, and informed consent and privacy rules and regulations in the U.S.A (2, 14).

When all the practical, legal, and scientific preparation for a project is completed, the logistic phase of transferring data and samples starts. The files of the project are encrypted prior to sending to the researcher or research group. Depending on the study design, after the analyses have been completed and a copy of the results has been sent to the biobank database, the information (codes) about case and controls are delivered to the scientists. The data delivered may not be copied or distributed to anyone that does not have permission to use the data from the biobank. Permission is only given for the purpose of accessing information for the specific study. It is also prohibited to develop separate databases from the information or data delivered from the biobank. It is important that the individual's right to privacy is guaranteed.

3.8. Material Transfer Agreements

MTAs are agreements that are established between biobanks that agree to provide specimens and the recipient scientist or organization. MTAs with scientists, companies, or other organizations should be formal documents that are signed by both the recipient (applicant) and the principal investigator of the biobank providing the specimens. The MTA form contains contact information and a description of the research project, and states the general terms and conditions for the transfer of specimens and information that the applicant must agree upon (2). These conditions may vary among biobanks and according to standards and practices in various countries (e.g., see NCI best practices), but generally adhere to the following conditions:

- The applicants may only conduct noncommercial basic research, clinical research, and epidemiologic research on the samples.
- The applicants agree to return all samples and unused portions of samples, including extracted DNA, as well as results of tests and samples analyses with regard to individual samples, as soon as the research project has been completed. Note that this provision may not apply in all countries.
- The applicants own the results of the research (results of analyses, register data, collected patient material, etc.). Other researchers or companies wishing to use the results must first obtain permission from the researchers.
- The applicants are forbidden to use the samples or information in research conducted in cooperation with, under the assignment of, under the license of, or under similar conditions in connection with a commercial company, unless prior written approval has been obtained from the provider biobank.
- If results of the project lead to a patent application, the applicants should make efforts to do so in cooperation with the biobank. The applicants are, however, forbidden to sell or transfer patent rights or rights related to patent applications for discoveries based on the results of the research project or

samples or to any other manner commercialize the results, without prior written approval from the biobank.

- The applicants agree not to transfer samples or information to persons not a part of the laboratory approved by the biobank for the analysis of the samples. The applicants are also forbidden to transfer their rights or obligations according to the MTA contract without prior written approval from the biobank.

The MTA form should be used for all projects with some exceptions concerning complicated commercial projects, where it can be replaced by agreements with special provisions for the handling of the samples and the legal issues associated with the development of industrial projects.

3.9. Informatics for Specimen Retrieval

The informatics plan is the heart of a biobank, and it coordinates all activities within the biobank. General informatics practices for biobanking are discussed in Chapter 21. The discussion in this section applies primarily to epidemiologic biobanks.

Standardized and modifiable software should be used for all data processing activities within the biobank, including specimen tracking, clinical annotation, and depending on the studies supported by the biobank may include case selection and control matching. In all data administration activities, it is generally very important to name all study data files, all working data files, and all variables in a standardized way. A set of common data elements should be agreed upon. This is especially crucial if the activities have grown to a certain point, where there is no possibility to keep track of the logistics in a manual system. Data entry should be controlled by double keying, and the entry invalid or extreme values should be controlled by editing protocols.

Specific programs are used for the retrieval of samples. Retrieval is authorized by the biobank coordinator or PI. The standard programs used for this purpose must be modified for the actual study, and a sample selection list printed if the samples are to be retrieved manually. When automated storage and retrieval systems are used, the robots are instructed in a corresponding manner. Automation speeds up the process, and it prevents human mistakes to some extent, but automated processes may be error prone. As noted in Subheading 2.1, automation is costly, and it is most easily applied when the biological samples are standardized, e.g., when only DNA in standard size vials is processed.

It is also a biobank's responsibility to sort the samples in an appropriate way before sending them to the analytic laboratory, since all codes that identify cases and controls must be hidden. Some studies require a randomization of the samples or some other case-control order. Sorting as well as other handling of samples must be performed without any risk of thawing or warming the samples. Different routines are applied depending on the amount of samples to be handled. Another important service is the preparation of a randomized set of internal controls for inter- and intrabatch validation. To blind these QC samples for the analytic laboratories, the samples must be prepared and labeled within the biobank. For example, control samples with a known concentration of DNA should be inserted and kept blinded to the analytic laboratory for QC purposes.

For all data handled in the biobank database, the software to be used should be defined and standardized. The database in a biobank may contain many different materials which puts specific demands on the handling of the database. Some of the materials may be the property of the biobank, others may belong to external scientists via a system of "biobank deposit boxes," where the biobank administers the biobank samples and questionnaires for the scientists or by adding clinical data collected and "owned" by the scientist/clinician. This part of the biobank data is covered by the system of GBP.

3.9.1. Information Security of the Biobank Database

Standard protocols that guarantee information security must be established. The managers of various biobank sections should have access to the database system but at different levels of security, depending on their roles and the need to access certain sensitive information. The security level is established on the basis of need within each area of the biobank work. Some managers and other employees are authorized to read, others to write and make changes in certain registries and databases. For users whose assignments change or who resign, authorization becomes invalid immediately.

Access to the system is recommended to be protected by a login process with password security, and a password is activated by the screensaver whenever there is too long absence of activity. All data should be stored in a common internal server with terminals connected to it, without any direct or indirect contacts with internet. The server is recommended to be placed in a fire and burglary secure closet with a secure coded lock. The door to the research department of the biobank is also recommended to be locked at all times. No electronic data with personal information are allowed to be taken out from the biobank office without being encrypted or made anonymous.

3.9.2. Privacy and Confidentiality of Biobank Information

The information policy is an important part of the quality work of a biobank. To ensure that all employees are aware of secrecy laws and information policies, an agreement form on information policy should be signed. This contract states that all information within a biobank is confidential. If specimens from hospital patients are involved, specific laws related to privacy within the healthcare system covers the confidentiality of patient information within the biobank. In the U.S.A., patient confidentiality is regulated by the HIPAA Privacy Rule (2).

Practices for contacts with the media are especially important to follow in order to protect both the donors of the samples and the unpublished results of the scientists. In contacts with the media or other external organizations, only issues concerning the biobank itself may be discussed. Questions that concern the clients of the biobank, usually scientists or companies, must be referred to the clients. Statements to the media are usually made by the PI, someone appointed by the PI or other authorized individuals. Issues that are covered by privacy laws or confidentiality agreements must always be handled with great care.

3.9.3. Data Sharing Policy

There are several systems for sharing data with other scientists, the general public, or with the donors of the biological samples. The following principles may be applied and individually evaluated in the different biobanks:

- Scientists using biobank samples should be encouraged to share their results with the media whenever new and important results are published, by issuing a press release. The biobank should be acknowledged in the press release with the purpose of getting information to the donors concerning how and why their donated specimens have been used. It is expected that this information will influence their willingness to participate in biobank research in the future. With the dramatic development of new techniques, biobank research is expected to increase in the future.
- Some results of biobank research may be directly put into the Web site of the biobank combined with general information of the biobank, including its rules, QA system, and individual questionnaires used in different studies and inventory of biological samples. Some parts of the Web site may be restricted to use only by authorized individuals.
- NCI on a regular basis posts the individual results of single nucleotide polymorphisms (SNPs) on a restricted Web site for the purpose of sharing data with other scientists, and promoting overall scientific development through this exchange of information. The present view in some European countries is that data that could be tracked back to the individuals via a code are to be regarded as personal information no matter how sophisticated the coding systems are, and therefore not allowed to be posted on a Web site.
- Data sharing among collaborators or with the scientific community in general is encouraged by the U.S. National Institutes of Health as part of it data sharing policy (2). This policy applies to NIH grants of more than 500,000 U.S. Dollars.

3.10. Guidelines for Evaluating the Value of Biobanks

Internationally, a huge number of biobanks are available. There are specific registries available for the identification of the biobanks (20). Within the GBP, a system for overall evaluation of individual biobanks is proposed, based on specific guidelines. The system contains the following information:

- The biobank impact factor (BIF);
- The prioritization process for the utilization of biobank samples;
- Documentation and evaluation of deviation reports within the QA system;
- The extent of automation for sample processing;
- Regular seminars within the biobank on quality issues;
- One or more systems for quality audits;
- A system for resolving conflicts, which cannot be resolved by a steering committee or expert groups, by independent and respected scientists attached to the biobank;
- A system for sharing of samples with other biobanks and external scientists;
- Effectiveness of communication of results to the general public, including the donors;
- An assessment of any of the biobank's research that has been the basis for new methods of prevention, diagnosis, or treatment of major diseases.

The BIF has been suggested by Anne Cambon-Thomsen to evaluate the quality and quantity of the scientific output of a biobank (21). Combined with other evaluation factors suggested by GBP, the BIF is expected to be a valuable tool for funding organizations and biobank owners in the future. Whenever there is a conflict between scientists or between scientists and authorities, the GBP system proposes two ways of resolving the conflict: (1) Assign the conflict to an expert group for resolution. This is usually a successful way of solving the problem when the conflict concerns a scientific matter. (2) When the conflict is more structural, it should be solved by the steering committee or by a group of independent and respected scientists, a system which has been most helpful within GBP.

Among the evaluation criteria stated above, the creation of an internal assessment of the most important results of an individual biobank that have been the basis for new methods of prevention, diagnosis, or treatment of major diseases needs a specific clarification. This assessment is very important for the funding organizations and the biobank owners, and it should be evaluated separately by independent experts.

3.11. International Biobanking Guidelines

There is considerable variation in national laws and local practices that are applied to the usage and storage of biological samples, personal information, and medical records among countries around the world. This variability complicates the conditions for collaboration between scientists from different countries, and it can to some extent inhibit future sharing of research data and samples, and the possibility to carry out collaborative research, if regulations are not harmonized between countries. A process in that direction has been initiated in the Nordic countries. In addition, several sets of standards and best practices have been published that will be helpful in establishing higher quality standards for biobank operations and biospecimens.

Some examples are provided below:

3.11.1. ISBER

"Best Practices for Repositories: Collection, Storage and Retrieval of Human Materials for Research" (3). These standards reflect the collective experience of members of the organization (see http://www.isber.org). ISBER was founded in 2000 as an educational forum for the discussion of repository to provide information and guidance on the safe and effective management of specimen collections.

3.11.2. CoE Guides

CoE Guides (Council of Europe, 22) has legally binding requirements concerning handling of tissues and cells. In other words, the same standards are supposed to exist throughout Europe when it comes to healthcare systems, and especially in cases of emergencies.

3.11.3. The First-Generation Guidelines for NCI-Supported Biorepositories (Now NCI Best Practices for Biospecimen Resources)

The First-Generation Guidelines for NCI-Supported Biorepositories (now NCI Best Practices for Biospecimen Resources) (The U.S. National Cancer Institute, (2)) were developed in order to optimize NCI-Supported Biorepositories through the adoption of best practices, and provide high-quality biospecimen resources for NCI's research programs. This process was initiated in 2002, and resulted from a report compiled that showed substantial heterogeneity in biorepository management practices across the Institute. The guidelines describe primarily the situation in the U.S.A.

3.11.4. EuroBioBank (23)

EuroBioBank (23) is a network of biological banks in Europe providing human biological material (DNA, tissue, cell) for research on rare diseases. This project is financed by the EU, under the Fifth Framework Programme for Research and Development (FP5) "Quality of Life and Management of Living Resources." The primary purposes are to harmonize and spread quality banking practices and to distribute quality material and associated data through the network. Some of the SOPs are publicized at the project's homepage (http://www.eurobiobank.org/index.htm)

3.11.5. Harmonizing Population-Based Biobanks and Cohort Studies to Strengthen the Foundation of European Biomedical Science (24)

This is a P3G-project (The Public Population Project in Genomics) which is a nonprofit international consortium to promote collaboration between researchers in the field of population genomics.

3.11.6. The Swedish National Biobank Program (25)

A collaborative program in consortium Swedish universities was established in the area of functional genomics and biobanks, with funding from the Wallenberg foundation. One of the projects is to establish a nationwide quality assurance system for collection, handling, storing, and documenting biological samples in biobanks based on the quality assurance standards known as "Good Biobanking Practise," created at a research biobank at Umeå University. This initiative has lead to a modified version of the Quality Assurance manual recommended by the Swedish County Council Association for *use in all biobanks throughout the Swedish healthcare system*. The program has also developed common quality standards for evaluating methods of handling samples with regards to DNA, RNA, and protein.

3.11.7. Cancer Control Using Population-Based Registries and Biobanks (CCPRB)

Cancer Control using Population-Based Registries and Biobanks (CCPRB, (26, 27)) is a Network of Excellence project within the sixth framework program of the EU that aims to improve control of cancer by facilitating research that links biobanks and cancer registries. One of the purposes is to provide the study base for uniquely large population-based prospective studies on cancer. Another aim is to establish a Europe-wide network for spreading the awareness of best practise quality standards for biobank-based research.

3.11.8. Comments of Future Perspectives for QA System of Biobanks: Biobanking as a Profession

As discussed above, the QA system covers all the activities of a biobank, and the QA activities must be the concern of the whole organization, as shown in the area of transfusion medicine (28). Regular seminars of important QA issues are needed in all major biobanks, and in some areas (e.g., transfusion medicine) academic programs have been suggested (28). The area of "biobanking" is becoming more and more complicated and many organizations are now involved in the development of laws, rules, and guidelines for the biobanks. A new profession, "Biobankers," has even been suggested for individuals who have a special competence and work within the area of biobank development and biobank research (29).

One important task for the future profession of "Biobankers" is to promote research on biobanks and to guide scientists to optimize biobank research according to QA and other rules. The "Biobankers" must always keep in mind that innovative research needs flexibility and freedom. The rules of the biobanks may not impair that crucial component of research.

4. Notes

1. Of crucial importance in shipment of samples is the communication between the sending and receiving laboratories. That communication must be direct and clear.
2. In selecting samples, thought must be given to the actual situation for the selected samples, e.g., sometimes matching must be performed to correct for the number of freeze-thaw cycles.
3. Concerning biosafety, all biological samples should be regarded as being potentially infectious (2).
4. It is unethical not to use the biobank samples for research purposes, but the utilization of the samples must also be proven to be of a specific value. If the samples are not used for projects that meet the criteria for high-quality research, innovative results will never occur. Therefore, in biobank research, the emphasis must always be to create systems for the efficient use of the samples after an internal evaluation process. This process is one of the most important parts of the QA system of the biobank.

References

1. Eiseman E, Haga SB, editors. (1999) Handbook of human tissue sources: A national resource of human tissue samples. Santa Monica, CA: RAND Corporation.
2. U.S. Department of Health and Human Services National Institutes of Health National Cancer Institute. First-Generation Guidelines for NCI-Supported Biorepositories. 2006. Available from: http://biospecimens.cancer.gov/bestpractices.
3. Campbell JD, Skubitz APN, Somiari SB, Sexton KC, Pugh RS (2008) International Society for Biological and Environmental Repositories (ISBER). (2008) Best practices for repositories I: Collection, storage, and retrieval of human biological materials for research. *Cell Preserv Technol* **3**, 5–48.
4. Vaught J. (2006) Blood collection, shipment, processing, and storage. *Cancer Epidemiol Biomarkers Prev* **15**(9), 1582–1584.
5. Ericsson C, Franzen B, Nister M. (2006) Frozen tissue biobanks. Tissue handling, cryopreservation, extraction, and use for proteomic analysis. *Acta Oncol* **45**, 643–661.
6. Tammen H, Schulte I, Hess R, Menzel C, Kellmann M, Mohring T, Schulz-Knappe P. (2005) Peptidomic analysis of human blood specimens: comparison between plasma specimens and serum by differential peptide display. *Proteomics* **5**, 3414–3422.
7. International Air Transport Association (IATA). Infectious substances and diagnostic specimens shipping guidelines 2006. 7th ed. Available from: http://www.iatabooks.com.
8. International Organization for Standardization. Available from: http://www.iso.ch/iso/en/ISOOnline.frontpage.
9. Eiseman E, Bloom G, Brower J. et al., editors. (2003) Case studies of existing human tissue repositories: "Best practices" for a biospecimen resource for the genomic and proteomic era. Santa Monica, CA: RAND Corporation.
10. Grizzle WE, Fredenburgh J. (2001) Avoiding biohazards in medical, veterinary, and research laboratories. *Biotechn Histochem* **76**, 183–206.
11. Occupational Safety and Health Administration (OSHA) Standards for Hazardous and Toxic Substances. Available from: http://www.osha.gov/SLTC/hazardoustoxicsubstances/standards.html.
12. Information about the Helsinki Declaration of 1964. Available from: http://www.cirp.org/library/ethics/helsinki/.

13. Godard B, Schmidtke J, Cassiman J-J, Ayme S. (2003) Data storage and DNA banking for biomedical research: Informed consent, confidentiality, quality issues, ownership, return of benefits. A professional perspective. *Eur J Hum Genet* **11** (Suppl 2), S88–S122.
14. Rose H. (2003) An ethical dilemma. The rise and fall of UmanGenomics – The model biotech company? *Nature* **425**, 123–124.
15. Ness R. (2007) Biospecimen ownership: Point. *Cancer Epidemiol Biomarkers Prev* **16**, 188–189.
16. Dressler L. (2007) Biospecimen ownership: Counterpoint. *Cancer Epidemiol Biomarkers Prev* **16**, 190–191.
17. Medical Research Council (2001) Human tissue and biological samples for use in research, operational and ethical guidelines.
18. Medical Research Council (1999) Human tissue and biological samples for use in research. Report of the Medical Research Council Working Group to develop operational and ethical guidelines. London: Medical Research Council.
19. Protocol EPIC Steering Committee. Lyon, October, 2003.
20. CARTaGENE Project. Available from: http://www.cartagene.qc.ca/index2.cfm?lang=1.
21. Cambon-Thomsen A. (2003) Assessing the impact of biobanks. *Nat Genet* **34**, 25–26.
22. Directive 95/46/EC. The European Union Data Protection (Official Journal L 281, 23/11/1995, pp 31–50).
23. Eurobiobank. Available from: http://www.eurobiobank.eu.
24. The Public Population Project in Genomics (P3G). Available from: http://www.p3gconsortium.org.
25. SWEGENE. The Postgenomic Research and Technology Programme in South Western Sweden. Available from: http://www.swegene.org/bioetik.
26. Cancer Control using Population Based Registries and Biobanks (CCPRB). Available from: http://www.cancerbiobank.org.
27. Andersson K, Bray F, Arbyn B. et al., (2010). The interface of population-based cancer and biobanks in etiological and clinical research – current and future perspectives. *Acta Oncologica*; Early online, 1–8.
28. Foss ML, Moore SB. (2003) Evolution of quality management: Integration of quality assurance functions into operations, or “quality is everyone’s responsibility”. *Transfusion* **43**, 1330–1336.
29. Hirtzlin I, Dubreuil C, Préaubert N. et al., (2003) An empirical survey on biobanking of human genetic material and data in six EU countries. *Eur J Hum Genet* **11**, 465–488.

Chapter 14

Extraction, Quantitation, and Evaluation of Function DNA from Various Sample Types

Malin Ivarsson and Joyce Carlson

Abstract

Two vital pre-requisites for genetic epidemiology have been fullfiled during the past decade and have led to a virtual explosion of knowledge concerning disease risks. Reliable databases over genetic variation derived from, e.g. the HUGO and HapMap projects, coupled with technological advances make large-scale genetic analyses and downstream bioinformatics suddenly affordable. Although recent prospective population-based biobanks have included DNA collection and purification in their planning, it is the older projects that currently are of greatest value due to the numbers of accumulated disease endpoints. In this chapter, methods to purify and use DNA derived from a variety of archival materials, including whole blood, formalin-fixed paraffin-embedded (FFPE) tissues, sera, dried blood spots (DBS), cervical cell suspensions, and mouthwash are presented and evaluated in a context of quality control guidelines to provide objective measure of the usefulness of various sample types for genetic epidemiology.

Key words: DNA extraction, Dried blood spots, Formalin-fixed paraffin-embedded tissue, Mouthwash samples, Serum, Plasma, Cervical cell suspension, Whole blood, Whole genome amplification, DNA quantification

1. Introduction

Many methods have been developed for the extraction of DNA from biological substances. Early methods were often manual and time consuming, but some produced large amounts of high-quality DNA in experienced hands (1). The goal of this chapter is to present guidelines for evaluating methods, creating a basis for comparison, and selection of new methods, as more and more advanced commercial alternatives become available. The basic essential guidelines involve the evaluation of *quantity*, *purity*, *structural integrity*, and *function*. One of the basic challenges involved in biobank-related research is that we must always assume

Joakim Dillner (ed.), *Methods in Biobanking*, Methods in Molecular Biology, vol. 675,
DOI 10.1007/978-1-59745-423-0_14, © Springer Science+Business Media, LLC 2011

that our future needs are unknown. We can, therefore, never demonstrate that a particular sample will function perfectly or that it is adequate. We can only describe its characteristics within the vocabulary of our available technology. The selection of methods for both DNA extraction and for qualitative evaluation is dictated by available sample volumes, sample DNA content, the local laboratory environment, and expected downstream applications. Our personal experience has included a total of >100,000 DNA extractions over the past decade.

In general, the collection and storage of biobank samples should be performed within "Good Biobanking Practice", including the concepts of sample identification codes, restricted access to the code key, complete tracking of samples and their derivatives, complete tracking of the chain of custody in sample handling, and reliable storage and retrieval of samples. Although our work has been done primarily on existing sample collections, some differences in yields due to collection tubes, methods or storage conditions can be noted and can guide prospective sample collection and storage.

As biobank samples frequently can be suboptimal in quality and are nearly always available in limited amounts, we try to start each new project by creating a fresh, abundant, homogeneous, and representative *control material* typical of the sample type to be handled for method development. Pilot DNA extractions and evaluations are then performed on this material before consuming the unique biobank samples. The inclusion of one or a few such control samples in each extraction batch and calculation of yield statistics over time enable the detection of changes in quality of reagent batches and technical problems within the chosen system.

2. Materials

All water is sterile and Millipore filtered and all pipette tips that are introduced into stock DNA solutions are disposable, sterile, and equipped with aerosol barriers to prevent contamination from one sample to another.

2.1. DNA Extraction from EDTA Whole Blood: Qiagen Mini-preparation Protocol

1. QIAamp DNA Mini kit (Qiagen).
2. 99.9% Ethanol (Kemetyl AB).

2.2. DNA Extraction from EDTA Whole Blood: Qiagen Autopure LS Maxi-preparation Protocol

1. Autopure LS Maxi-preparation kit (Qiagen).
2. 100% Isopropanol (Fisher Scientific).
3. 70% Ethanol (Kemetyl AB).
4. Proteinase K (Saveen Werner) Store at +4°C.
5. Autopure LS extraction robot (Qiagen, Hilden, Germany).

2.3. DNA Extraction from Dried Blood Spots

1. EZNA Forensic DNA reagent kit. (Omega Bio-Tek).
2. 70% Ethanol (Kemetyl AB).

2.4. DNA Extraction from Formalin-Fixed Paraffin-Embedded Tissues

1. Digestion buffer: 50 mM Tris–HCl, pH 8.3, 1 mM EDTA, 0.5% Tween 20 (Merck-Schuchardt).
2. Paraffin beads (make).
3. QIAquick Gel Extraction Kit (Qiagen).
4. 100% Isopropanol (Fisher Scientific).
5. Proteinase K (Saveen Werner).
6. QIAamp kit (Qiagen).
7. 70% Ethanol (Kemetyl AB).

2.5. DNA Extraction from Serum and Plasma: QIAamp MinElute Virus Spin Protocol

1. QIAamp MinElute Virus Spin kit (Qiagen).
2. tRNA (Sigma).
3. 70% Ethanol (Kemetyl AB).

2.6. DNA Extraction from Serum and Plasma: MagNA Pure LC Total Nucleic Acid Isolation Protocol

1. MagNA Pure LC instrument (Roche Diagnostics, Penzberg, Germany).
2. Total Nucleic Acid Isolation Kit (Roche Diagnostics, Penzberg, Germany).

2.7. DNA Extraction from Cervical Cell Suspensions

1. 154 mM NaCl.
2. 10 mM Tris–HCl, pH 7.4.

2.8. DNA Extraction from Mouthwash Samples

1. Oragene DNA purification kit (DNA Genotek, Inc., Ottawa, Ontario, Canada).
2. 70%, 95% Ethanol (Kemetyl AB).
3. TE-buffer (konc?).

2.9. Quantitation of DNA Yield: UV Absorbance at 260 and 280 nm

1. 96-well plastic micro-titre plates for dilution (Sarstedt).
2. TE buffer: 10 mM Tris–HCl and 1 mM EDTA, pH 8.0.
3. Fluostar Optima (BMG, LabVision).
4. NanoDrop™ (NanoDrop Technologies, Inc., Wilmington, USA).

2.10. Quantitation of DNA Yield: PicoGreen Fluorescence

1. 96-well plastic micro titre plates for dilution (Sarstedt).
2. 96-well black micro-titre plates for fluorescence measurements (Greiner nr 655076).
3. PicoGreen dsDNA quantitation kit (P7589 Molecular Probes, Eugene, USA). Store dark at –20°C.
4. Fluostar Optima (BMG, LabVision).

2.11. Quantitation of DNA Yield: OliGreen Fluorescence

1. 96-well plastic micro-titre plates for dilution (Sarstedt).
2. Heating block.
3. Ice.
4. 96-well black micro-titre plates for fluorescence measurements (Greiner nr 655076).
5. OliGreen ssDNA quantitation kit (P7589 Molecular Probes, Eugene, USA). Store dark at –20°C.
6. Fluostar Optima (BMG, LabVision).

2.12. Quantitation of DNA Yield: SYBR-Green QuantiTect PCR Kit

1. SYBR-Green QuantiTect PCR Kit (Qiagen, Hilden, Germany).
2. 7900HT sequence detection system (Applied Biosystems).

2.13. Quantitation of DNA Yield: TaqMan SNP Analysis

1. TaqMan Master Mix (Applied Biosystems).
2. F2 20210G>A 40× Assay Mix (dbSNP rs1799963, Applied Biosystems). Store all reagents dark at –20°C.
3. DNA standard with genotype F2 20210 G/G of known concentration (prepared at the laboratory).
4. 7900HT sequence detection system (Applied Biosystems).

2.14. Determination of Extracted DNA Fragments: Agarose Gel Electrophoresis

1. Gel electrophoresis apparatus (e.g. MGU-502T or SGU-2626T-02, C.B.S.Scientific Co., DelMar CA, USA), including tray and combs.
2. Voltage supply (e.g. EC105, E-C Apparatus Corp.).
3. UV-light board.
4. CCD camera with printer (optional for digital storage).
5. 10× TBE buffer: 0.9 M Tris, 0.9 M Boric acid and 10 mM EDTA pH 8.3. 1× working solution is prepared by dilution in water.
6. Ethidium bromide 10 mg/mL (AppliChem). Note carcinogenic! Wear gloves! Store dark at +4°C.
7. 10× Loading dye: 12.5% (w/v) Ficoll 400 and 0.0025% (w/v) bromphenol blue. Store at room temperature. Mix one part dye to nine parts sample at loading.
8. Marker XIII, 50 bp ladder (Boehering Mannheim).
9. NuSieve GTG Agarose (Cambrex Bio Science Rockland, Inc., Rockland, ME, USA).
10. SeaKem LE Agarose (Cambrex).
11. Erlenmeyer flask.
12. Micro-wave oven or a heating plate.

For 0.8% gel in a 130 × 100 mm gel tray.

1.6 g NuSieve GTG Agarose.

1.6 g SeaKem LE Agarose.

75 mL 1× TBE.

3.5 μL Ethidium bromide.

2.15. Usefulness of Extracted DNA: Automated SNP Analysis

1. SEQUENOM Mass Array (SEQUENOM MassARRAY, SanDiego, USA).
2. Appropriate oligonucleotides (Metabion) with sequences as determined by the SEQUENOM software for the SNPs of interest.
3. All other reagents and equipment from the SEQUENOM Mass Array manufacturer.
4. 7900HT sequence detection system (Applied Biosystems).
5. TaqMan MGB "assay by design" 40× AssayMix (Applied Biosystems).

   HFE H63D

   Forward 5′-GAT GAC CAG CTG TTC TGT TTG-3′.

   Reverse 5′-CCA CAT CTG GCT TGA AAT TCT ACT G-3′.

   Probe 1 5′-VIC-CGA CTC TCA TGA TCA TA-MGB-3′.

   Probe 2 5′-FAM-CGA CTC TCA TCA TCA TC-MGB-3′.

   HFE C282Y

   Forward 5′-GGC TGG ATA ACC TTG GCT GTA C-3′.

   Reverse 5′-TCC AGG CCT GGG TGC TC-3′.

   Probe 1 5′-VIC-CCT GGC ACG TAT AT-MGB-3′.

   Probe 2 5′-FAM-ACC TGG TAC GTA TAT C-MGB-3′.
6. No AmpErase UNG Master mix (Applied Biosystems).
7. Control DNA with known genotypes for the both polymorphisms.

2.16. Usefulness of Extracted DNA: SNP Analysis by RFLP

1. 10× PCR buffer, GenAmp PCR buffer II (Applied Biosystems).
2. 25 mM $MgCl_2$, GenAmp $MgCl_2$ solution (Applied Biosystems).
3. 100 mM dNTPs, GenAmp dNTP blend (Applied Biosystems).
4. Distilled water.
5. 10 μM Oligonucleotides (DNA technology).

   HFE H63D

   Forward 5′-GAC CTT GGT CTT TCC TTG TTT GAA GC-3′.

Reverse 5′-GGG CTC CAC ACG GCG AC-3′.

HFE C282Y

Forward 5′-CCA GGG CTG GAT AAC CTT GGC T-3′.

Reverse 5′-CCC AGA TCA CAA TGA GGG GCT G-3′.

6. AmpliTaq Gold DNA polymerase (Applied Biosystems).
7. Control DNA with known genotypes for the both polymorphisms.
8. Restriction enzyme *Bcl*I and 10× buffer G (Fermentas).
9. Restriction enzyme *Rsa*I and 10× buffer Tango (Fermentas).

2.17. Usefulness of Extracted DNA: Multiple Displacement Amplification

1. GenomiPhi DNA amplification kit (GE Healthcare). Store at −80°C.
2. TE buffer: 10 mM Tris–HCl and 1 mM EDTA, pH 8.0.

3. Methods

3.1. DNA Extraction from EDTA Whole Blood: Qiagen Mini-preparation Protocol

This method is used for extracting DNA from EDTA whole blood. Start with 200 μL EDTA whole blood according to the Purification of DNA from Whole Blood Protocol in the QIAamp 96 DNA Blood Handbook (see Note 1).

3.2. DNA Extraction from EDTA Whole Blood: Qiagen Autopure LS Maxi-preparation Protocol

This method is used for extracting DNA from EDTA whole blood. DNA is extracted from 4.5 mL EDTA whole blood using the Gentra Autopure LS robotic system protocols for fresh blood (<5 days-old, stored at +4 to +8°C), blood frozen on the day of collection at −80°C or blood that has been frozen at −20°C according to the manufacturers' instructions (see Note 2).

3.3. DNA Extraction from Dried Blood Spots

DNA is extracted from fresh or archival DBS samples with the EZNA kit according to the manufacturers' instructions.

Extend the proteolysis time from 45 min to 24 h (see Note 3).

3.4. DNA Extraction from Formalin-Fixed Paraffin-Embedded Tissues

DNA is extracted from 1 to 3 5 μm sections of Formalin-Fixed Paraffin-Embedded (FFPE) liver tissue using the QIAquick Gel Extraction Kit.

Incubate the sections at 80°C for 20 min in a 1.5 mL Eppendorf tube containing 180 μL digestion buffer (from QIAquick kit), mixing by vortex once during the incubation.

Cool briefly at +4°C.

Remove the paraffin plug at the top of the tube manually using a sterile needle (see Note 4).

Reheat samples to 56°C and add 20 μL protease solution.

Incubate at 56°C for at least 3 h.

Add buffer QG (600 μL) and isopropanol (200 μL) to the lysate and follow the gel extraction protocol from step 6.

Elute the DNA in 50 μL buffer EB.

Re-extract tissue sections of samples that function poorly on PCR using Proteinase K digestion (21) followed by QIAamp purification (22) (see Note 4).

3.5. DNA Extraction from Serum or Plasma: QIAamp MinElute Virus Spin Protocol

This method is used for the extraction of DNA from archival serum or plasma samples. DNA is extracted from 1 mL serum or plasma samples using the QIAamp MinElute Virus Spin kit and protocol.

Prepare 5 × 200 μL aliquots of each serum or plasma sample according to steps 1–5 of the protocol using an additional 30 μg tRNA in the AL buffer mixture.

Heat the samples for 5 min at 96°C after the protease step (see Note 5).

Follow the protocol from step 6.

Apply the first aliquot to the spin column, centrifuge and discard the filtrate according to step 8.

Repeat this procedure with the remaining four aliquots, and then follow the protocol from step 9.

Evaporate the samples to dryness after elution and dissolve in 10 μL H_2O (see Note 6).

3.6. DNA Extraction from Serum or Plasma: MagNA Pure LC Total Nucleic Acid Isolation Protocol

DNA is extracted from 100 to 200 μL serum or plasma according to the manufacturers' instructions using the MagNA Pure LC instrument and Total Nucleic Acid Isolation Kit.

Elute the DNA in 50 μL buffer.

Use aliquots of pooled fresh sera as positive controls and water as negative controls (see Note 7).

3.7. DNA Extraction from Cervical Cell Suspensions

This protocol is used for the extraction of both viral (human papilloma virus) and human DNA.

DNA is extracted from cervical cells that have been suspended in 1 mL 154 mM NaCl and stored at −20°C for 2–5 years.

Thaw the samples.

Centrifuge the samples at 3,000 × *g* for 10 min.

Resuspend the pellets in 1 mL 10 mM Tris–HCl, pH 7.4.

Refreeze samples at −20°C.

Thaw samples and boil for 10 min to release DNA (see Note 8).

3.8. DNA Extraction from Mouthwash Samples

DNA is extracted from mouthwash samples according to the manufacturers' instructions (see Note 9).

3.9. Quantitation of DNA Yield: UV Absorbance at 260 and 280 nm

Once DNA has been extracted, the quantity of DNA (genomic and mitochondrial) is generally determined by the measurement of UV absorption at 260 and 280 nm or fluorescence of an intercalating dye. The measurement of Abs 260/Abs 280 can be performed in a quartz cuvette in a high-quality UV spectrophotometer, but this frequently requires more material than one is willing to sacrifice in biobank contexts. Both UV absorption and fluorescence of medium to high throughput extractions can be measured in 96/384-well micro-titre plates which requires less material than a cuvette.

A NanoDrop™ performs spectrophotometric measurements on a single micro-litre of solution (see Note 10). This method determines both the concentration and purity of the DNA.

Dilute DNA extracts in TE-buffer (see Note 11).

Include TE blanks and positive controls with known DNA concentration with each analysis.

Determine the DNA concentration in each individual sample from the calibration curve (see Note 12).

3.10. Quantitation of DNA Yield: PicoGreen Fluorescence

This method determines the concentration of double-stranded DNA.

Use the PicoGreen dsDNA Quantitation Kit, following the manufacturer's instructions on a Fluostar Optima Fluorescence spectrophotometer (see Note 13).

Determine the linear measurement range for each spectrophotometer when using this kit, and adjust typical sample dilution protocols accordingly (see Note 11 and 14).

Include TE blanks and positive controls with each analysis.

Calculate the concentration of DNA in a sample from the standard curve.

3.11. Quantitation of DNA Yield: OliGreen Fluorescence

This method determines the concentration of single-stranded DNA.

Use the OliGreen ssDNA Quantitation Kit, following the manufacturers' instructions on a Fluostar Optima Fluorescence spectrophotometer with the following alterations.

Use the dsDNA standard provided in the PicoGreen kit to create a standard curve.

Dilute DNA extracts in TE-buffer (see Note 11).

Include TE blanks and positive controls with each analysis.

Heat-denature both the standard and the samples.

Place on ice.

Add OliGreen reagent and let stand for 10 min.

Read fluorescence and calculate the concentration of DNA in a sample using the standard curve (see Note 15).

3.12. Quantitation of DNA Yield: SYBR-Green QuantiTect PCR Kit

A real-time PCR method useful in quantifying extracted DNA is the SYBR-Green QuantiTect PCR Kit. It is used according to the manufacturers' instructions and the following PCR programme in absolute quantification mode on a 7900HT sequence detection system:

95°C for 1 min,

40 cycles of 95°C for 15 s and 60°C for 1 min.

Hold at 10°C (see Note 16).

3.13. Quantitation of DNA Yield: TaqMan SNP Assay

Another useful real-time PCR method for quantifying extracted DNA is the TaqMan SNP assay. Use the TaqMan SNP assay for the F2 20210 G>A SNP in the human genome on the Applied Biosystems 7900HT instrument in the absolute quantification mode according to the manufacturers' instructions. Read the plate in allelic discrimination mode to validate the absence of GA or AA genotypes.

Use a DNA standard of known concentration in a series of tenfold dilution from 10^7copies to a single copy of the DNA template to create a standard curve (see Note 17).

3.14. Size Determination of Extracted DNA Fragments: Agarose Gel Electrophoresis

Electrophoresis is performed as described (2) (see Note 18).

Mix agarose and TBE in an Erlenmeyer flask.

Cover and boil in a micro-wave oven or on a heating plate.

Mix and cook until agarose is completely dissolved.

Add approx. 0.04% Ethidium Bromide, mix and cool to 65°C.

Pour into a gel tray that has been taped at both ends.

Place combs and cool at room temperature for 30 min.

To run, lace gel in the electrophoresis apparatus, cover with 1× TBE buffer, and gently remove comb.

Mix nine parts sample with one part 10× loading dye and apply samples to wells.

Apply the marker to appropriate wells.

Close cover, and apply voltage.

Run small gels at 90 V for 15–60 min and large gels at 190 V from 45 to 120 min.

Photograph the finished gel exposed on a UV-light board, and determine fragment size by comparison to the marker bands (see Note 19).

3.15. Usefulness of Extracted DNA: Automated SNP Analyses

Within a specific project, it is always desirable to demonstrate that the DNA extract obtained will function for the intended use. In biobank contexts, we frequently wish to store aliquots of the sample for future (unknown and unanticipated) uses. Thus, ambition must be tempered with practicality and economy – both financial and regarding the wisest use of limited amounts of sample. The testing situations should be relevant for the intended use.

Analyse SNPs with the SEQUENOM Mass Array according to the manufacturers' instructions.

Mix 1–10 ng sample with 0.125 μL AsayMix and 12 μL MasterMix in a 25 μL reaction.

Run the following PCR programme 50°C for 2 min, 95°C for 10 min, 50 cycles of 95°C for 15 min, and 60°C for 1 min.

Analyse the samples in the 7900HT instrument set to allelic discrimination mode (see Note 20).

3.16. Usefulness of Extracted DNA: SNP Analysis by RFLP

Mix dNTPs and dilute to create a solution containing 1.25 mM of each dNTP.

Create a 50 μL PCR reaction mix using 5 μL buffer, 8 μL 1.25 mM dNTP, 3 μL 25 mM $MgCl_2$, 2 μL of each 10 mM primer, 1.25U Taq polymerase, and 5μL DNA sample or control.

Run the following PCR programme: 95°C for 9 min, 30 cycles of 95°C for 40 s, 60°C for 30 s, 72°C for 1 min, and 72°C for 10 min, hold at 10°C.

Cleave the HFE H63D PCR products using *Bcl*I in a 25 μl reaction containing 2.5 μL 10× buffer G, 0.4U enzyme and 9 μL PCR product. Incubate at 55°C over night.

Cleave the HFE C282Y PCR products using *Rsa*I in a 25 μL reaction containing 2.5 μL 10× buffer Tango, 0.4U enzyme, and 9 μL PCR product. Incubate at 37°C for 2 h.

Visualise the fragments on a 2% agarose gel by electrophoresis as described above (Subheading 3.3.1).

Compare the fragment lengths of the samples with the controls to determine genotypes (see Note 21).

3.17. Usefulness of Extracted DNA: Multiple Displacement Amplification

Use the GenomiPhi kit according to the manufacturers' instructions.

Dilute the Multiple Displacement Amplification (MDA) product tenfold in TE buffer.

Incubate on a shaker overnight to dissolve DNA (see Note 22).

3.18. Conclusions

DNA extracted from EDTA whole blood using mini- or maxi-preparations generally yields micro-gram amounts of high molecular weight DNA of a quality that is sufficient for successful downstream genotyping and MDA. DBS yield low amounts (ng) of DNA but of sufficient quality for successful genotyping and MDA. Storage of DBS samples at −20°C prevents DNA degradation and fragmentation. FFPE tissue usually yields relatively large amounts of DNA that can be successfully used for genotyping, preferably designed for short amplicons. However, the extraction is time-consuming and the DNA largely fragmented and unsuitable for MDA. Plasma and serum samples give low (ng) DNA yields. Plasma or serum DNA extracted with the QIAamp or MagNA Pure protocols can be successfully used for genotyping, whereas MagNA Pure extracts are not suitable for MDA. Cervical cell suspension extracts can be successfully used for the detection of viral DNA. DNA yield from mouthwash samples is generally less than 100 μg although some samples may yield no DNA, probably caused by incorrect sampling technique or leakage.

Among the quantitative methods; UV absorbance, PicoGreen fluorescence, and real-time PCR, UV absorbance at 260/280 nm is the least expensive and provides the added value of the 260/280 ratio – a measure of purity. It requires and consumes the largest amount of sample, and seems least reliable at truly low levels. In the presence of RNA and/or DNA from other species, this method overestimates the native DNA content.

The PicoGreen method is rapid, sensitive, and specific for dsDNA but involves the added cost of the PicoGreen reagent.

Real-time PCR is the most sensitive of the three methods, detecting down to a single copy of DNA (for the human genome, this is about 3.5 pg). Its cost in 384-format is comparable to that of PicoGreen in 96-format, and it provides the additional information of functionality in a PCR reaction as well as the potential for species discrimination.

In general, it is wise to consume as little sample as possible for high-quality results in all biobank-based research. Pilot studies using representative materials should be performed prior to all large projects, with the documentation of fragment size of extracted DNA, its quantity, purity, and function in project specific applications.

4. Notes

1. Prolonged heat incubation time prior to the evaluation of the DNA does not improve DNA yield (Table 1). A single freeze-thaw cycle (−20°C for ≥48 h) generally increases the yield, presumably due to cell lysis, however, the variation in yield also increases drastically (Table 1). 200 μL frozen (−20°C) whole blood samples can yield up to 100 μg high molecular weight DNA, but only about 1.5% of the samples have yields of >20 μg DNA when determined by PicoGreen fluorescence.
2. Fresh blood yields a mean of about 150 μg DNA, blood frozen on the day of collection at −80°C yields a mean of

Table 1
DNA yield from 200 μL whole blood according to length of heat incubation using the QIAamp whole blood protocol

	Qiagen 10 min	Qiagen 30 min	Qiagen 60 min	Qiagen frozen
N	96	96	96	96
DNA yield	pg/cell	pg/cell	pg/cell	pg/cell
Mean	4.96	4.58	1.88	5.99
Median	4.84	4.64	1.67	4.76
S.D.	1.25	1.93	1.23	4.08
C.V.	0.25	0.42	0.65	0.68
Minimum	1.06	0.47	0.00	0.84
Maximum	12.59	13.50	9.21	19.98

about 120 μg DNA and blood frozen at −20°C yields a mean of about 130 μg DNA. Large DNA pellets can be difficult to dissolve in less than 1.0 mL buffer. This difficulty can be noted as instability (CV > 10%) of repeated concentration measurements. Dissolving the DNA pellets in 2.0 mL buffer with rocking at room temperature overnight may improve solubility (CVs around 2%). Yields are generally lower if samples have been poorly mixed in the EDTA tubes at the time of collection so that small clots have formed in the tubes. Such clotting problems are rare if plastic tubes with lyophilised K_2EDTA powder are used rather than tubes containing K_3EDTA liquid solution (3). Very low can occur in some samples due to pellet loss by the Autopure LS instrument.

3. Commercial filtre paper products intended for nucleic acid collection and designed to bind PCR inhibitors while enabling the release of DNA without the need for proteolytic enzymes or incubation with chemicals are available (4). Using 100 μL of fresh control whole blood with known WBC on Schleicher and Schüll 2992 a 6 mm disc typically yields 13% of the theoretical DNA content with high molecular weight (5). The presence of PCR inhibitors can be minimised by diluting the samples by a factor of ≥20. Storage of samples at −20°C reduces DNA degradation and fragmentation and increases success rate in genetic analyses compared to samples stored at room temperature.
4. Addition of a paraffin bead prior to the 80°C incubation facilitates the removal of a solid paraffin disc after cooling. DNA obtained from FFPE tissue is generally <1,000 bp in size, with abundant fragments <200 bp (6). The use of the QIAquick spin column, removing the smallest fragments of DNA, may improve the

Table 2
Performance of DNA extracted from FFPE tissue in PCR reactions according to amplicon length

Amplicon size	Success rate in PCR
495 bp	0%
191 bp	14%
127 bp	86.7%
93 bp	95.9%
79 bp	92.5%
73 bp	98.1%

Table 3
Genotyping results of WGA products of plasma DNA according to WGA template

		Genotype		
WGA template (ng)	Genome copies	HFE H63D	HFE C282Y	AAT E342K
Sample 1		*H/H*	*C/Y*	*E/K*
2	500	H/H	C/Y	E/K
0.2	50	H/H	C/Y	E/K
0.02	5	H/H	Y/Y	K/K
Sample 2		*H/H*	*C/Y*	*E/E*
1.4	350	H/H	C/Y	E/E
0.2	50	H/H	C/Y	E/E
0.02	5	H/H	Y/Y	E/E
Sample 3		*H/D*	*C/Y*	*E/E*
0.1	25	H/D	C/Y	E/E
0.02	5	H/D	C/Y	E/E

efficiency of PCR reactions. Performance of DNA, in typical PCR reactions, increases with decreasing amplicon lengths (Table 2).

5. This ensures the destruction of all proteolytic activity.
6. Typical total DNA yield ranges from 0.2 to 70 ng DNA/mL serum or plasma and SNP genotyping is successful and accurate in 98% of the reactions. Using ≥50 genome copies as DNA template in a MDA reaction generally results in maintained heterozygosity (Table 3) and a MDA product that typically contains high molecular weight DNA (6).

7. These control pools can be spiked with known concentrations of plasmid DNA to determine the sensitivity of subsequent assays. Serial dilutions of the MagNA eluate can be used in real-time PCR for both a human specific sequence (the F2 20210 G>A assay) and for a microbial sequence to detect the presence of inhibitors to the PCR. In some instances, results are improved by the addition of 0.1% Bovine Serum Albumin (BSA, Sigma-Aldrich, Stockholm, Sweden) and 3.5 mM $MgCl_2$. DNA extracted by MagNA does not perform well in MDA (GenomiPhi) reactions (7).
8. Cervical cell suspensions obtained in this way are not homogeneous. Clinical considerations require the maintenance of sample aliquots for the validation of clinical diagnoses, so the entire sample cannot be converted into a homogeneous DNA solution. Nonetheless, successful and accurate genotyping analyses of viral DNA are generally obtained from samples extracted this way (8).
9. Mouthwash extracts show significantly greater estimates of DNA content by Abs 260/280 than by PicoGreen, perhaps due to the presence of RNA and ssDNA as well as dsDNA in the samples. The DNA content determined by real-time PCR is an average of about 70% of that determined by PicoGreen. This may indicate either the presence of dsDNA from microbial species or the presence of inhibitors to the PCR reaction, potentially including fragmented DNA. Some samples may produce no measurable DNA which is probably caused by incorrect sampling technique or leakage due to incorrectly screwed lids. DNA yield from mouthwash samples is generally less than 100 μg, as determined by PicoGreen fluorescence.
10. The precision and accuracy of this instrument is limited by the precision and accuracy of manually pipetting 1 μL of DNA solution – having highly variable viscosities over large concentration ranges – onto the photocell.
11. Dilute samples for UV measuring 1:10 and samples for PicoGreen measuring 1:10, mini-preparations, or 1:10 and 1:100, maxi-preparations, in TE-buffer. DNA extracts from plasma, serum or DBS usually contain low amounts of DNA and can be used undiluted or in larger volumes than the standard protocol by adjusting the volume of TE-buffer.
12. Measurement error on the most dilute samples frequently results in suboptimal A260/280 ratios, i.e. <<1.8 or >2.0. This can either be tolerated or repeat measurement can be performed on undiluted samples. The purity of DNA is generally assessed by the ratio of UV absorbance at 260 nm/280 nm. Samples with ratios >1.8–1.9 generally contain RNA species as impurity and ratios <1.8 are generally contaminated by protein.

13. Aqueous solutions of extracted nucleic acids may contain denatured single-stranded (ss) DNA, double-stranded (ds) DNA and ribosomal, transfer, and mRNA. The use of an intercalating fluorescent marker quantifies only dsDNA.
14. Use one series from 20 to 1,000 ng/mL for calibration when measuring DNA from samples with relatively high DNA yield such maxi- and mini-preparations. Use a series from 1 to 50 ng/mL when measuring DNA from samples with low DNA yield such as plasma and sera.
15. The OliGreen kit is designed for the quantitation of cDNAs or oligonucleotides, not for the quantitation of samples with large amounts of fragmented or denatured cellular dsDNA as is done here. In this situation, compared to the results of PicoGreen and real-time PCR, OliGreen has given consistently false high results by 30–60%.
16. An advantage of real-time PCR is that one can use a probe, specific for a single species-specific sequence in, for example the human genome, assuming equal representation of all parts of the genome.
17. One human genome is, roughly approximated, = 3.7 pg (i.e. 1 ng DNA = 270 copies of the human genome). When comparing the quantities of microbial and human genomes and evaluating the amount of functional DNA in preparations of questionable purity, it is of interest to probe a sequence either lacking genetic variation, or containing a known single nucleotide polymorphism (SNP), such as the TaqMan SNP assay for a rare SNP in the human genome (F2 20210 G>A). When performing pilot studies on new materials, it is wise to perform real-time PCR on serial tenfold dilutions of the DNA extract to evaluate the potential presence of inhibitors to the PCR. A plot of Cycle threshold (Ct) values versus log of copy number ideally has a slope of −3.32 and an intercept of about 39 if the PCR efficiency is one and no inhibitors are present (ABI Prism 7900 Course).
18. To view relatively intact DNA with large fragment size (1–21 kb), solutions of 0.8–1% agarose are appropriate. A 2% solution is suitable for fragments between 1,000 and 50 bp, and a 4% solution is suitable for 300–10 bp fragments.
19. Gels are about 6 mm deep and can be stored at +4°C up to 3 days covered with plastic foil. Samples migrate from the cathode (−) towards the anode (+). Due to very small amounts of DNA in some sample types, e.g. archival serum/plasma, the DNA fragments will not be visible. In such cases, fresh control materials can be used, e.g. pooled anonymized remnants of clinical samples, for method development. DNA extracts can also be pooled and run for size determination, if not needed more desperately for other applications.

20. Usefulness should be demonstrated on the intended analytical platform. The Applied Biosystem TaqMan 7900 HT and the SEQUENOM Mass Array systems design most assays for SNPs with amplicons <125 bp, placing low requirements for structurally intact DNA. Other automated assays – including pyrosequencing and mini-sequencing or single base extension strategies on capillary electrophoresis platforms may use similar or much larger amplicons. Nonetheless, PCR amplification is always inefficient in the presence of abundant fragments containing only part of the amplicon. Usefulness of the sample for many applications may be improved by the removal of small fragments, e.g. by using QIAquick Spin columns or chromatographic methods (9) to enrich larger fragments, at the cost of losing some total DNA.
21. If samples will be subjected to analysis using restriction fragment length polymorphisms (RFLP), long amplicon sizes (≥150 bp) are frequently required. The chemical composition of the DNA solution, including pH, buffering capacity, ionic strength, and the concentration of specific divalent ions as well as EDTA or other chelators, is relevant for successful enzymatic cleavage and subsequent analysis. In biobanking contexts, sufficient amounts of DNA are not always available for the purification by precipitation – and the logistics of such steps for thousands of samples are not desirable.
22. Samples containing small amounts of DNA can be evaporated to dryness and the pellet used as template in the MDA reaction. Diluting the MDA product in TE buffer dissolves the DNA product and increases the homogeneity of the sample. Successful and representative (equal replication over all parts of the genome) performance of a whole genome amplification procedure is dependent on the availability of primer binding sites (excellent throughout the genome when using random hexamer/heptamer sequences as with Phi29 MDA kits), long intact genomic sequences, and sufficient amounts of DNA template to assure equal bi-allelic representation. Numerous reports have evaluated the concordance in SNP genotyping results, demonstrating between 80% and 99.5% concordance, depending on the quality of starting material and the complexity of the SNP analysis platform (5, 10–14). It has been shown that whole genome amplification using the phi29 DNA polymerase and MDA technique may introduce large inversions and chimeric artefacts in a systematic fashion that must be considered when interpreting sequencing results of amplified material (15).

Acknowledgements

The authors thank the Knut and Alice Wallenberg Foundation who through SWEGENE have financed the laboratory equipment, and Joakim Dillner, PI for the Cancer Control and Prevention using Registries and Biobanks (EU FP6) for economic support of MILS. We thank Anna Söderlund-Strand for access to her data on cervical cell suspensions, Maria Sterner and Liselotte Hall for help with the SEQUENOM MassArray analyses, and Sophia Harlid and Christina Gerouda for DNA extractions from serum on the MagNAPure.

References

1. Maniatis, T., Fritsch, E. F., Sambrook, J. (1982) *in* "Molecular Cloning, A Laboratory Manual", pp. 280–81, Cold Spring Harbor Laboratory.
2. Maniatis, T., Fritsch, E. F., Sambrook, J. (1982) *in* "Molecular Cloning, A Laboratory Manual", pp. 150–63, Cold Spring Harbor Laboratory.
3. Arzoumanian, L. (2002) Becton Dickinson TechTalk (VS5998–1).
4. Makowski, G. S., Davis, E. L., and Hopfer, S. M. (1997) Amplification of Guthrie card DNA: effect of guanidine thiocyanate on binding of natural whole blood PCR inhibitors. *J Clin Lab Anal* **11**, 87–93.
5. Sjoholm, M. I., Dillner, J., and Carlson, J. (2007) Assessing quality and functionality of DNA from fresh and archival dried blood spots and recommendations for quality control guidelines. *Clin Chem* **53**, 1401–7.
6. Sjoholm, M. I., Hoffmann, G., Lindgren, S., Dillner, J., and Carlson, J. (2005) Comparison of archival plasma and formalin-fixed paraffin-embedded tissue for genotyping in hepatocellular carcinoma. *Cancer Epidemiol Biomarkers Prev* **14**, 251–5.
7. Ivarsson, M. I., Dillner, J., and Carlson, J. (2009) Validity of maternal genotypes in DNA from archival pregnancy serum samples. *Clin Chem* **55**, 842–3.
8. Söderlund-Strand, A., Dillner, J, and Carlson, J. (2008) High-throughput genotyping of oncogenic human papillomaviruses with MALDT-TOF mass spectrometry. *Clin Chem* **54**, 86–92.
9. Maniatis, T., Fritsch, E. F., Sambrook, J. (1982) *in* "Molecular Cloning, A Laboratory Manual", pp. 464–67, Cold Spring Harbor Laboratory.
10. Paez, J. G., Lin, M., Beroukhim, R., Lee, J. C., Zhao, X., Richter, D. J., Gabriel, S., Herman, P., Sasaki, H., Altshuler, D., Li, C., Meyerson, M., and Sellers, W. R. (2004) Genome coverage and sequence fidelity of phi29 polymerase-based multiple strand displacement whole genome amplification. *Nucleic Acids Res* **32**, e71.
11. Montgomery, G. W., Campbell, M. J., Dickson, P., Herbert, S., Siemering, K., Ewen-White, K. R., Visscher, P. M., and Martin, N. G. (2005) Estimation of the rate of SNP genotyping errors from DNA extracted from different tissues. *Twin Res Hum Genet* **8**, 346–52.
12. Dickson, P. A., Montgomery, G. W., Henders, A., Campbell, M. J., Martin, N. G., and James, M. R. (2005) Evaluation of multiple displacement amplification in a 5 cM STR genome wide scan. *Nucleic Acids Res* **33**, e119.
13. Tzvetkov, M. V., Becker, C., Kulle, B., Nurnberg, P., Brockmoller, J., and Wojnowski, L. (2005) Genome-wide single-nucleotide polymorphism arrays demonstrate high fidelity of multiple displacement-based whole-genome amplification. *Electrophoresis* **26**, 710–5.
14. Paynter, R. A., Skibola, D. R., Skibola, C. F., Buffler, P. A., Wiemels, J. L., and Smith, M. T. (2006) Accuracy of multiplexed Illumina platform-based single-nucleotide polymorphism genotyping compared between genomic and whole genome amplified DNA collected from multiple sources. *Cancer Epidemiol Biomarkers Prev* **15**, 2533–6.
15. Lasken, R. S., and Stockwell, T. B. (2007) Mechanism of chimera formation during the Multiple Displacement Amplification reaction. *BMC Biotechnol* 7, 19.

Chapter 15

Cervical Cytology Biobanks as a Resource for Molecular Epidemiology

Marc Arbyn, Kristin Andersson, Christine Bergeron, Johannes Bogers, Magnus von Knebel-Doebertitz, and Joakim Dillner

Abstract

A cervical cytology biobank (CCB) is an extension of current cytopathology laboratory practice consisting in the systematic storage of Pap smears or liquid-based cytology samples from women participating in cervical cancer screening with the explicit purpose of facilitating future scientific research and quality audit of preventive services. A CCB should use an internationally agreed uniform cytology terminology, be integrated in a national or regional screening registry, and linked to other registries (histology, cancer, and vaccination). Legal and ethical principles concerning personal integrity and data safety must be respected strictly. Biobank-based studies require approval from ethical review boards.

A CCB constitutes a nearly inexhaustible resource to perform fundamental and applied biologic research. In particular, it can contribute in answering questions on the natural history of HPV infection and HPV-induced lesions and cancers, screening effectiveness, exploration of new biomarkers, and surveillance of short- and long-term effects of the introduction of HPV vaccination.

To understand the limitations of CCB, more studies are needed on quality of samples in relation to sample type, storage procedures, and duration of storage.

Key words: Molecular epidemiology, Cervical cytology, Pap smears, Liquid-based cytology, Biobanking, Human papillomavirus, HPV

1. Introduction

European guidelines recommend that cytology-based cervical cancer screening should be offered once every 3–5 years to all women in the European Union (EU) aged 25–64 years (1–3).

Moreover, the European Council proposes that the member states should organize this screening, which includes that all steps of the screening process are registered at the individual level: attendance of women in screening, results of the screen tests, and

Joakim Dillner (ed.), *Methods in Biobanking*, Methods in Molecular Biology, vol. 675,
DOI 10.1007/978-1-59745-423-0_15, © Springer Science+Business Media, LLC 2011

follow-up findings of screen-positive women, such as repeat cytology, HPV (human papillomavirus) triage results, and/or histologic outcomes (3). The personal records from the cytologic registry should be linkable to histologic databases, with the cancer registry and population files, using unique identifiers in order to monitor the quality and to evaluate the impact of the program (4).

Every year, a few tens of millions of cervical cell samples are taken in EU. In countries with well-organized screening, the 10-year cumulative attendance rate is well above 90% of all women belonging to the target age range. The average number of smears planned over a life time varies among member states: between 7 in Finland and the Netherlands and more than 50 in Germany, Austria, and Luxembourg (5–7). Cervical cell samples are usually stored for 5–10 years because of medico-legal reasons or for quality control. Certain European laboratories have archived series of Pap smears over several decades. This makes archival of cervical samples an extremely valuable resource for research where genetic susceptibility, exposure to external etiologic factors and the outcome (cervical and other malignancies) can be documented through the link with cancer registries.

The knowledge that cervical cancer is caused by persistent infection with high-risk human papillomavirus (hrHPV) types has prompted the development of HPV tests usable for screening and even prophylactic vaccines. In the near future, these tools will be translated in new policies of cancer prevention. Series of multiple cytologic biospecimen offer an interesting framework to address essential pending questions on the optimal use of these new preventive tools (8).

A CCB linked to a screening registry, a cancer registry, and in the future with a HPV vaccination registry offers a unique tool for etiologic studies, studies on the natural history and molecular pathways of carcinogenesis, the evaluation of screening and vaccination effectiveness, and the evaluation of early biomarkers (viral DNA, RNA, or proteins or immunologic, genetic, or other biologic host factors) usable in cancer screening or triage of screen positives. Even when in the future detection of nucleic acids of HPVs would become the standard screening methodology, collection of cellular material will be needed to perform HPV tests.

Removing cover slips from archived conventional Pap smears, extraction of DNA from scraped cells or from micro-dissected dyskaryotic cells, DNA amplification of human or viral genomic sequences, and HPV genotyping were demonstrated to be feasible in previous studies (9–15). Liquid-based cytology (LBC) is a newer method of preparing cervical cytology samples, where collected cells are transferred in a fixative and evenly dispersed across slides. Importantly, the left-over sample contains large amounts of cells that in the methanol-based fixative contain high-quality DNA, RNA, proteins, and well-preserved cellular morphology.

In the current chapter, we propose practical and logistical guidelines on how to prepare, process, report, store, and handle cervical cytology samples in routine practice and discuss biomolecular standardized procedures, possible molecular and epidemiologic study objectives, and inherent ethical and legal issues involved in cytology biobanking.

2. Materials

2.1. Preparation of Cervical Cell Samples

1. Sampling material such as endocervical broom, spatula, endocervical brush, or extended tip spatula is used.

2.2. Processing and Staining of Liquid-Based Cytology

1. Hematoxylin–eosin (alcohol).
2. ThinPrep (Cytyc, Boxborough, USA).
3. SurePath (formerly AutoCyte PREP, TriPath Imaging Inc., Burlington, USA).
4. CYTOscreen® (Seroa Inc, Monaco, France).
5. Turbitec® (Labonord, Templemars, France).
6. PapSpin® or SpinThin® (Shandon, Spitzburg, USA).
7. Cytoslide® (Menarini Diagnostics, Firenze, Italy).

2.3. Reporting and Registration of Cytopathologic Results

1. Laboratory Information System (LIS).

2.4. Storage of Samples

1. 2 ml Ependorff tubes (Sarstedt Ref no. 72694006).

2.5. Extraction of Nucleic Acid from Archival Cervical Cytology Samples

2.5.1. Extraction of DNA from Pap Smears

1. 50 ml tubes (Sarstedt Ref no. 62547254), 2 ml tubes (Sarstedt Ref no. 72694006), 1.5 ml tubes (Sarstedt Ref no. 72692005).
2. Petri dishes.
3. Xylene (VWR, 24640-1), room temperature (see Note 2).
4. Ethanol 99.5%, 95%, 70%, room temperature.
5. TE buffer: 10 mM Tris–HCl, 1 mM EDTA, pH 8.0, store at 4°C.
6. Proteinase K (Roche, 745723) 20 mg/ml, store frozen at –20°C, do not freeze-thaw.

2.5.2. Extraction of DNA from Liquid-Based Cytology Samples

1. Qiamp DNA Mini Kit (QIAGEN, Hilden, Germany).
2. Magna Pure (Roche Molecular Diagnostics, Pleasanton, California, USA).
3. 2 ml tubes (Sarstedt Ref no. 72694006).
4. Specimen Transport Medium (STM; Digene Corp., Gaithersburg, Maryland, USA).
5. Lysis buffer: 10 mM Tris–HCl, 10 mM NaCl, 10 mM EDTA, 4% SDS, pH 7.8.
6. Proteinase K (Roche, 745723) 20 mg/ml, store frozen at -20°C, do not freeze-thaw.
7. Saturated NH_4Ac, room temperature.
8. Ethanol 99.5%, 70%, room temperature.
9. TE-buffer: 10 mM Tris–HCl, 1 mM EDTA, pH 8.0, store at 4°C.
10. Magna Pure LC Total Nucleic Acid Isolation Kit (Roche).

2.5.3. Extraction of RNA from Liquid-Based Cytology Samples

1. RNeasy (QIAGEN, Hilden, Gemany).
2. Masterpure Extraction kit (Epicentre, Madison, Wisconsin, USA).
3. TRIzol (Invitrogen, Carlsbad, California, USA).

2.6. Amplification, Identification of Nucleic Acid Sequences

1. Hybrid Capture II assay (Digene Corp., Gaithersburg, Maryland, USA).
2. Amplicor HPV Test (Roche Molecular Diagnostics, Pleasanton, California, USA).
3. PreTect HPV-Proofer (NorChip AS, Kokkastua, Norway).

2.7. Proteomics

1. CINtec cytology kit (MTM Labs, Heidelberg, Germany).

3. Methods

3.1. Preparation of Cervical Cell Samples

The precursors of cervical cancer arise mainly in the transformation zone (TZ) between the ectocervical multilayer squamous epithelium and the endocervical columnar epithelium (16–18). Therefore, it is important that cell material is sampled primarily from this zone. The presence of metaplastic squamous cells and endocervical cells, in addition to squamous cells, indicates that the TZ has been sampled but cannot provide assurance that the complete TZ has been targeted.

For conventional Pap smears, the procedure of collection of cellular material is well described in several training manuals for practitioners (19–21). Sample takers should scrape cells from the entire TZ with appropriate sampling devices, such an endocervical broom, a combination of a spatula and an endocervical brush, or an extended tip spatula. The collected cells should be spread quickly to a glass slide and fixed immediately to avoid drying artifacts (19).

For LBC, the same procedures should be followed as those for sampling cells for Pap smears, but only plastic devices may be used. The collected cells should be quickly transferred into a vial with fixative liquid according to the instructions of the manufacturer of the LBC system.

Subsequently, the slide or vial, labeled with identification data, and the completed request form are sent to the laboratory for processing, staining, and cytologic interpretation.

3.2. Processing and Staining of Pap smears or Liquid-Based Cytology Samples

Staining of conventional Pap smears facilitates the recognition of cell components. All stains have the same principles. The nuclei accept the basophilic components of the dye and stain blue with hematoxylin. The cytoplasm stains eosinophilic with eosin–alcohol preparations, the most commonly employed being EA36 (commercially available as EA50) (22). Processing and staining should be performed according procedures described in guidelines (4).

Several commercial systems have been developed in the last 15 years for staining of LBC samples, among which ThinPrep and SurePath (formerly, AutoCyte PREP) are the most well known. With the ThinPrep-2000 or the more fully automated ThinPrep-3000 processor, the liquid is aspirated through a membrane that detains the cellular material, which is then stamped onto a slide in the form of a very thin layer. The SurePath material is sedimented through a density gradient (23). Only ThinPrep and SurePath have so far been approved in USA by the Food and Drug Administration (FDA). Other fluid-based systems are manufactured and are used in Europe, such as CYTOscreen®, Turbitec®, PapSpin®, SpinThin®, and Cytoslide®.

Transfer via a fluid medium increases the likelihood of representative smears (24). The ThinPrep and SurePath systems produce circular areas that contain an average of 50,000–75,000 randomly selected cells, whereas the conventional Pap smear usually contains 100,000–250,000 cells, which is about one-fifth of the cellular material available on the sampling device (24).

Fixation of the cell material is optimal in LBC. The visualization of a calibrated thin layer of properly distributed epithelial cells, with reduced numbers of red and white blood cells, facilitates cytologic interpretation (25, 26).

Multiple smears can be made or additional investigations performed on the residual fluid (e.g., HPV DNA or RNA, or Chlamydia DNA testing) (27, 28).

The staining of LBC is part of the kit and is well described in the various packet inserts.

3.3. Reporting and Registration of Cytopathologic Results

Cytologic findings should be classified according to an internationally agreed system such as *The Bethesda System*, including a judgment of specimen adequacy and grading of squamous and glandular epithelial cells (29, 30). Patient identifiers, cytologic results, and results of ancillary molecular tests must be entered into the laboratory information system (LIS) using a standardized coded format. This LIS should allow transmission of data to the national or regional cancer screening registry and/or cancer registry. Information on the type of storage and retrievability of the cytologic biospecimen are essential components of LIS.

3.4. Storage of Samples

Conventional smears should be stored in a dry and dark surrounding. After years of storage, air bubbles may appear and the contact with air can affect the quality of the sample. If the sample needs to be reread, removal and renewal of the cover slip is recommended to ensure high-quality cytologic assessment.

The storage of whole vials with residual liquid is not performed routinely, and usually only the prepared fixed slides are kept for medico-legal reasons. Keeping cell pellets in Ependorff tubes reduces the volume of storage space substantially. One method for storage of LBC samples that is currently under evaluation is to transfer the well-mixed liquid to a 2-ml Ependorff tube, centrifuge the sample at 3,220 × *g*, and then carefully remove the supernatant. The pellet can then be stored in the freezer. The fixative fluid used after sample collection contains several alcohol components and other stabilizing agents which may interfere with biomolecular testing; thus, it is very important to remove the solution prior to storage. It has been shown that storage of liquid-based samples at ambient temperature decreases the stability of PCR amplifiable genes and nuclear preservation over time compared to frozen storage (31, 32).

Long-term storage of DNA/RNA extracted from LBC material is currently under investigation. Probably, the best way will be to store the extracted nucleic acids at –80°C. However, this involves substantial costs for the laboratory.

3.5. Extraction of Nucleic Acid from Archival Cervical Cytology Samples

Extraction of high-quality nucleic acid suitable for PCR amplification is dependent on adequate specimen collection and handling as well as on the absence of inhibitors of the polymerase reaction. Long-time storage of samples induces DNA degradation. However, studies on conventional Pap smears have shown that the method of fixation is more important for DNA recovery than the length of storage (33). Inclusion of acetic acid in the fixative solution, which gives an enhanced nuclear staining for microscopy, seems to reduce the length of amplifiable genomic DNA fragments. It has also been shown that the amount of cellular material in conventional Pap smears varies substantially, which requires normalization of DNA copy numbers to make the comparison of samples and studies possible (34).

Several studies have tested the extraction and stability of nucleic acid in LBC samples and it has been shown that the method of fixation used as well as the extraction procedure affects the quality and amount of nucleic acid that can be extracted. After storage in ThinPrep, recovery of both DNA and RNA is almost unaffected (35, 36), whereas storage in SurePath diminishes the recovery of both DNA and RNA rapidly (37).

3.5.1. Extraction of DNA from Pap smears

Extracting DNA from a Pap smear will inevitably involve destruction of at least a part of the Pap smear, which might be in conflict with the mandatory regulation to store the smears for medico-legal reasons. Electronic preservation by partial or complete scanning of the cytologic preparation before extraction could provide an interesting solution for the legal and medical safety of both patient and laboratory, and offer logistic advantages for central storage and exchange between researchers.

Extraction of DNA from Pap smears is a very laborious procedure and requires careful handling of the samples to avoid contamination between samples (see Note 1). A possible DNA extraction procedure is as follows, but commercial DNA isolation techniques have also been shown to be suitable for DNA extraction from Pap smears (38).

Removal of cover slips (see Note 1).

1. Put each slide in individual labeled 50-ml tubes (xylene and ethanol can remove the labeling of the slide).
2. Add enough xylene to cover the cover slip, approximately 35 ml.
3. Leave the tubes standing in a rack to soak for 5–6 days in the fume hood.
4. After 5–6 days, remove the cover slides using 5-ml pipette tips (if not possible, use a sterile scalpel) and leave the slides in the xylene for at least one more hour.

Destaining

5. Pour off the xylene in a waste bottle and put the slides in labeled Petri dishes.
6. Add 10 ml of ethanol, 99.5%, to the Petri dishes and soak for at least 1 h.
7. Remove the ethanol and replace it with 10 ml of 95% ethanol for 5 min.
8. Remove the 95% ethanol and replace it with 10 ml of 70% ethanol for 5 min.
9. Remove the 70% ethanol and let the slides dry in room temperature.

Digestion

10. Prepare digestion buffer; 25 μl of proteinase K (20 mg/ml) per ml of TE buffer.

11. Add 600 μl of the digestion buffer to the slide. Spread it over the slide with a sterile pipette tip and dislodge the cells from the glass surface with the tip.
12. Transfer the cell suspension into a labeled, sterile 1.5-mL tube with screw cap.
13. Add another 10 μl proteinase K (20 mg/ml) to each tube (final concentration of proteinase K 0.82 mg/ml).
14. Incubate at 37°C over night.
15. Inactivate proteinase K in a water bath at 100°C for 10 min.
16. Do a quick centrifugation (5 s) of the tubes to remove any liquid from the lid.
17. The sample is ready for PCR or can be frozen at -20°C until PCR is run.

3.5.2. Extraction of DNA from Liquid-Based Cytology Samples

For extraction of DNA from LBC samples, different commercially available kits have been used in research studies, for example, QIamp DNA Mini Kit. Here we describe one commercially available method (Magna Pure) and one non-commercial system that works well for LBC samples (39). More methods for extraction of total nucleic acid are found in sections 3.5.3 and 3.5.4 about RNA extraction.

Before extraction of the samples and preferably before storage, the buffer of the samples must be changed:

1. The sample in the original collection tube is vortexed and 10 ml is transferred to a suitable centrifugation tube. Centrifuge at 3,220 × *g* for 15 min and carefully pour of the supernatant (the pellet is rather loose).
2. Transfer the pellet to a 2-ml tube with screw cap and centrifuge at 3,220 × *g* for 10 min. Remove all of the supernatant and dissolve the pellet in 750 μl of STM and store at -80°C until extraction (if extracting with Magna Pure, store in aliquots).

Extraction with Magna Pure:

1. Thaw one 250-μl aliquot and add another 150 μl of STM to the sample.
2. Extract 200 μl of the dissolved sample with Magna Pure using the Magna Pure LC Total Nucleic Acid Isolation Kit according to the manufacturer's description. Extracted samples can be stored at -20°C until PCR analysis.

Manual extraction:

1. Mix 300 μl of lysis buffer with 1.6 μl of Proteinase K. Mix by inverting. Add the mixture to 100 μl of the sample and vortex.

2. Incubate in a heating block at 37°C over night.
3. Add 150 µl of saturated NH_4Ac and vortex.
4. Centrifuge the samples at 16,000 × *g* for 15 min.
5. Transfer the supernatant to a new tube.
6. Add 900 µl of absolute ethanol (99%) to the supernatant and vortex.
7. Precipitate the DNA for 30 min at –20°C.
8. Centrifuge the sample at 16,000 × *g* for 5 min.
9. Discard the supernatant carefully without disturbing the pellet.
10. Add 500 µl of 70% ethanol and vortex.
11. Centrifuge the sample at 16,000 × *g* for 5 min.
12. Discard the supernatant carefully without disturbing the pellet.
13. Dry the pellet in a heating block at 37°C for about 30 min or until the pellet is dry.
14. Dissolve the pellet in 100 µl of TE-buffer. The sample is then ready for PCR or can be frozen at –20°C until the PCR is run.

For methods to control the quality of extracted DNA, see Chapter 14.

3.5.3. Extraction of RNA from Liquid-Based Cytology

Extraction of RNA from LBC samples has been reported in several publications and RNeasy is a commonly used method (36, 37).

For extraction of total nucleic acid, the Masterpure Extraction kit (35) and TRIzol (37) have been described.

Residual material from fresh ThinPrep samples are suitable for RNA isolation and subsequent analysis by PCR, real-time PCR, or nucleic acid sequence-based amplification (NASBA)-based amplification. However, reduced RNA recovery and RNA degradation in cell lines and exfoliated cells fixed in SurePath have been reported (40).

3.5.4. Extraction of RNA from Pap Smears

Using RNA transcripts of known genes involved in the carcinogenesis as biomarker requires the ability to obtain high-quality RNA. The detection of RNA transcripts of viral genes (for instance, those coding for the oncoproteins E6 or E7 of certain hrHPV types) may assist in differentiating between progressive and non-progressive HPV infections or precursor lesions, therefore serving as a more specific biomarker for carcinogenesis (41, 42).

RNA in biologic specimens is rapidly degraded by RNAses unless adequately fixated. However, several studies have proven the feasibility of utilizing mRNA recovered from cytologic Pap specimens as a resource for ancillary molecular biologic studies (43, 44).

3.6. Amplification, Identification of Nucleic Acid Sequences

A multitude of HPV detection systems exist but the clinical accuracy (to predict prevalent or incident cervical cancer or its precursors) has been evaluated thoroughly only for a few PCR-based methods and the Hybrid Capture II assay. For a synthetic overview of test characteristics of currently used and novel methods, and a description of potential applications in clinical practice and research, we refer to published reviews (45–48).

PCRs, such as the GP5+/6+ or (PG)MY09/11 PCR, target a conserved region of the viral L1 gene, amplifying different HPV types in a single reaction (49–51). PCR systems targeting short DNA sequences, such SPF10 (52, 53) or the Amplicor HPV test (45), have a high analytic sensitivity and are considered to be particularly appropriate for HPV DNA testing in archived cytologic preparations, stored over long periods, in which parts of the viral genome may be fragmented. After PCR amplification, distinction of different genotypes can be achieved by hybridization with type-specific probes or reverse line blot assays. Type-specific PCRs often target specific sequences of the early genes. Several innovating micro-array chips, Luminex micro-arrays, and mass spectrometry technologies have been developed recently which allow high-throughput assessment and identification of multiple HPV types (48, 54). However, the performance of these newer methods on material from stored cervical samples has not yet been evaluated.

Viral mRNA can be detected by real-time PCR (55) or NASBA (56). A commercial kit, PreTect HPV-Proofer, has been developed.

For an overview of currently available tools to assess changes at the level of the human genome, which are associated or could be associated with the development of cervical cancer, or which are the consequence of the interaction with viral oncogenes, we refer to a recent review (57). In brief, we mention comparative genome hybridization (CGH) detecting chromosomal abnormalities, single nucleotide polymorphism (SNP) profiling to identify point mutations, and circular DNA micro-arrays to assess mRNA expression profiles.

3.7. Proteomics

The expression of certain cell cycle regulating proteins, such as cyclin-dependent kinases (CDKs), cyclins, CDK inhibitors, and others, is altered in cervical neoplastic lesions as a consequence of the interaction with viral oncogenes (58, 59). The degree of expression correlates with the severity of cervical disease and/or with progression of disease and, therefore, some are proposed as biomarkers for carcinogenesis (60).

These overexpressed proteins can be identified by immunocytochemistry on fresh or stored conventional or liquid cervical cytology preparations.

Most experience exists for the detection of the CDK inhibitor p16INK4a. Release of pRB function through binding with the HPV E7 protein results in overexpression of p16, which is a marker for progression of HPV infection to cervical cancer transformation (61).

High-quality detection of overexpressed p16 in cervical cells is possible by using a commercial staining kit for cytologic samples (CINtec cytology kit, MTM Labs, Heidelberg, Germany) on residual cytologic material stored for 5 years or more in ThinPrep solution, whereas immunostaining of processed ThinPrep slides reveals less favorable results (von Knebel-Doeberitz, personal communication).

For LBC with SurePath technology, it was shown that low cellularity and long-term storage in water negatively influence p16 immunoreactivity (62). Good results were obtained using the residual 2 ml of fluid left in the original SurePath vial stored for several months. Care should be taken to cover the hole in the cap, made by the processing machine.

The use of a standardized p16 nuclear scoring system including morphologic criteria is crucial for optimizing accuracy and inter-rater reproducibility (63, 64).

p16 immunocytochemistry is also possible on stored conventional Pap smears (65, 66).

Numerous other cell-regulating proteins, proliferation markers, or viral oncogene products (Ki67, p53, pRB, Cyclin E, Mcm5, Cdc6, c-myc, E6, E7, etc.) have been tested for using cytochemistry on Pap smears and liquid cervical preparations (57, 60).

Proteomics can also be used to screen profiles of large numbers of proteins using mass spectrometry technology and antibody micro-arrays, but until now none of these novel methods have been applied on cervical cytology specimen (67–69).

3.8. Quality Assurance in Clinical Cytology Biobanking

A biobank of archived cervical cytology samples should preferentially involve all Pap smears or liquid biospecimen of a well-circumscribed population, collected in whatever circumstances (screening, follow-up for previous abnormality, or clinically indicated). Completeness, expressed as the percentage of all cervical cell preparations collected in a given target area available for CCB, should be an important quality indicator.

A uniform standard, internationally agreed reporting system for cytologic grading, should be used, complete with relevant clinical data and linkable with external clinical data bases. Data should be available in a standard coded database format.

A CCB must respect procedures of data safety, and coding of identifying information.

It is important that the samples are stored under conditions that preserve them for future use and that prevent contamination between samples, and that the quality of the sample is checked regularly.

Research need to be done to define optimal storage procedures for LBC specimen, assuring high-quality material for biomolecular assays.

3.9. Study Base for Cytologic Biobanking

Biospecimens play a critical role in translating basic science discoveries into clinical applications (70). A cervical cytology biobank, embedded in a comprehensive cytology and HPV test registry, and linked with a cancer registry, constitutes a nearly inexhaustible resource to perform fundamental and applied biologic research, and also public health relevant studies.

3.9.1. Audit of Screening Histories After the Diagnosis of Cervical Cancer

Linkage between the cancer and screening registry, including cytologic reviewing of stored Pap smears, is extremely instructive to identify errors in the screening program. This type of audit can be extended by matching cancer cases to control women without cancer, which yields odds ratios that reflect effectiveness of screening (71).

3.9.2. Etiologic Studies

Women participating in cervical cancer screening constitute huge cohorts, providing the necessary statistic power to study rare outcomes, such as vagina, vulva, anus, or oro-pharyngeal cancer, which are also causally linked by HPV, but for which the HPV type-specific population attributed risk is not known precisely.

Moreover, nested case–control studies can be easily conceptualized where cases can be women, selected from the cancer registry, and where exposure can be assessed using archived cytology samples, taken at young age when risk for acquiring HPV infection is the highest. Such nested case–controls provide more realistic estimates of the relative risk associated with HPV infection (odds ratios for expressing association with cervical cancer are in the range >5 to <100) (9, 12, 14, 15) compared to conventional case–control studies, where HPV is assessed in cancer biopsies and in cervical samples of age-matched women (72). HPV prevalence in controls of similar age as cervical cancer cases is low, due to natural viral clearance resulting in inflated odds ratios (odds ratios in the range of 50–500) (72).

HPV DNA detection in well-conserved archived cytology samples could also be a more reliable measure of risk exposure than serologic assays, such as those applied in pioneering Nordic studies using maternity serum banks (73, 74). These serology-based cohort studies often have shown rather low odds ratios for association between HPV seropositivity and subsequent cervical cancer (in the range of 3–13).

The role of cofactors, such as *Chlamydia trachomatis* and possible other micro-organisms, as well as genetic susceptibility and molecular pathways of carcinogenesis can be assessed using cancer registry-linked cytobiobanks.

3.9.3. Secular Trends in HPV Exposure

HPV genotyping in historical series of smears can be used to understand cohort effects in incidence of and mortality from HPV-related cancers and to explain international differences in cervical cancer burden (75–77).

3.9.4. Biomarkers of Carcinogenesis

Introduction of HPV screening instead of cytology-based screening will necessitate the use of more specific triage strategies for instances based on molecular markers such as HPV type-specific persistence, evolution of HPV type-specific load, integration of viral DNA into the host genome, expression of viral oncogenes, and altered expression of cell cycle regulating proteins (48). Again, well-preserved cytologic specimen offers an excellent tool to explore the predictive value of specified progression markers or to explore undefined markers using micro-array technology.

An interesting and efficient study design to select useful biomarkers to triage women with low-grade squamous intraepithelial lesions is a case–control study where cases are women who had LSIL that progressed to more serious disease, and controls are women with LSIL that regressed spontaneously. Immunostaining or HPV nucleic acid assays applied on the archived index samples with LSIL, blinded to the outcome, can identify clinically useful markers.

Another fundamental type of case–control study involves archived smears from women with cancer matched to those without cancer. It is expected that genomics, RNAomics, and proteomics profiles from numerous case and control series will contribute in disentangling the molecular pathogenesis of cancer and that this could reveal new applications for diagnostic and targeted therapeutic interventions (57, 69, 78).

3.9.5. Monitoring of Early and Late Impact of HPV Vaccination

Cytologic biobank surveys involving HPV genotyping of archived cervical cell samples allows monitoring of changes in HPV type distribution subsequent to vaccination, at population level, and by linkage with vaccination registries, at individual level (8, 79). A primary indicator of interest is the prevalence trend and geographic spread of vaccine HPV types (HPV 6, HPV 11, HPV 16, and HPV18), vaccine related and unrelated types (80, 81). Occurrence of cervical lesions associated with type-specific HPV infections in relation to date of vaccination will provide early answers to pending questions regarding HPV vaccination, such as possible type replacement, waning immunity, cross-protection, and duration of this protection.

It would be of interest to integrate HPV vaccination monitoring activities (such as sentinel HPV genotyping at sexual or reproductive health services for teenage girls, self-sampling surveys, and genital wart registries) into the cytology biobank-based information system.

3.10. Ethical and Legal Issues

The possible objectives of cytology biobank research, outlined in the study base above or not yet conceptualized, fall beyond the immediate medical purpose of collecting a cervical cell sample and require, therefore, a particular ethical and legal framework (see Chapters 1 and 2). Restrictive consent procedures may be unpractical and could result in serious selection biases compromising the scientific validity of studies.

Broad consent to the use of biological human samples for *future medical research* is an ethically laudatory principle promoted by CCPRB (Cancer Control using Population-based Registries and Biobanking), a large biobank research network supported by the 6th Framework Programme of the European Union (82). Routine informed consent for biobanking of the cervical smears for medical research purposes was instituted in Sweden in 2004 and has worked exceptionally well, with more than 700,000 consents annually. In certain EU countries, a specific biobank law might regulate future use of stored human biospecimen, whereas in others existing legal regulations regarding patient rights, scientific research and privacy protection could be considered as sufficient. Whatever the legal regulation in place, the following general ethical conditions will have to be fulfilled (82, 83):

1. Personal information should be handled safely, involving adequate coding of identifying data in such a way that unauthorized access is impossible while at the same time enabling efficient matching of relevant data;
2. Donors of biologic samples should be granted the right to withdraw consent;
3. Every new study should be approved by an ethical review board.

Concerning consent for research on *previously collected samples*, the following cases can be distinguished: (a) no explicit consent was requested; (b) consent was requested and obtained for a specific research without explicit exclusion of other research; (c) consent was requested and obtained for a specific research explicitly excluding other research. In case (c), no samples should be included in biobank research, unless sample providers are contacted for new consent, whereas in cases (a) and (b), biobank-based research without consent can be considered as ethically acceptable under the condition that personal safety is guaranteed and the study is approved by an ethical review board (84). Individual feedback of study results to sample providers should be considered as unnecessary (84). However, information about studies and their results via public media is recommended since it increases the fundamental trust of the community in research.

A particular case is the re-examination or assessment of stored samples, in the framework of an established quality audit or planned surveillance activity, such as review of previous Pap smears of a woman who developed cervical cancer, or HPV genotyping in cervical samples of vaccinated or non-vaccinated women (85). Such cases should not require consent of the sample provider if the official public health authority has defined these activities as a legal duty to verify effectiveness and security of specific preventive services.

3.11. Conclusion

A cervical cytology biobank, embedded in a comprehensive cytology and HPV test registry, and linked with a cancer registry, constitutes a nearly inexhaustible resource to perform fundamental and applied biologic research and to address possible implications of the introduction of new strategies of cancer prevention, reducing the dependency on expensive and long-lasting prospective studies.

Cervical cytologic biobanking is nothing more than a structural extension of current established practice, but it requires substantial logistic efforts to make it a high-quality tool for operational research. A solid medico-legal framework should be created not only guaranteeing the integrity and autonomy of sample providers, but also enabling the availability of data and specimen for well-designed studies. In particular, HPV genotyping on archived cytologic samples will assist in monitoring the early and late impact of the introduction of prophylactic HPV vaccination.

4. Notes

1. To check for contaminations, including a blank, clean slide in each extraction round is recommended.
2. Carefully follow the safety regulations when working with xylene. A fume hood and gloves that do not leak xylene are minimum protection.

Acknowledgments

We thank Prof. M.G. Hansson (Centre for Bioethics at Karolinska Institutet and Uppsala University, Uppsala Science Park, SE-75185 Uppsala, Sweden) and Don Chalmers (Faculty of Law School, University of Tasmania, Australia) for the review of subchapter 9 on legal and ethical issues. We are also grateful to Cindy Simoens (Scientific Institute of Public Health, Brussels) for editorial assistance.

Funding was received from the 6th Framework Programme (European Commission, DG Research, Brussels, Belgium) through the CCPRB Network (University of Lund, Malmö, Sweden), the Institute for the Promotion of Innovation by Science and Technology in Flanders (IWT-Vlaanderen, refnum 060081), and the Belgian Cancer Foundation.

References

1. Council of the European Union (2003) Council Recommendation of 2 December 2003 on Cancer Screening. *Off. J. Eur. Union* **878**: 34–38.
2. Advisory Committee on Cancer Prevention (2000) Recommendations on cancer screening in the European Union. Advisory Committee on Cancer Prevention. *Eur. J. Cancer* **36**: 1473–1478.
3. European Commission. European Guidelines for Quality Assurance in Cervical Cancer Screening – 2nd Edition. Arbyn M., Anttila A., Jordan J., Schenck U., Ronco G., Segnan N., and Wiener H., eds. (2008), 2nd edn. Office for Official Publications of the European Communities, Luxembourg.
4. Wiener H.G., Klinkhamer P., Schenck U., Arbyn M., Bulten J., Bergeron C., and Herbert A. (2007) European guidelines for quality assurance in cervical cancer screening: Recommendations for cytology laboratories. *Cytopathology* **18**: 67–78.
5. Anttila A., Ronco G., Clifford G., Bray F., Hakama M., Arbyn M., and Weiderpass E. (2004) Cervical cancer screening programmes and policies in 18 European countries. *Br. J. Cancer* **91**: 935–941.
6. Linos A. and Riza E. (2000) Comparisons of cervical cancer screening programmes in the European Union. *Eur. J. Cancer* **36**: 2260–2265.
7. Ballegooijen M., van Marle M.E., Patnick J., Lynge E., Arbyn M., Anttila A., Ronco G., and Habbema D.F. (2000) Overview of important cervical cancer screening process values in EU-countries, and tentative predictions of the corresponding effectiveness and cost-effectiveness. *Eur. J. Cancer* **36**: 2177–2188.
8. Arbyn M. and Dillner J. (2007) Review of current knowledge on HPV vaccination: An appendix to the European Guidelines for Quality Assurance in Cervical Cancer Screening. *J. Clin. Virol.* **38**: 189–197.
9. Wallin K.L., Wiklund F., Angstrom T., Bergman F., Stendahl U., Wadell G., Hallmans G., and Dillner J. (1999) Type-specific persistence of human papillomavirus DNA before the development of invase cervical cancer. *N. Engl. J. Med.* **341**: 1633–1638.
10. Ylitalo N., Sorensen P., Josefsson A.M., Magnusson P.K.E., Andersen P.K., Ponten J., Adami H.-O., Gyllensten U.B., and Melbye M. (2000) Consistent high viral load of human papillomavirus 16 and risk of cervical carcinoma in situ: A nested case-control study. *Lancet* **355**: 2194–2198.
11. Ylitalo N., Josefsson A., Melbye M., Sorensen P., Frisch M., Andersen P.K., Sparen P., Gustafsson M., Magnusson P., Ponten J., Gyllensten U., and Adami H.O. (2000) A prospective study showing long-term infection with human papillomavirus 16 before the development of cervical carcinoma in situ. *Cancer Res.* **60**: 6027–6032.
12. Zielinski G.D., Snijders P.J., Rozendaal L., Voorhorst F.J., van der Linden H.C., Runsink A.P., De Schipper F.A., and Meijer C.J.L.M. (2001) HPV presence precedes abnormal cytology in women developing cervical cancer and signals false negative smears. *Br. J. Cancer* **85**: 398–404.
13. Hamidi A.E., Liu H., Zhang Y., Hamoudi R., Kocjan G., and Du M.Q. (2002) Archival cervical smears: A versatile resource for molecular investigations. *Cytopathology* **13**: 291–299.
14. van der Graaf Y., Molijn A., Doornewaard H., Quint W., van Doorn L.J., and van den Tweel J. (2002) Human papillomavirus and the long-term risk of cervical neoplasia. *Am. J. Epidemiol.* **156**: 158–164.
15. Gunnell A.S., Tran T.N., Torrang A., Dickman P.W., Sparen P., Palmgren J., and Ylitalo N. (2006) Synergy between cigarette smoking and human papillomavirus type 16 in cervical cancer in situ development. *Cancer Epidemiol. Biomarkers Prev.* **15**: 2141–2147.
16. Burghardt E. (1970) Latest aspects of precancerous lesions in squamous and columnar epithelium of the cervix. *Int. J. Gynecol. Obstet.* **8**: 573–580.

17. Burghardt E., Pickel H., and Girardi F. Colposcopy Cervical Pathology. eds. (1998), 3rd revised and enlarged edition. Georg Thieme Verlag, Stuttgart.
18. Boon M.E. and Suurmeijer A.J.H. The Pap Smear. eds. (1993), 2nd edition. Coulomb Press Leyden.
19. Arbyn M., Herbert A., Schenck U., Nieminen P., Jordan J., McGoogan E., Patnick J., Bergeron C., Baldauf J.J., Klinkhamer P., Bulten J., and Martin-Hirsch P. (2007) European guidelines for quality assurance in cervical cancer screening: Recommendations for collecting samples for conventional and liquid-based cytology. *Cytopathology* **18**: 133–139.
20. BSCC. How to take a cervical smear (3rd edition). 2003. Uxbridge, British Society of Clinical Cytology
21. NCCLS (1994) Papanicolaou technique; approved guideline. National Comity for Clinical Laboratory Standards, Pensylvania.
22. Koss L. and Gompel C. Introduction to gynecologic cytopathology with histologic and clinical correlations. Mitchell C.W., eds. (1999), Pradel edn.
23. Howell L.P., Davis R.L., Belk T.I., Agdigos R., and Lowe J. (1998) The AutoCyte preparation system for gynecologic cytology. *Acta Cytol.* **42**: 171–177.
24. Hutchinson M.L., Isenstein L.M., Goodman A., Hurley A., Douglas K.L., Mui K.K., Patten F.W., and Zahniser D.J. (1994) Homogeneous sampling accounts for the increased diagnostic accuracy using the thinprep processor. *Am. J. Clin. Pathol.* **101**: 215–219.
25. Austin R.M. and Ramzy I. (1998) Increased detection of epithelial cell abnormalities by liquid-based gynecologic cytology preparations. *Acta Cytol.* **42**: 178–184.
26. Linder J. and Zahniser D. (1997) The ThinPrep Pap test. A review of clinical studies. *Acta Cytol.* **41**: 30–38.
27. Ferenczy A. and Franco E.L. (1997) Human Papillomavirus DNA testing using liquid-based cytology, in: *New Developments in Cervical Cancer Screening and Prevention* (Franco, E. L. and Monsonego, J., eds.), Blackwell Science, Cambridge, pp. 343–353.
28. Sherman M.E., Schiffman M.H., Lorincz A.T., Herrero R., Hutchinson M.L., Bratti C., Zahniser D., Morales J., Hildesheim A., Helgesen K., Kelly D., Alfaro M., Mena F., Balmaceda I., Mango L., and Greenberg M. (1997) Cervical specimens collected in liquid buffer are suitable for both cytologic screening and ancillary human papillomavirus testing. *Cancer* **81**: 89–97.
29. Solomon D., Davey D., Kurman R., Moriarty A., O'Connor D., Prey M., Raab S., Sherman M.E., Wilbur D., Wright T.C., and Young N. (2002) The 2001 Bethesda System: Terminology for reporting results of cervical cytology. *JAMA* **287**: 2114–2119.
30. Herbert A., Bergeron C., Wiener H., Schenck U., Klinkhamer P.J., Bulten J., and Arbyn M. (2007) European guidelines for quality assurance in cervical cancer screening: Recommendations for cervical cytology terminology. *Cytopathology* **18**: 213–219.
31. Castle P.E., Solomon D., Hildesheim A., Herrero R., Concepcion B.M., Sherman M.E., Rodriguez A.C., Alfaro M., Hutchinson M.L., Terence D.S., Kuypers J., and Schiffman M.A. (2003) Stability of archived liquid-based cervical cytologic specimens. *Cancer* **99**: 89–96.
32. Bergeron C., Cas F., Fagnani F., Didailler-Lambert F., and Poveda J.D. (2006) Human papillomavirus testing with a liquid-based system: Feasibility and comparison with reference diagnoses. *Acta Cytol.* **50**: 16–22.
33. Canfell K., Gray W., Snijders P.J., Murray C., Tipper S., Drinkwater K., and Beral V. (2004) Factors predicting successful DNA recovery from archival cervical smear samples. *Cytopathology* **15**: 276–282.
34. Moberg M., Gustavsson I., and Gyllensten U. (2004) Type-specific associations of human papillomavirus load with risk of developing cervical carcinoma in situ. *Int. J. Cancer* **112**: 854–859.
35. Tarkowski T.A., Rajeevan M.S., Lee D.R., and Unger E.R. (2001) Improved detection of viral RNA isolated from liquid-based cytology samples. *Mol. Diagn.* **6**: 125–130.
36. Cuschieri K.S., Beattie G., Hassan S., Robertson K., and Cubie H. (2005) Assessment of human papillomavirus mRNA detection over time in cervical specimens collected in liquid based cytology medium. *J. Virol. Methods* **124**: 211–215.
37. Powell N., Smith K., and Fiander A. (2006) Recovery of human papillomavirus nucleic acids from liquid-based cytology media. *J. Virol. Methods* **137**: 58–62.
38. Boulet G.A.V., Horvath C.A.J., Berghmans S., Moeneclaey L.M., Duys I.S.M., Arbyn M., Depuydt C.E., Vereecken A.J., Sahebali S., and Bogers J.J. (2008) Cervical cytology biobanking: Quality of DNA from archival cervical Pap-stained smears. *J. Clin. Pathol.* **61**(5), 637–641.
39. Depuydt C., Vereecken A.J., Salembier G.M., Vanbrabant A.S., Boels L.A., van Herck E., Arbyn M., Segers K., and Bogers J.J. (2003)

Thin-layer liquid-based cervical cytology and PCR for detecting and typing human papillomavirus DNA in Flemish women. *Br. J. Cancer* **88**: 560–566.

40. Horvath C.A., Boulet G., Sahebali S., Depuydt C., Vermeulen T., Vanden Broeck D., Vereecken A., and Bogers J. (2007) Effects of fixation on RNA integrity in a liquid-based cytology setting. *J. Clin. Pathol.*
41. Cuschieri K.S., Whitley M.J., and Cubie H.A. (2004) Human papillomavirus type specific DNA and RNA persistence – Implications for cervical disease progression and monitoring. *J. Med. Virol.* **73**: 65–70.
42. Molden T., Nygard J.F., Kraus I., Karlsen F., Nygard M., Skare G.B., Skomedal H., Thoresen S.O., and Hagmar B. (2005) Predicting CIN2+ when detecting HPV mRNA and DNA by PreTect HPV-proofer and consensus PCR: A 2-year follow-up of women with ASCUS or LSIL pap smear. *Int. J. Cancer* **114**: 973–976.
43. Chuaqui R., Cole K., Cuello M., Silva M., Quintana M.E., and Emmert-Buck M.R. (1999) Analysis of mRNA quality in freshly prepared and archival Papanicolaou samples. *Acta Cytol.* **43**: 831–836.
44. Liu H., Huang X., Zhang Y., Ye H., El Hamidi A., Kocjan G., Dogan A., Isaacson P.G., and Du M.Q. (2002) Archival fixed histologic and cytologic specimens including stained and unstained materials are amenable to RT-PCR. *Diagn. Mol. Pathol.* **11**: 222–227.
45. Iftner T. and Villa L.L. (2003) Chapter 12: Human papillomavirus technologies. *J. Natl. Cancer Inst. Monogr.* 80–88.
46. Arbyn M., Sasieni P., Meijer C.J., Clavel C., Koliopoulos G., and Dillner J. (2006) Chapter 9: Clinical applications of HPV testing: A summary of meta-analyses. *Vaccine* **24 S3**: 78–89.
47. Cuzick J., Mayrand M.H., Ronco G., Snijders P., and Wardle J. (2006) Chapter 10: New dimensions in cervical cancer screening. *Vaccine* **24**: 90–97.
48. Arbyn M., Dillner J., Schenck U., Nieminen P., Weiderpass E., Da Silva D. et al. (2007) Chapter 3: Methods for screening and diagnosis, in: *European Guidelines for Quality Assurance in Cervical Cancer Screening* (Arbyn, M., Anttila, A., Jordan, J., Ronco, G., Schenck, U., Segnan, N., Wiener, H. and European Commission, eds.), Office for Official Publications of the European Communities, Luxembourg, pp. 1–69.
49. Coutlee F., Gravitt P., Kornegay J., Hankins C., Richardson H., Lapointe N., Voyer H., and Franco E. (2002) Use of PGMY primers in L1 consensus PCR improves detection of human papillomavirus DNA in genital samples. *J. Clin. Microbiol.* **40**: 902–907.
50. de Roda Husman A.M., Walboomers J.M., van den Brule A.J., Meijer C.J.L.M., and Snijders P.J. (1995) The use of general primers GP5 and GP6 elongated at their 3' end with adjacent highly conserved sequences improves human papillomavirus detection by PCR. *J. Gen. Virol.* **76**: 1057–1062.
51. Söderlund-Strand A., Carlson J., and Dillner J. (2009) Modified general primer PCR system for sensitive detection of multiple types of oncogenic human papillomaviruses. *J. Clin. Microbiol.* **47**: 541–546.
52. Kleter B., van Doorn L.J., Schrauwen L., van Krimpen K., Burger M., ter Harmsel B., and Quint W. (1998) Novel short-fragment PCR assey for highly sensitive broad-spectrum detection of anogenital human papillomaviruses. *Am. J. Pathol.* **153**: 1731–1739.
53. Kleter B., van Doorn L.J., Schrauwen L., Molijn A., Sastrowijoto S., ter Schegget J., Lindeman J., ter Harmsel B., Burger M., and Quint W. (1999) Development and clinical evaluation of a highly sensitive PCR-reverse hybridization line probe assay for detection and identification of anogenital human papillomavirus. *J. Clin. Microbiol.* **37**: 2508–2517.
54. Söderlund-Strand A., Dillner J., and Carlson J. (2008) High-throughput genotyping of oncogenic human papillomaviruses with MALDI-TOF mass spectrometry. *Clin. Chem.* **54**: 86–92.
55. Sotlar K., Selinka H.C., Menton M., Kandolf R., and Bultmann B. (1998) Detection of human papillomavirus type 16 E6/E7 oncogene transcripts in dysplastic and nondysplastic cervical scrapes by nested RT-PCR. *Gynecol. Oncol.* **69**: 114–121.
56. Smits H.L., van Gemen B., Schukkink R., Van der Velden J., Tjong A.H., Jebbink M.F., and ter Schegget J. (1995) Application of the NASBA nucleic acid amplification method for the detection of human papillomavirus type 16 E6-E7 transcripts. *J. Virol. Methods* **54**: 75–81.
57. Martin C.M., Astbury K., and O'Leary J.J. (2006) Molecular profiling of cervical neoplasia. *Expert. Rev. Mol. Diagn.* **6**: 217–229.
58. Keating J.T., Ince T., and Crum C.P. (2001) Surrogate biomarkers of HPV infection in cervical neoplasia screening and diagnosis. *Adv. Anat. Pathol.* **8**: 83–92.
59. zur Hausen H. (2002) Papillomaviruses and cancer: From basic studies to clinical application. *Nat. Rev. Cancer* **2**: 342–350.

60. IARC. Cervix Cancer Screening. IARC Handbooks of Cancer Prevention. Vol. 10. eds. (2005). IARCPress, Lyon.
61. von Knebel D.M. (2002) New markers for cervical dysplasia to visualise the genomic chaos created by aberrant oncogenic papillomavirus infections. *Eur. J. Cancer* **38**: 2229–2242.
62. Sahebali S., Depuydt C., Boulet G.A., Arbyn M., Moeneclaey L.M., Vereecken A.J., van Marck E.A., and Bogers J.J. (2006) Immunocytochemistry in liquid-based cervical cytology: Analysis of clinical use following a cross-sectional study. *Int. J. Cancer* **118**: 1254–1260.
63. Wentzensen N., Bergeron C., Cas F., Eschenbach D., Vinokurova S., and von Knebel D.M. (2005) Evaluation of a nuclear score for p16(INK4a)-stained cervical squamous cells in liquid-based cytology samples. *Cancer* 461–467.
64. Bergeron C., Wentzensen N., Cas F., and Doeberitz M.V. (2006) The p16INK4a protein: A cytological marker for detecting high grade intraepithelial neoplasia of the uterine cervix. *Ann. Pathol.* **26**: 397–402.
65. Klaes R., Friedrich T., Spitkovsky D., Ridder R., Rudy W., Petry U., Dallenbach-Hellweg G., Schmidt D., and von Knebel D.M. (2001) Overexpression of p16(INK4A) as a specific marker for dysplastic and neoplastic epithelial cells of the cervix uteri. *Int. J. Cancer* **92**: 276–284.
66. Nieh S., Chen S.-F., Chu T.-Y., Lai H.-C., and Fu E. (2003) Expression of p16INK4A in Papanicolaou smears containing atypical squamous cells of undetermined significance from the uterine cervix. *Gynecol. Oncol.* **91**: 201–208.
67. Wulfkuhle J.D., Liotta L.A., and Petricoin E.F. (2003) Proteomic applications for the early detection of cancer. *Nat. Rev. Cancer* **3**: 267–275.
68. Kuramitsu Y. and Nakamura K. (2006) Proteomic analysis of cancer tissues: Shedding light on carcinogenesis and possible biomarkers. *Proteomics* **6**: 5650–5661.
69. Yim E.K. and Park J.S. (2006) Role of proteomics in translational research in cervical cancer. *Expert. Rev. Proteomics.* **3**: 21–36.
70. Ozols R.F., Herbst R.S., Colson Y.L., Gralow J., Bonner J., Curran W.J., Jr., Eisenberg B.L., Ganz P.A., Kramer B.S., Kris M.G., Markman M., Mayer R.J., Raghavan D., Reaman G.H., Sawaya R., Schilsky R.L., Schuchter L.M., Sweetenham J.W., Vahdat L.T., and Winn R.J. (2007) Clinical cancer advances 2006: Major research advances in cancer treatment, prevention, and screening – A report from the American Society of Clinical Oncology. *J. Clin. Oncol.* **25**: 146–162.
71. Sasieni P., Adams J., and Cuzick J. (2003) Benefit of cervical screening at different ages: Evidence from the UK audit of screening histories. *Br. J. Cancer* **89**: 88–93.
72. Munoz N., Bosch F.X., de Sanjose S., Herrero R., Castellsague X., Shah K.V., Snijders P.J., and Meijer C.J. (2003) Epidemiologic classification of human papillomavirus types associated with cervical cancer. *N. Engl. J. Med.* **348**: 518–527.
73. Lehtinen M., Dillner J., Knekt P., Luostarinen T., Aromaa A., Kirnbauer R., Koskela P., Paavonen J., Peto R., Schiller J.T., and Hakama M. (1996) Serologically diagnosed infection with human papillomavirus type 16 and risk for subsequent development of cervical carcinoma: Nested case-control study. *BMJ* **312**: 537–539.
74. Dillner J., Lehtinen M., Bjorge T., Luostarinen T., Youngman L., Jellum E., Koskela P., Gislefoss R.E., Hallmans G., Paavonen J., Sapp M., Schiller J.T., Hakulinen T., Thoresen S., and Hakama M. (1997) Prospective seroepidemiologic study of human papillomavirus infection as a risk factor for invasive cervical cancer. *J. Natl. Cancer Inst.* **89**: 1293–1299.
75. Laukkanen P., Koskela P., Pukkala E., Dillner J., Laara E., Knekt P., and Lehtinen M. (2003) Time trends in incidence and prevalence of human papillomavirus type 6, 11 and 16 infections in Finland. *J. Gen. Virol.* **84**: 2105–2109.
76. Bray F., Loos A.H., McCarron P., Weiderpass E., Arbyn M., Moller H., Hakama M., and Parkin D.M. (2005) Trends in cervical squamous cell carcinoma incidence in 13 European countries: Changing risk and the effects of screening. *Cancer Epidemiol. Biomarkers Prev.* **14**: 677–686.
77. Arbyn M., Raifu A.O., Autier P., and Ferlay J. (2007) Burden of cervical cancer in Europe: Estimates for 2004. *Ann. Oncol.* **18**: 1708–1715.
78. Cheung A.N. (2007) Molecular targets in gynaecological cancers. *Pathology* **39**: 26–45.
79. Dillner J., Arbyn M., and Dillner L. (2007) Translational mini-review series on vaccines: Monitoring of human papillomavirus vaccination. *Clin. Exp. Immunol.* **148**: 199–207.
80. Lehtinen M., Kaasila M., Pasanen K., Patama T., Palmroth J., Laukkanen P., Pukkala E., and Koskela P. (2006) Seroprevalence atlas of infections with oncogenic and non-oncogenic human papillomaviruses in Finland in the 1980s and 1990s. *Int. J. Cancer* **119**: 2612–2619.

81. Lehtinen M., Herrero R., Mayaud P., Barnabas R., Dillner J., Paavonen J., and Smith P.G. (2006) Chapter 28: Studies to assess the long-term efficacy and effectiveness of HPV vaccination in developed and developing countries. *Vaccine* **24**: 233–241.
82. Hansson M.G., Dillner J., Bartram C.R., Carlson J.A., and Helgesson G. (2006) Should donors be allowed to give broad consent to future biobank research? *Lancet Oncol.* **7**: 266–269.
83. Hansson M.G. (2005) Building on relationships of trust in biobank research. *J. Med. Ethics* **31**: 415–418.
84. Helgesson G., Dillner J., Carlsson J., Bartram C.R., and Hansson M.G. (2007) Ethical framework for previously collected biobank samples. *Nat. Biotechnol.* **25**: 973–975.
85. Arbyn M., Wallyn S., Van Oyen H., Nys H., Dhont J., and Seutin B. (1999) The new privacy law in Belgium: A legal basis for organised cancer screening. *Eur. J. Health Law* **6**: 401–407.

Chapter 16

Biobanking of Fresh Frozen Tissue from Clinical Surgical Specimens: Transport Logistics, Sample Selection, and Histologic Characterization

Johan Botling and Patrick Micke

Abstract

Access to high-quality fresh frozen tissue is critical for translational cancer research and molecular diagnostics. Here we describe a workflow for the collection of frozen solid tissue samples derived from fresh human patient specimens after surgery. The routines have been in operation at Uppsala University Hospital since 2001. We have integrated cryosection and histopathologic examination of each biobank sample into the biobank manual. In this way, even small, macroscopically ill-defined lesions can be procured without a diagnostic hazard due to the removal of uncharacterized tissue from a clinical specimen. Also, knowledge of the histomorphology of the frozen tissue sample – tumor cell content, stromal components, and presence of necrosis – is pivotal before entering a biobank case into costly molecular profiling studies.

Key words: Biobank, Molecular pathology, Fresh frozen tissue, Cryosection, Quality control

1. Introduction

Most pathology departments at surgical hospital centers have routines for the reception and handling of fresh tissue specimens for frozen section diagnostics. In our experience, a biobanking manual can be based on this established structure to allow collection of unfixed tissue samples from all types of surgical lesions. The procedures were primarily designed for tumor specimens but other lesions, such as vessels affected by atherosclerosis, inflammatory bowel disease specimens, or rejected transplanted organs, can also be processed in the same way. Presently at our hospital, the majority of surgical specimens submitted for histopathologic diagnosis are transported fresh to the department of pathology

Joakim Dillner (ed.), *Methods in Biobanking*, Methods in Molecular Biology, vol. 675,
DOI 10.1007/978-1-59745-423-0_16, © Springer Science+Business Media, LLC 2011

for potential biobanking of fresh frozen tissue. The protocol described here is compatible with the general recommendations on frozen tissue biobanking published by the TuBaFrost and EORTC networks (1, 2).

Diagnostic security is an important concern for all clinical biobanks. Here, the decision to cut and remove tissue from a fresh specimen is placed on the pathologist who finally will report the diagnosis. A well-known limiting factor in frozen tissue biobanking is the appropriate hesitance among pathologists to remove uncharacterized tissue for biobanking from clinical specimens, as this might jeopardize a correct diagnosis and potentially lead to wrong treatment decisions for the patient. To address this problem, cryosection and histologic examination of the tissue samples are performed up-front at the time of entry into the biobank. This leads to sampling of a wider range of specimens from a cohort of operated patients. In most frozen tissue biobanks, certainly in our own historical collections, there is a strong bias toward large, well-demarcated tumors. The possibility to sample macroscopically small and poorly demarcated diffusely growing tumors without a diagnostic hazard will result in less selection bias in translational cancer research imposed by biobanking routines. The frozen sections can also be used to assess the histologic parameters of a biobank sample before entry into a molecular study. The size of the sample, tumor cell content, type of stroma, presence of normal tissue, level of inflammation, and presence of necrosis will decide if a sample is apt for a certain research project or not.

Potential tissue degradation during transport of a specimen from the surgical theater to the biobank facility is an important concern and the subject of much debate. In our experience, most tissue types are stable for hours when transported on wet ice in terms of histomorphology, DNA quality, protein quality, and RNA quality, and work fine in most research applications (3). The lack of published hard data on degradation kinetics during specimen transport regarding DNA, proteins, and RNA for most organ types under relevant experimental conditions argues for a generous attitude from a prospective biobank perspective. That is, samples that can turn out to be valuable in the future should not be excluded from biobanking just because of a long transport time in fresh state. This generous attitude at the input end of a biobank should be coupled to strict quality control at the output end. Tissue quality criteria should be defined for each research project and only samples that meet these criteria should be used. Therefore, assays for DNA, protein, and RNA quality are important elements in a biobanking infrastructure. Issues regarding degradation and quality control assays are discussed in detail in Chapters 10, 17, and 25.

2. Materials

2.1. Surgical Theater

Handling of fresh specimens should be contained to a designated area. To avoid contamination between specimens, blood, tissue remnants, tissue fluids, microorganisms, etc., must be removed from the bench between each handled case. Gloves and instruments used to handle specimens should be changed between each case.

2.2. Department of Pathology: Tissue Selection

"Fresh tissue bench": A stainless steel bench and sink equipped with cutting board, filter paper, scalpel (handle and disposable blades), scissors, forceps, ruler, biopsy punches (3 and 8 mm in diameter, Kai Medical, Japan), tissue ink (Bradley Products, Inc., USA), cryomolds (25 × 20 × 5 mm, Tissue-Tek, Sakura Finetek, USA.), water proof marker pen, cryogel (OCT, Tissue-Tek, Sakura Finetek), thermos filled with blocks of dry ice and isopentane (2-methyl butane), surface disinfectant solution, and instrument disinfectant solution. After handling a case on the bench, the filter papers and instruments should be changed and all surfaces cleaned with soap and water, and disinfected by wiping with a surface disinfectant.

Refrigerator (+4°C). Low-temperature freezer (–80°C). Biobank log book and folder.

2.3. Department of Pathology: Fresh Tissue Biobank

Cryostat equipped with tape-transfer section module (CryoJane, Instrumedics Inc, USA, http://www.instrumedics.com). A CryoJane tape-transfer module can be fitted onto most standard cryostats.

Computer with database software.

Standard histotechnology materials – not listed.

3. Methods

3.1. Surgical Theater: Handling and Transport of Fresh Tissue

1. On the pathology referral chart, note the time point when the specimen is devitalized, i.e., removed from the patient (see Note 1).
2. Place the specimen in a clean plastic bag or wrap it in a clean surgical cloth. Small samples and biopsies can be put in jars or test tubes and immersed in cold saline solution (NaCl 0.9%).
3. Transfer the wrapped specimen, jar, or tube to a bucket partly filled with wet ice for transport at ±0°C. The specimen should not be in direct contact with ice or water during transport.
4. Call the Department of Pathology to inform the technician responsible for the reception of fresh specimens that a specimen

is ready for transport, and if the purpose is biobanking and/or frozen section diagnostics.

5. Send the specimen to the "reception desk" at the Department of Pathology. (a) Use the internal hospital transportation service and ask for "urgent delivery." (b) Delivery from external hospitals is made by external courier. The specimen must arrive the same day during office hours.

3.2. Department of Pathology: Tissue Selection

Upon arrival at the Department of Pathology, the fresh specimens are handled at the designated "fresh tissue bench." The procedure is illustrated in Fig. 1 (see Note 2).

1. Note the time of arrival of the specimen on the referral chart. Register the case in the clinical laboratory management system. Label specimen container and referral chart with the pathology case number.
2. Page the pathologist on call. If there is a delay, place the specimen container in the refrigerator.
3. Unpack the specimen at the "fresh tissue bench." Remove the sample from the container and place it on a clean sheet of filter paper on the cutting board. On the referral chart, note the macroscopic features such as measurements, weight, and description of lesion and surrounding tissue.
4. Cut out tissue pieces representing the lesion and surrounding normal tissue. The cryomold can hold a tissue piece of maximum 15 × 10 × 5 mm, allowing some space for complete cryogel coverage. Avoid contamination between tumor sample and benign tissue sample. Small lesions can be sampled by a punch biopsy probe. If a resection margin is included in the biobank sample, it should be inked for identification in the frozen section.
5. Mark cryomolds with biobank numbers (see Note 2). Place each piece of tissue into a separate cryomold. Cover with cryogel (see Note 3). Pick up the mold with a forceps and snap-freeze in isopentane/dry ice.
6. Note the time of freezing and biobank numbers on the referral chart. Also note the signatures of the pathologist and technician.
7. Transfer the snap-frozen tissue blocks to the low-temperature freezer (–80°C).
8. In the biobank log book, note the pathology case number and the corresponding biobank numbers for the frozen tissue molds. Make a copy of the referral chart and put it in the biobank folder for subsequent registration in the biobank database.

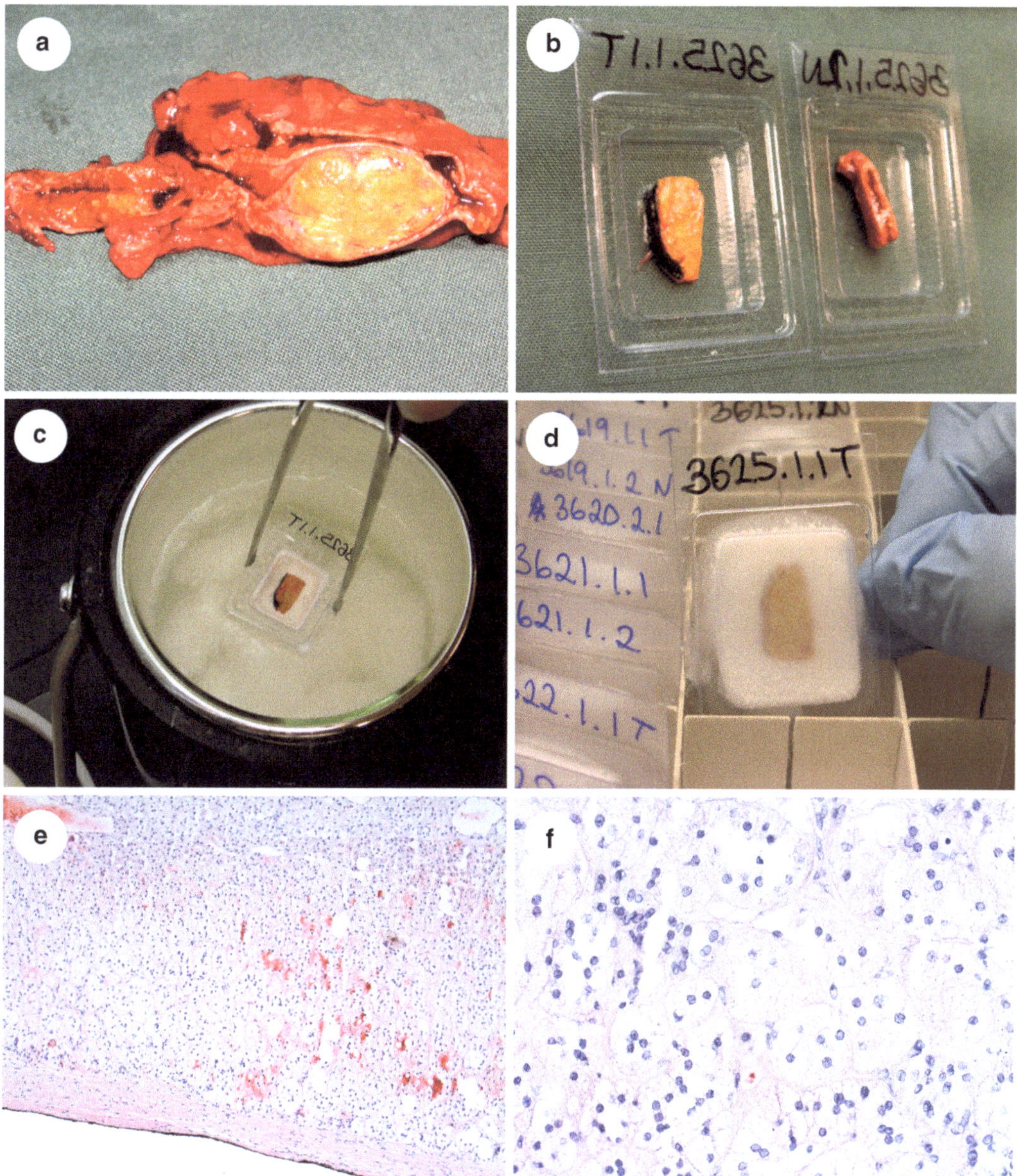

Fig. 1. A fresh adrenalectomy specimen with a yellow soft solid tumor, 2 cm in diameter, was submitted for biobanking (**a**). Samples representing the tumor and normal adrenal gland tissue were removed and put in cryomolds. The resection margin of the tumor was marked by *black ink* (**b**). The cryomolds were filled with cryogel and snap-frozen in isopentane cooled by crushed dry ice (**c**) and transferred to the −80°C freezer (**d**). HE-stained tape-transfer frozen sections were prepared from the frozen tissue blocks. The inked resection margin of the tumor sample is visible at the bottom of the section (**e**) original magnification 10x. The tumor is composed of nests of cells with uniform small nuclei and abundant pale cytoplasm typical of an adrenal cortical adenoma (**f**) original magnification 40x.

9. Return the remaining part of the specimen into the appropriate container and immerse it in buffered formalin for further handling according to standard protocols for fixed tissues.

3.3. Fresh Frozen Tissue Biobank: Cryosection, Histologic Characterization, and Database Registration

1. On one of the following days, the biobank technician makes cryosections (4 μm) of the frozen biobank samples. The CryoJane tape-transfer system is used to produce high-quality frozen sections. For detailed information about this procedure, see the manufacturer's instructions. In short, adhesive tape is used to support the section during cutting and transfer in order to avoid tears and folds. The section is cross-linked onto a coated glass slide using UV light and the tape can then be peeled off.
2. The frozen tape-transfer sections are fixed (Aqueous buffer salt mix, Instrumedics, and 10% final concentration of glutaraldehyde), stained with Hematoxylin–Eosin, mounted, and labeled with pathology case number and biobank number.
3. These biobank slides are then delivered to the pathologist in charge of the case together with the slides produced from the formalin-fixed part of the same specimen.
4. After the histopathologic diagnosis has been reported, the biobank technician registers the case in the biobank database. The entries include patient name, personal identification number, biobank number, age, sex, histopathologic diagnosis, topography and morphology codes according to SNOMED or ICD-O, timepoints for devitalization/arrival at pathology department/freezing, and position in low-temperature freezer. The biobank number serves as the primary code when data and samples are delivered for research use.
5. Optional: Digital microscopic photomicrographs (at 10× and 40× magnification) of the biobank slides can be made. These files can then be linked to the sample entry in the biobank database.

4. Notes

1. The responsibilities of the surgeon, nurse, and other operating room staff members regarding the different steps of specimen handling need to be clearly agreed upon and documented.

   Communication routines between surgical theater and pathology department must be clear and safe.

   After office hours, specimen should be fixed in buffered formalin and handled accordingly.
2. All biobanking protocols relating to clinical specimens need to be integrated with established local and national clinical and diagnostic procedures and must be authorized by relevant medical officials.

If there are other authorized protocols describing the handling and biobanking of specific organ specimens, these should be followed.

Marking of biobank blocks: Each frozen tissue block should have a unique biobank number as follows: biobank case nr: fraction nr: mold nr. Additional letters are optional; for example, T=tumor, N=normal, R=right, and L=left. Example: Case A12345/2005 (pathology case number) was given the biobank number 135. The case arrived in two separate fractions: (1) resected tumor with a margin of surrounding normal tissue and (2) lymph nodes. Tissue samples were frozen as follows:

135:1:1 T = tumor sample
135:1:2 T = another tumor sample
135:1:3 N = normal adjacent tissue
135:2:1 T = lymph node metastasis
135:2:2 N = normal lymph node

As molecular components are extracted from these blocks, traceability is important. For example, if tumor DNA is prepared from case 135 above, it should be documented if block 135:1:1, 135:1:2, or 135:2:1 was used.

3. Storage in cryogel prevents lyophilization of the samples. Many of our own historical samples stored for 10–20 years at –80°C in cryotubes have become freeze-dried and their annotated diagnosis can, therefore, not be confirmed by cryosection. Still, in our experience, the DNA and RNA in these samples are usually intact. Moreover, the orientation of a sample is maintained in a cryogel block and this facilitates downstream applications in microdissection projects where the sectioning plane through the sample needs to be preserved through several rounds of cutting (4). Commercial cryogels are supposedly inert and, in our experience, do not affect the DNA and RNA quality of the samples (3, 5). Similarly, we have not experienced any problems with cryogels in protein detection by immunohistochemistry or immunoflourescence (6). Concerns have been raised at biobank meetings regarding polyethylene glycol containing cryogels and interference with mass spectrometry protocols.

   Sample thawing during the life span of a fresh frozen biobank sample is probably the most important risk factor for tissue degradation (5). To avoid thawing when delivering material for research, a cryostat can be used to slice off an appropriate number of sections for DNA, RNA, or protein extraction (see Chapter 25). Alternatively, if larger portion of a sample is needed, the frozen tissue block can be cracked apart on a cutting board cooled on dry ice. Tissue disruption by cryosectioning is compatible with most extraction protocols. Other methods for tissue homogenization are reviewed by Schmitt et al. (7).

Acknowledgments

The skillful assistance of Simin Tahmasebpoor is gratefully acknowledged. The Fresh Tissue Biobank at Uppsala University was supported by the Swedish National Biobank Platform funded by Wallenberg Consortium North and Swegene.

References

1. Mager, S.R., Oomen, M.H., Morente, M.M., Ratcliffe, C., Knox, K., Kerr, D.J., Pezzella, F., and Riegman, P.H. (2007) Standard operating procedure for the collection of fresh frozen tissue samples. *Eur J Cancer* **43**, 828–34.
2. Morente, M.M., Mager, R., Alonso, S., Pezzella, F., Spatz, A., Knox, K., Kerr, D., Dinjens, W.N., Oosterhuis, J.W., Lam, K.H., Oomen, M.H., van Damme, B., van de Vijver, M., van Boven, H., Kerjaschki, D., Pammer, J., Lopez-Guerrero, J.A., Llombart Bosch, A., Carbone, A., Gloghini, A., Teodorovic, I., Isabelle, M., Passioukov, A., Lejeune, S., Therasse, P., van Veen, E.B., Ratcliffe, C., and Riegman, P.H. (2006) TuBaFrost 2: Standardising tissue collection and quality control procedures for a European virtual frozen tissue bank network. *Eur J Cancer* **42**, 2684–91.
3. Micke, P., Ohshima, M., Tahmasebpoor, S., Zhi-Ping, R., Östman, A., Ponten, F., and Botling, J. (2006) Biobanking of fresh frozen tissue: RNA is stable in nonfixed surgical specimens. *Lab Invest* **86**, 202–11.
4. Micke, P., Kappert, K., Ohshima, M., Sundquist, C., Scheidl, S., Lindahl, P., Heldin, C.H., Botling, J., Ponten, F., and Östman, A. (2007) In situ identification of genes regulated specifically in fibroblasts of human basal cell carcinoma. *J Invest Dermatol* **127**, 1516–23.
5. Botling, J., Edlund, K., Segersten, U., Tahmasebpoor, S., Engström, M., Sundström, M., Malmström, P.U., and Micke, P. (2009) Impact of thawing on RNA integrity and gene expression analysis in fresh frozen tissue. *Diagn Mol Pathol* **18**, 44–52.
6. Dimberg, A., Rylova, S., Dieterich, L.C., Olsson, A.K., Schiller, P., Wikner, C., Bohman, S., Botling, J., Lukinius, A., Wawrousek, E.F., and Claesson-Welsh, L. (2008) alphaB-crystallin promotes tumor angiogenesis by increasing vascular survival during tube morphogenesis. *Blood* **111**, 2015–23.
7. Schmitt, M., Mengele, K., Schueren, E., Sweep, F.C., Foekens, J.A., Brünner, N., Laabs, J., Malik, A., and Harbeck, N. (2007) European Organisation for Research and Treatment of Cancer (EORTC) Pathobiology Group standard operating procedure for the preparation of human tumour tissue extracts suited for the quantitative analysis of tissue-associated biomarkers. *Eur J Cancer* **43**, 835–44.

Chapter 17

Protein Extraction from Solid Tissue

Christer Ericsson and Monica Nistér

Abstract

Maximal extraction and solubilization of protein from diseased or healthy tissue is important to make the whole protein complement available for proteomic analysis. It also helps to maximize reproducibility and to minimize waste. Minimal degradation of the protein amino acid backbone or dephosphorylation is essential to preserve the analytical utility of the extract. Containment of the sample is important to minimize the risk of contamination to and from the sample. The proposed standard protocol for protein extraction and solubilization can result in 98% solubilization of brain tissue, corresponding to about 100 μg protein per mg tissue wet weight, by a frozen disintegration/SDS-based solubilization method: Tissue is crushed in the frozen state in a cryotube by shaking with a sterile steel ball. The crushing is followed by the extraction and solubilization in 2% SDS for 10 min, at 70°C, in a volume corresponding to ten times the tissue wet weight, with shaking. The containment in a cryotube helps to prevent contamination. The treatment with SDS sample buffer can inhibit protease and phosphatase activity. The resulting protein extracts can be used for SDS PAGE, 2-D PAGE, Western blotting, ESI-MS, and ELISA. The proposed standard protocol has the potential to find wide application where protein extraction, solubilization, identification, and quantitation from cryopreserved clinical samples are desirable.

Key words: Protein, Proteome, Phosphorylation, Extraction, Solubilization, Tissue, Analysis, Identification, Quantitation

1. Introduction

Immunohistochemistry and direct mass spectrometry are currently the only methods that can be used to analyze the protein complement of intact tissue. Other types of protein analyses require that the proteins first be released from the tissue (1).

Efficient protein release requires mincing, or homogenizing, the tissue so that the surface exposed to solvent is enlarged. A variety of homogenization methods exist, including handheld or motor-driven homogenizers, sonication, French pressure cell, grinding with alumina or sand, glass bead vortexing, detergent

Joakim Dillner (ed.), *Methods in Biobanking*, Methods in Molecular Biology, vol. 675,
DOI 10.1007/978-1-59745-423-0_17, © Springer Science+Business Media, LLC 2011

lysis, or frozen tissue crushing (2). Based on our studies, we favor freeze crushing in a frozen ball mill apparatus, such as the Mikro-dismembrator (S. B. Braun International). Freeze crushing involves shaking the frozen sample in a capped cryotube with a frozen steel ball. The containment in a cryotube during this phase reduces the risk of sample loss and the risk of contamination. Freeze crushing was as efficient in solubilization as was sonication, the other very efficient homogenization method, and has the added benefit of virtually eliminating the possibility of sample degradation during homogenization (3, 4), since the sample stays frozen during the homogenization.

As the crushed cells are thawed following homogenizing, for further processing, the endogenous proteases and other degrading enzymes gain access to proteins they do not normally access, resulting in the potential for protein degradation (5). This risk is minimized at this stage by using an as low a temperature above freezing as possible and by working quickly. The use of protease and other enzymatic inhibitors should also be considered at this point, but only as a last resort for specific purposes, as they do have drawbacks. The drawbacks include that covalent protease inhibitors, such as AEBSF, alter the migration pattern of some of the proteins they are intended to preserve, and that peptide-like protease inhibitors may interfere with peptide analysis (6). A guide to rational protease and phosphatase inhibitor selection is available (1).

Two of the most efficient ways to solubilize water-insoluble protein subunits is to use either the ionic detergent SDS (7, 8) or the chaotrope urea (9). Both work by denaturing the proteins (although by different mechanisms) and setting them free in solution. Since denaturation prevents proteins from folding into functional structures, the proteins become nonfunctional. Thus, functional studies are not normally possible on material solubilized this way. Instead identification and quantitation benefits from the more complete solubilization that then can be achieved compared to hydrophilic extraction. For functional studies, it is preferred to extract and solubilize the material with hydrophilic buffers (1).

Since there is currently no way to quantitate the initial amount of protein in a piece of tissue, we use percent solubilization as a measure of completeness of extraction and solubilization. The total amount of protein is then measured in the extract.

The very efficiency of SDS (or urea) solubilization suggests that it may be used as a standardized, "universal," protein solubilization for the identification and quantitation (but not for functional studies). While several major proteomic techniques, such as 2-D PAGE, LC-MS and antibody array (10–12), do not tolerate the 2% SDS present in SDS-PAGE sample buffer, and they do tolerate low amounts of SDS. Procedures to dilute the SDS or buffer exchange procedures to reduce the SDS concentration may be used to make SDS extracts compatible with sensitive downstream analytical methods. An added benefit to SDS extraction is that SDS denatures and inactivates many degrading enzymes, thus helping to preserve an intact proteome. We used brain tumor and mouse brain material to evaluate the potential of an SDS-based protein extraction and solubilization procedure (3).

2. Materials

1. Freeze crusher (Mikro-dismembrator S. B. Braun International).
2. Cryo tubes (Corning Incorporated, #2028).
3. Eppendorf tubes (Treff Lab AG, #96.8668.9.01).
4. Shaking heat block (Eppendorf, Thermomixer Compact, #5350 000.013).
5. Eppendorf centrifuge (Centrifuge 5415 D, #5425 000.219).
6. Sample preparation (Amersham Biosciences 2-D Clean-Up Kit: product code 80-6484-51.).
7. Protein concentration assay: BIO-RAD, RC DC Protein Assay #500-0119.
8. Shimadzu BioSpec-1601 E spectrophotometer.
9. 1× SDS sample buffer (8): 0.125 M Tris–HCl (pH 6.8), 2% sodium dodecyl sulfate, 10% glycerol, 0.001% bromophenol blue, 5% 2-mercaptoethanol.

3. Methods

1. A pathologist should inspect all human samples, and secure representative parts for diagnosis. Surplus material is used for this work (see Note 1). Sample cubes, with a side about 5 mm, should be placed in sealed plastic bags and subsequently be snap frozen by immersion in liquid nitrogen (N_2 (l)) and stored at –80°C until used.
2. Sample disintegration is performed by using frozen ball mill grinding (Mikro-dismembrator S. B. Braun International) in a cryogenic tube with a steel ball, followed by shaking in SDS sample buffer in an Eppendorf tube. The samples should be cooled in N_2 (l) prior to, and after, the sample disintegration, and remain frozen during the entire disintegration procedure (see Notes 2–5).

3. Solubilization of the protein content is done in Eppendorf tubes by the addition of 1× SDS buffer ten times the weight of the frozen tissue and subsequent incubation at 70°C and shaking at 1,400 rpm for 10 min (see Notes 6–8). Any remaining solid residue is sedimented at $13.2 \times 10^3 \times g$ for 5 min at room temperature. The supernatant is then transferred to a new labeled tube and the protein concentration measured.
4. Following heating, the sample is spun down so as to concentrate the liquid and sediment any remaining solid material. There should be no visible sediment at this point, indicating complete solubilization. The fraction solubilized can be determined by weighing the tube before and after solubilizing the sample and removing the supernatant. Any added weight would correspond to a solid tissue residue. Divide this weight by the initial wet weight to get the fraction NOT solubilized.
5. The protein concentration is determined, using a commercial assay and using bovine serum albumin as the standard (see Note 9).
6. The solubilized protein samples should be frozen and stored at –20°C until used (see Note 10).
7. Protein samples may in some cases be either diluted, or purified by precipitation followed by resolubilization, in the appropriate buffer for the downstream application.

4. Notes

1. Clinical samples should be handled with the assumption that they may contain known or unknown contagions. Therefore, use barrier protection, as formalized by, for example, the NIH Universal Precautions (http://www.niehs.nih.gov/odhsb/biosafe/univers.htm).
2. The samples can be weighed while frozen so that the correct amount of solubilization buffer can be added after freeze crushing. Work rapidly and use sterile equipment.
3. It is important that the samples be kept frozen at all times, since freezing and thawing would rupture the cell membranes and give degrading enzymes access to substrates they may not see in vivo.
4. Precool the cryotubes and steel balls for the freeze crushing in liquid nitrogen.
5. It will likely be necessary to titer the frequency and time of crushing. The sample should become a fine powder, like powdered sugar. White color is one indication that the sample has

been sufficiently crushed, followed by careful visual inspection of the grain size. Too high frequency or time results in damage to the tube, and the potential to loose or contaminate the sample. A typical protocol is for a frequency of 2,000 Hz and a time of 60 s.

6. After crushing, either centrifuge the samples or tap them, to collect the powder on the bottom of the cryotube.
7. Add 1× SDS sample buffer to the sample to initiate extraction and solubilization. Immediately vortex and confirm that the sample has been dispersed by visual inspection. Continue vortexing and inspecting until the sample is dispersed. This should take less than 1 min. If not, find and remedy the problem.
8. Heat and agitate the dispersed sample in a shaking heat block.
9. It is best to measure the protein concentration in the SDS solubilized sample itself, rather than in a parallel sample without SDS. Few such, SDS-tolerant, procedures exist. We suggest using the RC DC protocol given, as one alternative.
10. Store the sample frozen at –20°C or below. There does not seem to exist any systematic universally recognized studies of the integrity of proteins and posttranslational modifications in SDS and at –20°C. Pending such studies we recommend the above storage conditions as being cost-effective and in accordance with established experience.

Acknowledgment

The authors would like to thank the Wallenberg Consortium North for support.

References

1. Ericsson, C., B. Franzén, and M. Nistér, (2006) Frozen tissue biobanks; Tissue handling, cryopreservation, extraction and use for proteomic analysis. *Acta Oncologica*, **45**, 643–61.
2. Scopes, R.K., (1987) Making an Extract, in Protein Purification: Principles and Practice. Springer-Verlag: New York.
3. Ericsson, C., I. Peredo, and M. Nistér, (2007) Optimized protein extraction from cryopreserved brain tissue samples. *Acta Oncologica*, **46**(1), 10–20.
4. Butt, R.H. and J.R. Coorssen, (2006) Preextraction sample handling by automated frozen disruption significantly improves subsequent proteomic analyses. *Journal of Proteome Research*, **5**(2), 437–48.
5. Olivieri, E., B. Herbert, and P.G. Righetti, (2001) The effect of protease inhibitors on the two-dimensional electrophoresis pattern of red blood cell membranes. *Electrophoresis*, **22**(3), 560–5.
6. Rai, A.J., C.A. Gelfand, B.C. Haywood, D.J. Warunek, J. Yi, M.D. Schuchard, et al., (2005) HUPO Plasma Proteome Project specimen collection and handling: towards the standardization of parameters for plasma proteome samples. *Proteomics*, **5**(13), 3262–77.

7. Laemmli, U.K., (1970) Cleavage of structural proteins during the assembly of the head of bacteriophage T4. *Nature*, **227**(259), 680–5.
8. Laemmli, U.K. and M. Favre, (1973) Maturation of the head of bacteriophage T4. I. DNA packaging events. *Journal of Molecular Biology*, **80**(4), 575–99.
9. Bjellqvist, B., K. Ek, P.G. Righetti, E. Gianazza, A. Gorg, R. Westermeier, et al., (1982) Isoelectric focusing in immobilized pH gradients: principle, methodology and some applications. *Journal of Biochemical and Biophysical Methods*, **6**(4), 317–39.
10. Peng, J., J.E. Elias, C.C. Thoreen, L.J. Licklider, and S.P. Gygi, (2003) Evaluation of multidimensional chromatography coupled with tandem mass spectrometry (LC/LC-MS/MS) for large-scale protein analysis: the yeast proteome. *Journal of Proteome Research*, **2**(1), 43–50.
11. Nielsen, U.B., M.H. Cardone, A.J. Sinskey, G. MacBeath, and P.K. Sorger, (2003) Profiling receptor tyrosine kinase activation by using Ab microarrays. *Proceedings of the National Academy of Sciences of the United States of America*, **100**(16), 9330–5.
12. Harder, A., R. Wildgruber, A. Nawrocki, S.J. Fey, P.M. Larsen, and A. Gorg, (1999) Comparison of yeast cell protein solubilization procedures for two-dimensional electrophoresis. *Electrophoresis*, **20**(4–5), 826–9.

Chapter 18

Collection and Preservation of Frozen Microorganisms

Rosamaria Tedeschi and Paolo De Paoli

Abstract

The storage of the different microorganisms over long periods is necessary to ensure reproducible results and continuity in research and in biomedical processes and also for commercial purposes. Effective storage means that a microorganism is maintained in a viable state free of contamination or genetic drift and must be easily restored without genotypic or phenotypic alterations to its original characteristics and properties. To this end, different techniques have been described and advances in cryopreservation technology have led to methods that allow low-temperature maintenance of a variety of cell types, minimizing the risks of genetic change and are now recommended for long-term storage of most microorganisms.

This chapter summarizes the most important steps and components in the process of low- and ultra-low temperatures freezing of bacteria, parasites, yeasts and fungi, viruses, and recombinant microorganisms.

Key words: Cryogenic preservation, Long-term storage

1. Introduction

Cryogenic preservation is the act of freezing and storing cells at very low temperatures. The effects of the freezing and thawing process on living cells are not fully understood: when water changes from liquid to solid state, cellular metabolism ceases and, when cells are warmed and water returns to liquid, cellular function resumes. Cell cryopreservation process remains the main method of cell preservation to date, and the high survival rates achieved by this method are of interest from both the biophysical and the practical points of view. Storing over long periods of bacterial or fungal strains, parasites, and viruses allows future research study, and it is essential for clinical, epidemiological, educational, microbiological, and commercial reasons. A special approach is required for the proper storage of recombinant microorganisms.

Joakim Dillner (ed.), *Methods in Biobanking*, Methods in Molecular Biology, vol. 675,
DOI 10.1007/978-1-59745-423-0_18, © Springer Science+Business Media, LLC 2011

Effective storage means that the organism is being maintained in a viable state, free of contamination, and without alterations in genotypic or phenotypic characteristics, easily restorable to the same condition it was before preservation. Several researchers attempted to develop methods for 100% preservation of freezing–thawing of diverse cellular specimens (1, 2). This literature also includes detailed information on those microorganisms, which remain recalcitrant to cryopreservation methods or that cannot be preserved by freezing (3–6).

This chapter presents simple and broadly applied methods, from the clinical-microbiology laboratory point of view, which can be used for long period storage of bacteria, protozoa, fungi and viruses, and genetically modified microorganisms. The chapter includes several important steps or components involved in the process of cryopreservation at low- or ultra-low temperatures.

2. Safety Procedures

The infectious and often pathogenic nature of most of the microorganisms, viruses in particular, means that they must be carefully handled by experienced personnel in purpose designed and approved laboratories and following the appropriate safety conditions. Special precautions must be taken during collection, processing, and storage of biological samples. The personnel must be trained and the laboratory must adopt biological safety level II or III, according to the handled pathogens. Guidelines related to biosafety issues are available from the Centre of Disease Control and from the World Health Organization ((7, 8), http://www.cdc.gov; http://www.who.int).

3. Storage Temperature

Liquid nitrogen provides the lowest practical temperatures (-196°C) for storing different sorts of microorganisms and, because viability is preserved so well, it is used extensively. However, because of its relatively expensive cost, there have been many studies on the effect of higher temperatures on survival of various groups or kinds of microorganisms. The critical temperature is dependent on a number of factors, but –70°C appears to be sufficiently low to preserve most organisms (5). Freezing at –20°C is used to keep a few organisms for as long as 1–2 years, most have a poorer survival than at higher or lower temperatures because of the damage caused by ice crystal formation (5) and electrolyte fluctuations (4) at this temperature. At this temperature of storage, the preservation times vary depending on the medium used (5, 9). For long storage periods, microorganisms must be kept at temperatures of −70°C or lower in electric freezers and in liquid nitrogen storage containers. Close observation of the system that must have an adequate alarm mechanism is essential since any increase in temperature will reduce viability and, if thawing occurs, there are no guidelines for rapid restoration of the storage condition.

4. Storage Vials

Storage vials, supplied in a variety of sizes and from different commercial Companies, must withstand very low temperatures and keep their contents sterile. Plastic (polypropylene) vials are easier to use as compared to glass tubes. The most common sizes used are 1.0–1.8 ml tubes; generally, 0.5–1.0 ml of cell suspension is placed into each tube. The most important factors to be considered include cryotolerance (particularly when material has to be stored at liquid nitrogen temperatures), storage conditions, type of cells, and safety conditions. Usually, vials are conveniently packaged in boxes with grid to allow an easy recovery and picking up.

5. Age of Culture

To ensure the genetic stability of a culture, the number of passages from the original must be minimized. It is generally accepted that cells from the maximum stationary phase cultures are more resistant to damage by freezing and thawing than cells from the early or mid-log phase of growth, and the percentage of cells surviving is also increased by an increase in cell density (5).

6. Cryoprotective Agents

Cryoprotective agents, often added to the cellular suspension, serve to protect microorganisms from damage during the freezing process, storage, and thawing. There are two types of cryoprotective agents: ones that enter the cells, delay intracellular freezing, and minimize the solution effects (glycerol, DMSO) and others that protect the external milieu of the organism (sucrose lactose, glucose, mannitol, sorbitol, dextran, polyglycol) (10). Although there are no absolute rules in cryopreservation and many compounds have been tested and reviewed (11), glycerol and DMSO have been widely used and seem to be most effective; generally, the choice of a cryoprotective agent is dependent upon the type of micoorganisms to be preserved (Table 1). Glycerol and DMSO are generally used, after sterilization by autoclaving or filtration, respectively, at a concentration of 5–10% (vol/vol) (12, 13) and are not used together; the optimum concentration varies with the cell type and the highest concentration the cell can tolerate should be used. At the ATCC, mycoplasma and fungi that do not survive lyophilization are frozen in 10% glycerol. Rapid freezing without additives is allowed for long-term survival of protozoa. Of the external products, skim milk is most commonly used. It is purchased from medical product suppliers (e.g. Becton Dickinson, Oxoid) and used, after autoclaving, at a final concentration of 20% (wt/vol) in distilled water, the equivalent of regular milk (3, 14).

7. Preparation of Microorganisms for Freezing

Several factors must be considered when preparing microorganisms for cryopreservation: type of cell, viability, growth conditions, physiological state, and amount of cells. Microbial cells, in particular bacteria and yeasts, grown under aerated conditions show a greater effect of cooling and freezing than nonaerated cells (3). It was demonstrated that cell permeability is greater in aerated cultures and aerated cells dehydrate faster during cooling than nonaerated cells (15).

Most bacteria and yeasts are inoculated in a medium that adequately supports maximal growth; cultures are allowed to mature to late growth or stationary phase before being harvested, generally considering that, the greater the number of cells initially present the greater the recovery. Concentrations of at least 10^7 CFU/ml or higher are recommended (3, 5, 14). Microorganisms can be conveniently harvested from agar slants or plates, or when greater quantities are required, grown in broth cultures and thereafter harvested by centrifugation. In both cases, cells are generally suspended in fresh growth medium containing the cryoprotectant agent and aliquots frozen in volume of 0.2–0.5 ml. *Equilibration* is the period of time between mixing the cryoprotectant with the cell suspension and the beginning of the cooling process. It should take place at room temperature, and it allows time for the cryoprotective agent to penetrate the cells. For most cells, equilibration should occur for at least 15 min but no longer than 45–60 min (as the cryprotective agent may become toxic to the cells). The optimal equilibration time should be determined empirically for the cells being cryopreserved to maximize later recovery. Once the cells and the cryoprotectant have been combined and dispensed into vials, the next step is to cool the

Table 1
Common procedures for ultra-low temperatures preservation of microorganisms (adaptation from ref. 17)

Organism group	Cell amount	Cryopreservative	Storage temperature (°C)	Storage duration (years)
Gram-positive bacteria	10^7/ml	Sucrose, glycerol Skim milk, sucrose, glycerol	–20 –70 to –196	1–3 1–30
Streptococci	10^7/ml	Skim milk Skim milk	–20 –70 to –196	0.2 0.2–1
Mycobacteria	10^7/ml	Skim milk Skim milk	–20 –70 to –196	3–5 3–5
Gram-negative bacteria	10^7/ml	Sucrose, lactose Sucrose, lactose, glycerol	–20 –70 to –196	1–2 2–30
Spore forming bacteria	10^7/ml	Glucose Skim milk, glucose	–20 –70 to –196	1–2 2–30
Fungi hyphae spores	–[a] 10^6/ml	Glycerol, DMSO	–70 to –196	2–30
Protozoa	10^5 to 10^7/ml	Blood, nutrient broth + DMSO + sucrose DMSO or glycerol or blood + nutrient media	–20 to –40 –70 to –196	
Viruses cell free infected cells	–[b] 10^6/ml	DMSO + FCS (10%)	–70 to –196	1–30

[a] Mycelial masses are prepared for freezing of the hyphae of fungi regardless of cell amounts
[b] The number of infectious particles has little effect on the recovery of viruses and bacteriophage

suspension. *The rate of cooling* is another important parameter: it affects the rate of formation and size of ice crystals, as well as the solution effects that occur during freezing. For a wide variety of cells, a uniform cooling rate of 1°C per min from room temperature is effective. An easy-to-use system, designed to achieve a rate of cooling very close to what recommended, is the Nalgene "Mr Frosty" 1°C freezing container. Furthermore, nowadays, different companies provide freezers with computer-driven programmed cooling/thawing rates that allow a strict and effective control within an optimization of the cooling process. Uniform rates of cooling are required by particularly fastidious bacteria and nonsporulating fungi, while most bacteria and spore-forming fungi will tolerate less-than-ideal cooling rates and can be placed and stored at –60°C. Protists often require even more accurate cooling rates to minimize the detrimental effects of under cooling and the heat released during freezing.

8. Freezing Methods

The temperatures at which frozen microorganisms are stored affect the length of time after which they can be recovered. In general, a lower storage temperature implies a longer viable storage period (5). As a good method, it is recommended a slow, controlled-rate freezing at a rate of 1°C per min until the vials cool to temperature of at least –30°C, followed by more rapid cooling until the final storage temperature is achieved (3). When organisms must be stored permanently at –60 to –70°C, the vials can be placed directly into this freezer. However, when organisms are to be stored in liquid nitrogen, it is still recommended that vials be placed initially in a –60°C freezer for 1 h and then transferred to the liquid nitrogen.

9. Thawing and Reconstitution

The critical temperatures for the frozen microrganisms to get damaged occur between –40°C and –5°C. The rapid warming through these temperatures improves recovery rates. Recommendations are to rapidly warm stored material frozen in vials by placing them in a 37°C water bath until ice has disappeared (3, 5). As a general rule, rapid thawing is preferable to slow thawing (http://www.cryobiosystem-imv.com). Once the vial is thawed, the organism should be transferred immediately to an appropriate growth medium, to minimize exposure to the cryoprotective agent.

10. Procedures for Specific Organisms

Procedures for long-term storage of specific microorganims are described below and summarized in Table 1.

11. Storage of Bacteria

Low temperature storage greatly reduces genotypic and phenotypic drift for living bacteria and enables cultures to be used as standards, helping to ensure reproducible results in a series of tests or experiments. After storage and thawing, some bacterial species, characterized by a potentially reduced viability and/or stability of antigenic, molecular, and biochemical properties are defined "fastidious microorganisms" (as *Haemophilus* spp., *Neiseria gonorrhoeae*, *Campylobacter* spp.).

Several factors are critical to the stability and viability of a bacterial culture undergoing cryopreservation, as cell type, age growth conditions, population size, cryoprotectant used, cooling and storage conditions.

11.1. Growth and Preparation for Bacteria Cryopreservation

Bacteria being cryopreserved should be grown under optimal conditions on the recommended media (ATCC website: http://www.atcc.org). Cells can be harvested from broth culture (as pellet after centrifugation and by removing the supernatant), agar plates or slants, and suspended in fresh growth medium containing a cryoprotectant. For most bacteria, a concentration of 10^7 cells/ml is required for good recovery (14); the most commonly used vials are plastic cryovials with volume between 1.2 and 2.0 ml. Growing condition and at which point the bacteria are harvested are the major factors to be considered. Bacteria grown under aerate conditions tolerate the stress of freezing better than statically grown or nonaerated cultures, as reported for *E. coli* (15). Microbial cells harvested in late log or early stationary phase are generally more resistant to the stresses of freezing than younger or older harvested cells (5).

11.2. Cryoprotectant

Glycerol at 10% in culture medium (e.g. Trypticase soy broth, TSB) is recommended by ATCC for freezing nonfastidious bacterial cultures. It is important to respect the equilibration time, leaving the cells at room temperature for a minimum of 15 min, but no longer than 60 min, to ensure that the cryoprotectant has enough time to penetrate (13); during this time the operator can proceed to aliquot cell suspension into the cryovials.

11.3. Cooling and Storage Conditions

Once the cryoprotecant is added, the cultures are ready for cooling. The ideal cooling rate for bacteria is approximately –1°C per min (16) and a programmable-rate freezing apparatus can be used. Most bacteria will withstand less than ideal cooling rates and can be frozen by placing the vials on the bottom of the –60°C to –80°C freezer for 1 h, whereas for "fastidious strains" of bacteria more uniform cooling rates are required. In general, bacteria can be stored at –20°C for 1–3 years, at –70°C for 1–10 years, while freezing in liquid nitrogen at –196°C allows bacteria storage for up to 30 years (Table 1) (17). A simple and useful method based on the use of glass or plastic as carriers to support microorganisms (e.g. Microbank™, by Pro-Lab Diagnostics) offers a practical solution for long-term storage of frozen bacterial suspensions (18, 19). The protocol requires that a sterile vial with the cryopreservative fluid, containing approximately 25 porous beads, is open under aseptic conditions and inoculated with young colonial growth (18–24 h) picked from a pure culture to approximately a 3–4 Mcfarland standard; once closed, the vial should be inverted four to five times to emulsify suspension, without using vortex and, at this point, the microorganisms will be bound to the porous beads; the excess medium should be aspirated leaving the inoculated beads as free of liquid as possible. After inoculation, the cryovials are kept at –70°C for extended storage and when a fresh culture is required, a single bead is easily removed from the vial and used to directly inoculate a suitable bacteriological medium, while the vial returns as soon as possible to low storage temperature. By this approach, frozen bacteria may also survive a freezer failure for one or several days (12). The glass capillary method has proved to be another suitable method for preserving strict anaerobes bacteria (20). Considering the group of "fastidious microorganisms", comparative studies of preservation and storage of *Haemophilus influenzae* evaluated the influence of different suspension media, with or without cryoprotective agents, the supporting material, the initial inoculum concentration, and thawing time at room temperature of stored strains at –20°C and –70°C (21, 22). Recommendations from the results were: (a) starting with a concentrated inoculum; (b) suspending viable cells in MGY, BHI, or TSB supplemented with 25% glycerol; (c) thawing at room temperature for not more than 3 h. Gelatin-based media or TSB added with 10% glycerol was used in preserving at –20°C *N. gonorrhoeae* for at least 1 year (9) and a successive evaluation (23) of similar medium with *Campylobacter jejuni* also shown 85% recovery after 12 months of storage at –20°C, suggesting that this medium may thus be useful for the preservation of a variety of fastidious strains for at least 1 year. *Mycoplasmas* and *Chlamydiae*, despite their peculiar growth and metabolic requirements, can be successfully stored by cryopreservation for up to 10 years (24). *Helicobacter pylori* is also among the bacterial species "difficult to growth" and sensitive to storage conditions as compared to other intestinal bacteria (25), but previous studies reported about comparison of different cryopreservative media and use of particular procedures that allowed recovery of about 60% of *Helicobacter* strains after more than 3 years at –70°C (26, 27). The survival rate of bacteria after storage varied significantly according to species, being for Gram-positive higher than that of Gram-negative bacteria, probably because of their greater resistance of the structure and surface components (28). A wide range of aerobic and facultative anaerobic bacteria may survive long-time storage at –70°C to –80°C, especially if repeated freezing and thawing is prevented.

12. Storage of Fungi, Yeasts and Actinomycetes

There are many preservation protocols suitable for fungi as no individual preservation technique was ever applied to all fungi. Although lyophilization is widely used for preserving fungi and has been the only method used in some laboratories, several years ago ATCC began storing cultures in liquid nitrogen, as some fungi in their collection failed to survive the freeze-drying, in particular fungal cultures without spores (5). This storage method appears to approach ideal, although changes in physiology and genetic stability may occur in some isolates (29) requiring optimal and targeted protocols (30). In particular, fungi that cannot be preserved using traditional preservation methods, termed "preservation recalcitrant fungi", include those that do not readily sporulate in culture (e.g. some members of the Oomycota, some Basidiomycota) and others that are difficult to maintain in culture (e.g. Diplocarpon) or are facultative pathogens (6). Spore-forming fungi require harvesting of spores and suspension of the spores in fresh growth medium with the cryoprotectant. It is very important, during the process of freezing fungal spores, to take care in not delaying the freezing process too long, avoiding germination to occur prior to freezing. For fungi that do not form spores, special procedures for harvesting mycelia prior to freezing must be adopted (13). For fungi with tough mycelia, the culture is harvested from agar growth by cutting and removing agar plugs containing the mycelia and preserving it into fresh growth medium added with the cryoprotectant. Tough mycelia that do not adhere well to agar cultures are grown in broth culture, and the mycelial mass is blended prior to freezing (14). The two approaches aimed at optimizing the cryopreservation of fungi are, first, the use of light cryomicroscope to directly observe the effects of cooling and thawing on the microorganism, and, second, the comparison of the effects of defined preservation regimens on the viability, pathogenicity and morphological, physiological, and genomic stability of replicates before and after preservation (6, 31).

Regarding yeasts in particular, freezing at –80°C with 10% glycerol or 5% DMSO was the recommended method for proper and long storage of *Malassezia* spp. (32). Particular studies were also performed analyzing influence of cooling rate and freeze thaw cycles on yeasts viability during freezing (33).

13. Storage of Parasites

Cryopreservation at ultra-low temperatures has been applied to several species of parasites, such as *Toxoplasma gondii*, *Entamoeba histolytica*, *Trypanosoma cruzi*, *Giardia lamblia*, and *Trichomonas vaginalis*, and represents the method of choice for long-term storage, mostly considering the disadvantages due to propagation and preservation of parasites in vivo or in vitro, the need for labor, initial isolation, and loss of strains, bacterial an fungal contamination, and changes in the original biological and metabolic characteristics (34). The technical procedures for cryopreservation do not differ significantly from those used for the other microorganisms (3, 14). In general, type and concentration of the cryoprotectant and the rates of cooling and thawing are important factors affecting viability after cryopreservation. DMSO showed the highest cryoprotective effects for many species of protozoa at concentrations between 5 and 12.5%; glycerol is also considered an effective cryoprotectant, even if it permeates the cell more slowly than DMSO, requiring a period of equilibration (34); in particular, its use remarkably improves the survival rate of helmints that have never been successfully cryopreserved. Both rapid and slow cooling methods have been used depending on the parasitic species having high or low freezing tolerance, whereas exposure to a rapid thawing method using a water bath at 35–40°C produce better motility or infectivity for all parasites (34). Additional information on cryopreservation and detailed protocols of selected parasitic species may be found (e.g. *E. histolytica* (35), *Blastocystis hominis* (36)). Cryopreservation by vitrification represents another method for long-term storage of parasites, in particular for several species of helminths (37). Blood protozoa, such as *Plasmodium* spp. or *Trypanosoma* spp., can be stored from infected blood samples collected at the peak of parasitemia in liquid nitrogen, using 10% glycerol or 5% DMSO, respectively (17).

14. Storage of Viruses

Most viruses can be frozen as cell-free preparations without difficulties and do not require controlled cooling (14), whereas those viruses in viable infected cells do require it. Viruses are noncellular forms of life with small size, simple structure and absence of free water, and consequently are more stable than other microorganisms. Virus infectivity is retained well at temperature below –60°C, and it is significantly reduced in presence of a slow rise in temperature (more than 5°C). In general, larger viruses tend to be less stable than smaller ones; DNA viruses are more stable than RNA viruses (38). Viruses with envelopes are often less stable than nonenveloped viruses at room temperature, although this is not always evident at low- or ultra-low temperatures. Storage of virus suspension at –20°C is not recommended (5), but can be chosen if retention of virus infectivity is not essential and when the sample stored will be used for serodiagnostic purposes, e.g. as antigen in an enzyme-linked immunosorbent assay (ELISA) test, considering that the antigenic activity at this temperature is maintained. Storage in liquid nitrogen allows viruses to survive almost indefinitely but is not the most convenient and cost-effective method. Furthermore, viral stocks stored in liquid nitrogen containers may be exposed to cross-contamination if not preserved adequately in heat shrinkable tubes, and vapor phase containers are best indicated. Proteins are effective protectants for virus cryopreservation and the suspending medium used is, in general, tissue culture medium containing 10% or greater amount of serum or other proteins. Even the exact mechanisms are not deeply known, proteins keep virus infectivity during freezing, buffering capacity against pH changes (being the optimal pH for virus storage between 7.0 and 8.0), assist in colloidal dispersion of the virus particle and reduce or inhibit other processes that damage nucleic acids (38). About the use of particular cryoprotectants, sucrose-phosphate-glutamate containing 1% bovine albumin (SPGA) and hypertonic sucrose has been reported for storing labile viruses, as Respiratory Syncytial Virus (39). In general, it is good practice to use a high titer virus suspension and preserve it in small aliquots of 0.1–0.5 ml in cryotubes that should be frozen rapidly. Thawing frozen virus samples should be carried out just before the virus is to be used and rapidly by placing the cryotubes in a water bath at 37°C. Specimens known to contain viral pathogens may be kept at –70°C to –85°C for several years with reasonable recovery and without changes into the morphological characteristics as documented for enteric viruses in stool specimens (40). Viruses can be stored at ultra-low temperatures without particular treatment also in tissues or blood derived-specimens (serum, plasma, lymphocytes). Storage of peripheral blood lymphocytes in RPMI 1640 added with 10% fetal bovine serum and 10% DMSO is a commonly used method for HIV isolation (41). The effects of multiple freezing and thawing of serum specimens on acid nucleic stability, however, must be taken into account. Plasma stored for up to one year at –70°C showed quite stable levels of HIV RNA (42, 43) and viral DNA from TT virus, and HBV was still valuable by qualitative PCR after seven cycles of freezing and thawing (44). Viruses can be also preserved for very long period as nucleic acids. RNA must be precipitated with ethanol, which inhibit the enzymes that breakdown RNA, while DNA can be stored either under ethanol or as dried pellet. This method has several advantages since virus as nucleic acid can be frozen in extremely small volumes, in many aliquots, without the need for large volumes of storage capacity and almost indefinitely, but is not widely used.

15. Storage of Genetically Modified Microorganisms

Genetically modified organisms are widely used for both research and industrial production purposes and consequently regular programs of long-term viability, good preservation, and plasmid retention testing are important requirements. In general, such microorganisms can be preserved in a manner similar to the unmodified host cell, in most of cases (45, 46). In a recent study, cultures of recombinant *E. coli* strains, cryogenically frozen and stored at −80°C, showed stable viability and high plasmid retention over a period of up to 11 years; lower viability and/or plasmid retention may be likely due to the improper selection of initial colonies (47). The absence of gross structural instability of transfected sequences might be verified at least by restriction patterns analysis of the strains. Different parameters for the preparation and cryopreservation of recombinant microorganisms need to be considered, as suggested for recombinant cultures of *Saccharomyces cerevisiae* (48). Cryoprotectant concentrations of 2–5% glycerol in water resulted in optimal pre- and post-freezing recovery rates; the addition of the amino acids in the cryoprotectant media appeared to have a protective effect during deep freeze storage. Storage temperatures of −70°C or below and rapid thawing allows good recovery rates.

References

1. Albrecht, R.M., Orndorff, G.R., MacKenzie, A.P. (1973) Survival of certain microorganisms subjected to rapid and very rapid freezing on membrane filters. *Cryobiology* **10**, 233–9.
2. Dumont, F., Marechal, P.A., Gervais, P. (2003) Influence of cooling rate on *Saccharomyces cerevisiae* destruction during freezing: unexpected viability at ultra-rapid cooling rates. *Cryobiology* **46**, 33–42.
3. Alexander, M., Daggett, P.M., Gherna, R., Jong, J., Simione, F. (1980) American Type Collection Methods. Laboratory manual on preservation, freezing, and freeze-drying as applied to algae, bacteria, fungi and protozoa. American Type Culture Collection, Rockville, MD, p. 1–46.
4. Gherna, R.L. Preservation, p. 208–217. In: P. Gerhardt, R.G.E. Murray, R.N. Costilow, E.W. Nester, W.A. Wood, N.R. Krieg, and G.B. Phillips, Eds. Manual of Methods for General Bacteriology. ASM Press, Washington, DC (1981).

5. Heckly, R.J. (1978). Preservation of microorganisms. *Adv. Appl. Microbiol.* **24**, 1–53.
6. Smith, D., Ryan, M.J. Current status of fungal collections and their role in biotechnology. In: Handbook of Fungal Biotechnology. Marcel Dekker Published, 271 Madison Ave, NY (2003).
7. Biosafety in Microbiological and Biomedical Laboratories. U.S. Dept. of Health and Human Services and National Institutes of Health. Ed. J.Y. Richmond and R. McKinney, IV Edition. (1999).
8. Laboratory Biosafety Manual. II Edition (revised). WHO, Geneva (2003).
9. Harbec, P.S., Turcotte, P. (1996). Preservation of *Neisseria gonorrhoeae* at –20 degrees C. *J. Clin. Microbiol.* **34**, 1143–6.
10. Farrant, J. General observations on cell preservation. In: M.J. Ashwood-Smith and J. Farrant, Eds. Low Temperatures Preservation in Medicine and Biology. pp. 1–18. Pitman Medical Limited, Kent, England (1980).
11. Hubalek, Z. (2003) Protectants used in the cryopreservation of microorganisms. *Cryobiology* **46**, 205–29. Review.
12. Pell, P.A., Sneath, P.H. (1984) A note on the survival of bacteria in cryoprotectant medium at temperatures above 0 degrees C. *J. Appl. Bacteriol.* **157**, 165–7.
13. Simione, F.P. Cryopreservation Manual, Nunc Company, Rochester, New York (1992).
14. ATCC Preservation Methods: Freezing and Freeze Drying. Ed. F.P. Simione and E.M. Brown, American Type Culture Collection, Rockville, Maryland (1991).
15. Nei, T., Araki, T., Matsusaka, T. Freezing and injury to aerated and non aerated cultures of *Escherichia coli*. In: T. Nei, Ed. Freezing and Drying of Microorganisms. University of Tokyo Press, Tokyo, Japan (1969).
16. Mazur, P., Leibo, S., Chu, E. (1972) A two-factor hypothesis of freezing injury. *Exp. Cell Res.* **71**, 345.
17. Reimer, L., Carroll, K. Procedure for the storage of microorganisms. In: E. Murray, E. Baron., M. Pfaller, F. Tenover, and R. Yolken, Eds. Manual of Clinical Microbiology. pp. 67–73. ASM Press, Washington, DC (2004).
18. Feltham, R.K.A., Power, A.K., Power, Pell, P.A., Sneath, P.H.A. (1978) A simple method for storage of bacteria at –76 degrees C. *J. Appl. Bacteriol.* **44**, 313–316.
19. Jones, D. et al. Maintenance of bacteria on glass beads at –60°C to –76°C. In: Kirsop/Doyle, Eds. Maintenance of Microorganisms and Cultured Cells, II Edition. pp. 45–50. Academic Press, London (1991).
20. Hippe, H. Maintenance of methanogenic bacteria. In: Kirsop/Doyle, Eds. Maintenance of Microorganisms and Cultured Cells, II Edition. Academic Press, London (1991).
21. Aulet de Saab, O.C., de Castillo, M.C., de Ruiz Holgado, A.P., de Nader, O.M. (2001) A comparative study of preservation and storage of *Haemophilus influenzae*. *Mem. Inst. Oswaldo Cruz.* **96**, 583–6.
22. Votava, M., Stritecka, M. (2001) Preservation of *Haemophilus influenzae* and *Haemophilus parainfluenzae* at –70 degrees C. *Cryobiology* **43**, 85–7.
23. Gorman, R., Adley, C.C. (2004) An evaluation of five preservation techniques and conventional freezing temperatures of –20 degrees C and –85 degrees C for long-term preservation of *Campylobacter jejuni*. *Lett. Appl. Microbiol.* **38**, 306–10.
24. Furr, P.M., Taylor-Robinson, D. (1990) Long-term viability of stored mycoplasmas and ureaplasmas. *J. Med. Microbiol.* **31**, 203–6.
25. Ohkusa, T., Miwa, H., Endo, S., Okayasu, I., Sato, N. (2004) *Helicobacter pylori* is a fragile bacteria when stored at low and ultra-low temperatures. *J. Gastroenterol. Hepatol.* **19**, 200–4.
26. Shahamat, M., Paszko-Kolva, C., Mai, U.E., Yamamoto, H., Colwell, R.R. (1992) Selected cryopreservatives for long term storage of *Helicobacter pylori* at low temperatures. *J. Clin. Pathol.* **45**, 735–6.
27. Spengler, A., Gross, A., Kaltwasser, H. (1992) Successful freeze storage and lyophilisation for preservation of *Helicobacter pylori*. *J. Clin. Pathol.* **45**, 737.
28. Miyamoto-Shinohara, Y., Imaizumi, T., Sukenobe, J., Murakami, Y., Kawamura, S., Komatsu, Y. (2000) Survival rate of microbes after freeze-drying and long-term storage. *Cryobiology* **41**, 251–5.
29. Ryan, M.J., Jeffries, P., Bridge, P.D., Smith, D. (2001) Developing cryopreservation protocols to secure fungal gene function. *Cryo Letters* **22**, 115–24.
30. Smith, D., Thomas, V.E. (1998) Cryogenic light microscopy and the development of cooling protocols for the cryopreservation of filamentous fungi. *World J. Microbiol. Biotechnol.* **14**, 49–57.
31. Smith, D. (2001) Provision and maintenance of micro-organisms for industry and international research networks. *Cryo Letters* **22**, 91–6.
32. Crespo, M.J., Abarca, M.L., Cabanes, F.J. (2000) Evaluation of different preservation

and storage methods for *Malassezia* spp. *J. Clin. Microbiol.* **38**, 3872–5.

33. Dumont, F., Marechal, P.A., Gervais, P. (2006) Involvement of two specific causes of cell mortality in freeze-thaw cycles with freezing to -196 degrees C. *Appl. Environ. Microbiol.* **72**, 1330–5.
34. Miyake, Y., Karanis, P., Uga, S. (2004) Cryopreservation of protozoan parasites. *Cryobiology* **48**, 1–7.
35. Samarawickrema, N.A., Upcroft, J.A., Thammapalerd, N., Upcroft, P. (2001) A rapid-cooling method for cryopreserving *Entamoeba histolytica. Ann. Trop. Med. Parasitol.* **95**, 853–5.
36. Suresh, K., Init, I., Reuel, P.A., Rajah, S., Lokman, H., Khairul Anuar, A. (1998) Glycerol with fetal calf serum – a better cryoprotectant for *Blastocystis hominis. Parasitol. Res.* **84**, 321–2.
37. James, E.R. (2004) Parasite cryopreservation by vitrification. *Cryobiology* **49**, 201–10. Review.
38. Gould, E.A. (1999) Methods for long-term virus preservation. *Mol. Biotechnol.* **13**, 57–66.
39. Law, T.J., Hull, R.N. (1968) The stabilizing effect of sucrose upon respiratory syncytial virus infectivity. *Proc. Soc. Exp. Biol. Med.* **128**, 15–518.
40. Williams, F.P. Jr. (1989) Electron microscopy of stool-shed viruses: retention of characteristic morphologies after long-term storage at ultralow temperatures. *J. Med. Virol.* **29**, 192–5.
41. Gallo, D., Kimpton, J.S., Johnson, P.J. (1989) Isolation of human immunodeficiency virus from peripheral blood lymphocytes stored in various transport media and frozen at -60 degrees C. *J. Clin. Microbiol.* **27**, 88–90.
42. Sebire, K., McGavin, K., Land, S., Middleton, T., Birch, C. (1998) Stability of human immunodeficiency virus RNA in blood specimens as measured by a commercial PCR-based assay. *J. Clin. Microbiol.* **36**, 493–8.
43. Winters, M.A., Tan, L.B., Katzenstein, D.A., Merigan, T.C. (1993) Biological variation and quality control of plasma human immunodeficiency virus type 1 RNA quantitation by reverse transcriptase polymerase chain reaction. *J. Clin. Microbiol.* **31**, 2960–6.
44. Durmaz, R., Otlu, B., Direkel, S. (2002) Effect of multiple freezing and thawing of serum on TT virus and hepatitis B virus DNA positivity. *Arch. Virol.* **147**, 515–8.
45. Nierman, W.C., Feldblyum, T. (1985) Cryopreservation of cultures that contain plasmids. *Dev. Ind. Microbiol.* **26**, 423–34.
46. Nierman, W.C., Trypus, C., Deaven, L.L. (1987) Preservation and stability of bacteriophage lambda libraries by freezing in liquid nitrogen. *Biotechniques* **5**, 724–27.
47. Koenig, G.L. (2003) Viability of and plasmid retention in frozen recombinant *Escherichia coli* over time: a ten-year prospective study. *Appl. Environ. Microbiol.* **69**, 6605–9.
48. Schu, P., Reith, M. (1995) Evaluation of different preparation parameters for the production and cryopreservation of seed cultures with recombinant *Saccharomyces cerevisiae. Cryobiology* **32**, 379–88.

Chapter 19

Handling of Solid Brain Tumor Tissue for Protein Analysis

Christer Ericsson and Monica Nistér

Abstract

Optimal protein analysis requires unfixed tissue samples. We suggest handling the brain tumor tissue sterilely and coldly (on ice) for as short time as possible prior to processing, but for no more than 8 h. This simple protocol results in apparently intact morphology, immunoreactivity, protein integrity, and protein phosphorylation with the criteria we apply. Sample handling for Pathological Anatomical Diagnosis (PAD) and for protein analysis can be one and the same.

Key words: Brain, Tumor, Handling, Protein, Analysis, Proteomics, Biobank

1. Introduction

Traditional clinical pathological sample handling involves immersing the samples in formaldehyde cross-linking solution that fixes the macromolecules in place adequately for light microscopical and immunohistochemical examination. Experience has shown this to give excellent results for clinical Pathological Anatomical Diagnosis (PAD). With the recent developments in protein identification technology, particularly mass spectrometry, protein analysis has acquired an increased relevance for translational research, and has a large potential to help provide a more precise diagnosis and to suggest new therapy targets (1). Protein analysis is currently best performed on tissue that has not been fixed. Optimal handling of tissue for protein preservation has, therefore, become a priority.

The exact time a tissue can be maintained after removal from the body without significant alterations in the proteome is still an outstanding question. The possibilities of transplantation for some organs demonstrate that those organs can survive functionally for at least several hours ex vivo. In the survival process ex vivo,

Joakim Dillner (ed.), *Methods in Biobanking*, Methods in Molecular Biology, vol. 675,
DOI 10.1007/978-1-59745-423-0_19, © Springer Science+Business Media, LLC 2011

the cells would be expected to initially attempt protective reactions, including alterations in gene transcription, translation, and protein activity, before potentially succumbing to cell death and finally tissue lysis. As soon as devascularization starts, the tissue is potentially subject to artifactual alteration and then degradation of its metabolite pools. Complete preservation of metabolite levels would require "instant" enzyme inhibition, by microwaving live tissue (2) or using similar quick methods, and cannot currently be incorporated into a standard human tissue biobank scheme. Changes in gene transcription and translation would be expected to occur in response to the altered metabolite levels. But the extent of the response may be limited by the declining availability of metabolites necessary for the transcription and translation. Finally, the tissue will undergo lysis as a result of one or more of the cell death processes (3). An investigation into optimal handling conditions would include a decision on what criteria are relevant for protein analysis and then to perform time course experiments with those chosen criteria as readouts. It should be kept in mind that an optimal handling protocol is only as good as the criteria applied, and that future investigations may well necessitate more stringent handling paradigms. On the contrary, the handling paradigm should not require more stringent handling than is supported by the evidence. The best scenario would be to find a window of time, temperature, and other conditions that preserve the proteome apparently intact and constant, compared to what can be deduced to be the status in vivo. This scenario would allow valid comparisons of tissues from different donors, after compensating for any interindividual variability, to identify physiological or pathophysiological differences to better understand the mechanisms of a disease. A somewhat lesser scenario would be to accept certain alterations since they cannot at present be prevented with reasonable measures, but to document them and their time course, so they can be compensated for in the analysis.

In a surgical setting, the tissue will need to be transported to a pathology department where a pathologist will inspect the tissue, cut out suitable samples for histopathological diagnosis, and then freeze the remaining tissue for future use. It is, therefore, inevitable that some time will pass between surgery and sample freezing. The maximum allowable time before freezing would, in our opinion, logically follow from the maximum time that results in intact proteins by the criteria of undetectable proteolysis of marker proteins, undetectable abundant protein degradation, undetectable degradation of posttranslational modifications, and apparently intact morphology and immunoreactivity of the tissue.

Indications from human autopsy material show that samples can be taken *postmortem* and usefully studied by light microscopical and immunological staining. Human brain myelin basic protein is only subjected to minimal loss after 48 h (4, 5). The level of carboxymethylation of proteins from human post-mortem brain obtained within 24 h of death is not significantly different to the level present at the time of death (6). Recent studies show that mRNA is remarkably well preserved in intact tissue (7). Further studies of the temperature and time-dependence of the parameters listed appear warranted. One such study is forthcoming and is the basis for the below recommendation (Ericsson, Orrego, Peredo and Nistér).

We have studied the effect of clinically relevant times and temperatures on the integrity of normal brain and of glial brain tumor material. Our goal was to establish a handling routine that would accommodate clinical requirements of both neurosurgeons and neuropathologists and at the same time provide samples with a high degree of preservation of protein sequence, posttranslational modifications, and tissue morphology. To this end we performed during morphological, immunohistochemical, SDS-PAGE, Western blot, phosphoprotein, and 2-D PAGE experiments. Based on these studies, we provide a suggestion for a Brain Tumor Handling Standard Operating Procedure for clinical diagnosis and for protein analysis. All patient samples should have the appropriate ethical permission and informed consent. In the absence of evidence that more relaxed freezing routines are adequate, tissue for proteomic analysis should be snap-frozen in liquid nitrogen at –196°C or in isopentane cooled with dry ice at –80°C as quickly as possible, and stored at –80°C or below.

2. Materials

1. Wet icemaker.
2. Thermal insulation container.
3. Sterile disposables.
4. Liquid nitrogen.
5. Cryotubes (Nunc 375418), freezer boxes for cryotubes, –80°C freezers.
6. Freezer-safe, bar-coded, labels.
7. Sample database.

3. Methods

1. Unfixed tissue should be regarded as potentially infectious and be handled with barrier protection, according to, for example, the NIH universal precautions (http://www.niehs.nih.gov/odhsb/biosafe/univers.htm) (see Note 1).

2. The time of the start of resection should be noted, to allow calculation of the total devascularized time.
3. The tissue samples should be placed in sterile plastic containers under sterile conditions in the operating room, without delay, and the time, or time span, noted (see Note 2).
4. The sterile containers should be chosen to fit the sample size (see Note 3). The container should be fully immersed in wet ice.
5. The cold container should be transported to the department of Pathology without delay.
6. The samples should be filed for diagnosis.
7. From this point, a pathologist is responsible for the material. The pathologist should inspect the patient samples and remove representative parts for diagnosis (see Notes 4 and 5). Handling should be performed under sterile conditions and on a cold (4°C) surface.
8. The surplus representative material allocated by the pathologist can be used for protein analysis. This material should be cut into cubes with a side of about 5 mm to be suited for storage in cryotubes and for frozen disintegration for protein extraction and solubilization. As many samples as is deemed appropriate should be cryopreserved.
9. The samples should subsequently be snap-frozen by immersion in liquid nitrogen (N_2 (l)).
10. The samples should be frozen within 8 h of devascularization (see Note 6).
11. Make a note of the freezing time, so that the *actual* time between resection and freezing can be calculated.
12. The frozen samples should be transferred without thawing to a 1.8-ml prelabeled, precooled cryotube (see Note 7).
13. The sample and cryotube are to be held in liquid nitrogen until they can be transferred to –80°C storage.
14. The cryopreserved material is to be stored at –80°C in labeled boxes until used (see Note 8).
15. Make entries into sample database to document sample, handling parameters, and the location of the sample.
16. Maintain adequate security for samples and documentation.

4. Notes

1. Barrier protection will minimize contamination both from and to the sample.

2. The tissue should be handled sterilely to preserve the possibility to culture or FACS sort from the material (before freezing), in addition to biochemical analysis.
3. The sterile transportation container size should be matched to the resected sample size. It should not be so small as to deform the tissue, and not so big so as to impede the heat transfer from sample to ice.
4. The material for diagnosis should be treated according to the locally established protocol and is not further commented upon here.
5. It is essential that a trained pathologist is in charge of selecting and distributing available material for the different uses. This is the only way to safeguard the diagnostic material and to assure that representative material gets distributed.
6. It is important to note the actual time between resection and freezing, even if the target time is exceeded. The samples may still be useful for a range of analyses that do not require optimal handling.
7. It is convenient to use freezer-safe barcode labels on the cryotubes.
8. While the given handling protocol should provide adequate preservation, it remains important to verify the integrity of the material. For protein analysis of glioblastoma tissue, we suggest to analyze the ratio of apparent Glial Fibrillary Acidic Protein (GFAP) degradation products to the apparent intact protein as a sensitive indication of degradation. Other tissue will need other sensitive markers.

Acknowledgments

The authors would like to thank the Wallenberg Consortium North for financial support.

References

1. Ericsson, C., B. Franzén, and M. Nistér, (2006) Frozen tissue biobanks; Tissue handling, cryopreservation, extraction and use for proteomic analysis. *Acta Oncologica*, **45**, 643–61.
2. Cosi, C. and M. Marien, (1998) Decreases in mouse brain NAD+ and ATP induced by 1-methyl-4-phenyl-1, 2,3,6-tetrahydropyridine (MPTP): prevention by the poly(ADP-ribose) polymerase inhibitor, benzamide. *Brain Res*, **809**(1), 58–67.
3. Kroemer, G., W.S. El-Deiry, P. Golstein, M.E. Peter, D. Vaux, P. Vandenabeele, et al., (2005) Classification of cell death: recommendations of the Nomenclature Committee on Cell Death. *Cell Death Differ*, **12**(2), 1463–7.
4. Ansari, K.A., A. Rand, H. Hendrickson, and M.D. Bentley, (1976) Qualitative and quantitative studies on human myelin basic protein in situ with respect to time interval between

death and autopsy. *J Neuropathol Exp Neurol*, **35**(2), 180–90.

5. Berlet, H.H. and B. Volk, (1980) Studies of human myelin proteins during old age. *Mech Ageing Dev*, **14**(1–2), 211–22.
6. Goggins, M., J.M. Scott, and D.G. Weir, (1998) Regional differences in protein carboxymethylation in post-mortem human brain. *Clin Sci (Lond)*, **94**(6), 677–85.
7. Micke, P., M. Ohshima, S. Tahmasebpoor, Z.P. Ren, A. Ostman, F. Ponten, et al., (2006) Biobanking of fresh frozen tissue: RNA is stable in nonfixed susrgical specimens. *Lab Invest*, **86**(2), 202–11.

Chapter 20

Blood Plasma Handling for Protein Analysis

Christer Ericsson and Monica Nistér

Abstract

Blood handling routines have been worked out that result in consistent protein analytic results in clinical practice. It would seem reasonable to build on this experience when devising handling routines for new protein biomarker discovery. Consequently, normal blood sample handling precautions apply to blood sample handling for new biomarker discovery. The blood sample handling protocol mentioned below describes room temperature, or 4°C, platelet poor EDTA plasma collected within 90 min of venipuncture, handled, and screened to eliminate hemolysis. DNA can be isolated from the "buffy coat" that results as blood cells are sedimented to isolate the plasma.

Key words: Blood, Plasma, Handling, Protein, Analysis, Proteomics

1. Introduction

1.1. Functional Protein Integrity in Blood

The possibility of blood transfusion is one indication that the functional protein properties of blood can be maintained ex vivo for a considerable amount of time. The functionality, however, degrades with time. Changes occurring in whole blood during storage result in significant deterioration of clotting factors and ultimately in a total loss of function of granulocytes and platelets (1). Even so, a red cell concentrate may be stored for up to 21 days if kept at 4°C, without loss of apparent function of the red cells. Platelets may be stored for up to 72 h at 22°C without apparent loss of hemostatic function, but not at 4°C, since survival time at 4°C is significantly lower than that at 22°C. Furthermore, the stability of the hemostatic function, and platelet morphology, is enhanced in citrate coagulation-inhibited plasma compared to that in EDTA coagulation-inhibited plasma. Preserving active clotting factors requires processing within 6 h of collection (1). The intracellular ion potassium increases in plasma over time,

Joakim Dillner (ed.), *Methods in Biobanking*, Methods in Molecular Biology, vol. 675,
DOI 10.1007/978-1-59745-423-0_20, © Springer Science+Business Media, LLC 2011

indicating an inability of cells to maintain ion gradients across their cell membranes and a leakage of intracellular K^+. A typical clinical requirement, therefore, is for the blood sample to be less than 4 h old, and not to be hemolytic, to measure K^+ ions in plasma accurately (http://provtagningsanvisningar.karolinska.se/). Thus indications are that the proteins that are functional in blood remain so for at least 4–6 h ex vivo. This functionality would be a sensitive and broad-based bioassay for blood plasma functional protein integrity. Rapid freezing of plasma preserves the labile coagulation factors V and VIII (1). Rapid cryopreservation of plasma in less than 4–6 h following blood collection would, therefore, seem like a reasonable guideline for *functional* protein integrity.

1.2. Current Protein Biomarkers in Blood

A protein biomarker is a protein, or fragment thereof, whose altered quantity indicates a particular disease state. Presence of a protein biomarker indicates a change in expression, or state, of a protein that correlates with the risk of acute or chronic morbidity, with progression, or with susceptibility to a treatment. These proteins would be expected to vary physiologically in concentration between individuals. The reference interval is a measure of the variation that includes 95% of the variation in normal individuals, i.e., some apparently normal individuals show concentrations outside the reference interval. A separate measure, the discriminator value, or cut-off, indicates a concentration that discriminates healthy from sick individuals.

Albumin is the single most abundant protein in blood plasma, constituting about half of the cell-free protein by weight. The 22 most abundant proteins constitute about 99% of the dry weight of cell-free proteins in plasma. The difference in abundance between the most abundant and the least abundant known functional proteins of current clinical importance, albumin and the cytokines, is about 10–11 orders of magnitude (2). Changes in abundant plasma protein concentrations largely result from alterations in synthesis by hepatocytes in response to circulating inflammation-associated cytokines, the acute phase response (3). Other systemic changes include a tendency toward cachexia and thromboembolism in cancer (4, 5). Given that blood proteins physiologically vary in concentration and that a change of approximately 25% in plasma concentration has been suggested as a definition of an acute phase protein (3), it is clearly sometimes difficult to say if a difference in abundance corresponds to normal variation or can be used as a disease marker. This difficulty may have contributed to the significant number of initially suggested biomarkers that subsequently fail to be validated (6). It should be clear that the relatively nonspecific, systemic, alterations in the blood proteins that correspond to the acute phase response, cachexia, or increased risk of thromboembolism may well, in combination with more specific markers, ultimately find their way into a multiplexed disease-specific diagnostic assay.

Normal tissue turnover results in the formation of tissue degradation products that are subsequently eliminated through the blood. The abundance of the proteins released by pathologic tissue turnover would be expected to depend on the type of disease process, the total mass of pathologic tissue that turns over per unit time, and the abundance of the protein in the tissue (7). The larger tissue mass that turns over per unit time, the higher the concentration of degradation products in blood, all else being equal. Proteins that are released from the pathologic tissue that are characteristic of that tissue and that are abundant enough to be detected in blood would constitute potential biomarkers. It would be preferable that the biomarkers used for diagnosis are causal of the disease or the response.

Quantitation of several protein biomarkers of disease from blood samples is used in clinical practice (http://provtagningsanvisningar.karolinska.se/). The techniques employed include plasma protein electropherograms that assess any systemic inflammatory response among the most abundant plasma proteins. The increased appearance of specific characteristic proteins in blood can also serve as an indicator of specific tissue damage. Examples include aspartataminotransferase (ASAT) and alaninaminotransferase (ALAT) to assess liver damage, and creatinkinase (CK), troponin I and T (TnI and TnT), and myoglobin to assess heart tissue damage as an indication of heart infarction. These particular analytes are assessed in Li-heparin coagulation-inhibited plasma and require processing within 2–8 h. Other protein biomarkers for cancer, such as alpha fetoprotein, cancer-associated antigen 15-3 (CA 15-3), CA 19-9, CA 72-4, CA 125, S100 B, carcinoembryonic antigen (CEA), and prostate specific antigen (PSA), are measured in postcoagulation serum or Li-heparin plasma without specified time limit to analysis. EDTA or heparin plasma, or serum processed in 2–8 h or less, following collection would, therefore, seem like a reasonable guideline for current biomarker protein integrity.

1.3. Establishing Optimal Conditions for Screening for New Biomarkers in Blood

The emergence of high-throughput, multiplexed protein analysis is expected to provide a platform for discovering new protein biomarkers in a time-efficient and comprehensive fashion. This potential, however, largely remains to be realized. The reasons may include that not enough time has passed for the early biomarker candidates to be validated on a large scale, the limited sensitivity of the current high-throughput technologies (8, 9), physiologic variability, and the variability in handling and analysis of blood samples.

There seems to be a good reason to use the existing experience from clinical chemical practice and blood transfusion in devising optimal conditions for screening for new biomarkers in blood. That experience would indicate that a preanalytic lag of 2 h at room temperature would be acceptable. It should, however, be clear that multiplex analysis potentially is more demanding than analysis of single or a few analytes, since the possibility exists that individual critical analytes degrade with a unique time course. In addition, the sample collection should be as economical on personnel and financial resources as possible in order to make the collection of large sample volumes possible. This would specifically mean not to require more stringent collection requirements than is supported by evidence. This review discusses the current state-of-the-art protocol for standardized blood sample handling in maintaining consistent analytic utility of blood plasma for protein analysis.

It would be preferable if blood samples could be collected continuously in a healthcare system according to reasoned and standardized protocols, in order to obtain large numbers of representative and comparable samples with known patient outcomes. As discussed above, there exist protocols for clinical assays for protein biomarkers that specify the use of either (postcoagulation) serum, EDTA plasma, citrate plasma, or heparin plasma. Citrate and EDTA inhibit coagulation by chelating divalent cations, which subsequently inhibit enzymes involved in blood clotting. Heparin functions through the activation of antithrombin III. One of the current challenges is to determine which *one* blood derivative would be optimal in most cases. The top choice may not need to be optimal for all conceivable analytes, as long as the artifacts introduced can be documented. The Plasma proteome Project (PPP) of the Human Proteome Organization (HUPO) has determined that for protein analysis, EDTA coagulation-inhibited plasma is preferable to citrate- or heparin-inhibited plasma and to serum (10, 11). Platelet depletion was found to be beneficial in reducing contamination with platelet-derived proteins (12). Serum production causes reduction in proteins involved with clot formation and an increase in the number of detectable peptides by about 40% (12) and has been shown to be difficult to standardize. Nevertheless, it should be recalled that several of the current cancer biomarkers are measured in serum. Therefore, serum should probably not be recommended for biomarker screening, but may, after validation studies, be well suited for targeted assays of individual analytes.

The original assessment of a preference for EDTA plasma left open the maximum allowable time before separation of plasma and blood cells for optimal results, and also which temperature is optimal for most studies. It would seem preferable that a standardized blood plasma handling paradigm be based on systematic studies of the influence of time, temperature, and coagulation status and any other relevant variable on defined, identified, analytes, and that any changes observed fitted into a reasonable hypothesis of cause and effect. Specifically, higher temperatures may conceptually favor degradation, while low temperature may cause cold-induced activation of platelets in certain buffers and potentially subsequent release of platelet-derived proteins.

Several recent studies have addressed the question of changes in the protein composition of serum and plasma during processing, some looking at time before separation of plasma or serum (13–15) and some looking at effects of incubation after separation (13, 16). They have used chromatographic fractionation of the more abundant blood proteins and recorded any differences between various handling parameters by mass spectrometry. The methods used generate mass-to-charge spectra of peptides and proteins, where the identity of the corresponding protein is initially unknown. Clearly, the findings in those cases need to be validated with other technologies and sample sets to identify individual analytes and fit any changes into reasonable hypotheses, especially since it is noted that it can be difficult to obtain stable, reproducible SELDI-TOF MS results (17).

Serum and plasma samples show differences in their protein spectra, supporting the HUPO PPP conclusion that a choice between blood derivatives is necessary (13, 14, 16). The effect of visibly detectable hemolysis was readily detectable in the protein pattern, but overnight fasting, or not, had no apparent effect on the observed proteome (16).

In one study, many of the changes in the pre-centrifugation serum samples were readily apparent within 30 min of venipuncture, whereas virtually all significant changes in the plasma samples did not occur until 4 h after venipuncture (14). Many of the observed serum peaks arose directly from platelets or during coagulation, as determined by comparisons. In contrast, another study concluded that serum quality is compromised only if it is left to clot at room temperature for more than 3 h, or more than 24 h at 4°C (15). The reason for the discrepancy is not clear, but may be methodological. In a third study, it is concluded that keeping blood at room temperature for 1, 6, or 24 h changes the proteome profile considerably (13).

Postcentrifugation serum samples showed profound time-dependent changes in proteome profiles compared with EDTA or heparin plasma samples. Most changes within MS spectra occurred after a storage time of 4 h at room temperature (16). In contrast, another study found that only minimal changes of the serum proteome were noted within 6 h of incubation, while the changes became observable after 8 h of incubation. For serum and plasma samples stored up to 24 h at 4°C, the proteomes did not present with significant changes (13).

While it would seem like these data need to be followed up using identified proteins and specific reagents for each potential marker, they generally support the conclusion that serum preparation is difficult to standardize and that there are smaller changes in the plasma proteomes, than in the serum proteome, in the first hours. It will be interesting to test the hypothesis originally derived from the experience from transfusion and existing biomarker analysis that a preanalytic lag of about 2 h at room temperature would be acceptable, with respect to protein integrity, cellular integrity, and platelet activation.

1.4. Future Developments in Blood Handling and Validation

Unless new physical or chemical principles are brought to bear on blood plasma handling, we cannot at present reasonably expect any major improvements in existing blood handling paradigms. On the contrary, with the current state of information, we cannot even say that they unequivocally are needed. What we can expect is an examination and a validation of the current framework, and an optimization based on those findings. It currently remains to be determined what the optimal time and temperature would be for blood processing, and to develop validation markers for blood handling. One such study is underway (Ismail et al., manuscript in preparation). Subsequently, we can expect studies of the stability of individual analytes under the conditions of that optimal handling protocol, to serve as a reference for the study of alterations in disease.

2. Materials

1. Personal barrier protection, gloves, face shield, protective clothing
2. EDTA blood sample tubes
3. Crushed wet icemaker
4. Centrifuge capable of achieving 2500 RCF at 4°C or room temperature
5. One milliliter handheld pipette with disposable sterile tips
6. Sarstedt Filtropur S 0.2-μm filter (No./REF 83.1826.001)
7. Cryotube 1.8 ml (NUNC 375418)
8. Freezer-safe barcoded labels
9. Liquid nitrogen
10. –80°C freezer
11. Sample database
12. Container for safely disposing of potential biohazardous materials

3. Method

Based on (12).

1. The blood sample should be taken as a venous sample, without clenching or pumping of the fist (see Note 1).
2. Blood samples are collected into four 10-ml K_2EDTA plastic tubes, and inverted carefully ten times to distribute the anticoagulant (see Note 2).
3. The tubes should *either* be precooled on ice *or* remain at room temperature at all times.
4. Note the time of collection.
5. Allow the blood samples taken at room temperature to cool to room temperature for 20 min (see Note 3).
6. Note the time of cooling, if applicable.
7. The plasma is separated from the cells by centrifugation for 10 min at 2000 RCF at either 4°C or at room temperature (see Notes 1, 4, 5).
8. From each tube, 2 ml of the supernatant is removed and subjected to filtration through a low protein-binding 0.2-μm filter to remove any remaining platelets or other cells.
9. The "buffy coat" containing white cells, at the interface between the plasma supernatant and the erythrocyte pellet, can be saved separately as a source of DNA (see Notes 6, 7, 8).
10. The plasma samples are aliquoted into 1.8 ml cryovials and frozen by immersion in liquid nitrogen, without delay (see Notes 8, 9).
11. All aliquoting and freezing should be complete within 90 min.
12. Note the actual time of freezing (even if target times are exceeded), so actual processing time can be calculated. Note the temperature: 4°C or room temperature.
13. Enter data into database in accordance with legal and contractual requirements.
14. The plasma and "buffy coat" should be stored at –80°C.
15. Dispose of biological material waste in accordance with local requirements.

4. Notes

1. Apply NIH universal precautions to avoid contamination from or to the sample.

2. Sample collection at room temperature is less demanding than that at 4°C, and so should be used consistently if a reliable cold collection is not practically possible. For the sake of consistency, it is important to use one protocol or the other, and not to mix protocols.
3. Slow cooling avoids water condensation that may cause osmotic hemolysis.
4. Centrifugation is only partially effective in sedimenting platelets. Platelets are a major source of released proteins. Decant the plasma at the top of the tube to minimize contamination with platelets.
5. For an even more complete removal of platelets, filtration is added as a second step. The filtration should be performed with filters having sufficiently small pores to remove platelets and made from a low protein-binding material.
6. Carefully observe the buffy coat layer atop the erythrocytes as you aspirate it, in order to optimize the yield and purity. Buffy coat can be collected in the same kind of cryotubes as plasma.
7. We recommend snap freezing the buffy coat in liquid nitrogen.
8. Snap-freeze in liquid nitrogen while holding the tubes upright with forceps to keep the cap free of frozen liquid.
9. Apply local precautions when handling liquid nitrogen.

References

1. Blajchman, M.A., F.A. Shepherd, and R.A. Perrault, (1979) Clinical use of blood, blood components and blood products. *Can Med Assoc J*, **121**(1): 33–42.
2. Anderson, N.L. and N.G. Anderson, (2002) The human plasma proteome: history, character, and diagnostic prospects. *Mol Cell Proteomics*, **1**(11): 845–67.
3. Gabay, C. and I. Kushner, (1999) Acute-phase proteins and other systemic responses to inflammation. *N Engl J Med*, **340**(6): 448–54.
4. Boddaert, M.S., W.R. Gerritsen, and H.M. Pinedo, (2006) On our way to targeted therapy for cachexia in cancer? *Curr Opin Oncol*, **18**(4): 335–40.
5. Naschitz, J.E., D. Yeshurun, S. Eldar, and L.M. Lev, (1996) Diagnosis of cancer-associated vascular disorders. *Cancer*, 77(9): 1759–67.
6. Finlay, J.A., E.W. Klee, C. McDonald, J.R. Attewell, D. Hebrink, R. Dyer, B. Love, G. Vasmatzis, T.M. Li, J.M. Beechem, and G.G. Klee, (2006) A Systematic method for selection of promising serum protein biomarkers to improve prostate cancer (PCa1) detection. *Clinical Chemistry*, **52**(11): 2159-2162.
7. Diehl, F., M. Li, D. Dressman, Y. He, D. Shen, S. Szabo, et al., (2005) Detection and quantification of mutations in the plasma of patients with colorectal tumors. *Proc Natl Acad Sci U S A*, **102**(45): 16368–73.
8. Ericsson, C., Franzén, B. and Nistér, M., (2006) Frozen Tissue Biobanks; Tissue Handling, cryopreservation, extraction and use for proteomic analysis. *Acta Oncologica*, **45**: 643.
9. Hortin, G.L., (2006) The maldi-tof mass spectrometric view of the plasma proteome and peptidome. *Clin Chem*, **52**(7): 1223-37.
10. Omenn, G.S., D.J. States, M. Adamski, T.W. Blackwell, R. Menon, H. Hermjakob, et al., (2005) Overview of the HUPO Plasma Proteome Project: results from the pilot phase with 35 collaborating laboratories and multiple analytical groups, generating a core dataset of 3020 proteins and a publicly-available database. *Proteomics*, **5**(13): 3226–45.

11. Rai, A.J., C.A. Gelfand, B.C. Haywood, D.J. Warunek, J. Yi, M.D. Schuchard, et al., (2005) HUPO Plasma Proteome Project specimen collection and handling: towards the standardization of parameters for plasma proteome samples. *Proteomics*, **5**(13): 3262–77.
12. Tammen, H., I. Schulte, R. Hess, C. Menzel, M. Kellmann, T. Mohring, et al., (2005) Peptidomic analysis of human blood specimens: comparison between plasma specimens and serum by differential peptide display. *Proteomics*, **5**(13): 3414–22.
13. Hsieh, S.Y., R.K. Chen, Y.H. Pan, and H.L. Lee, (2006) Systematical evaluation of the effects of sample collection procedures on low-molecular-weight serum/plasma proteome profiling. *Proteomics*, **6**(10): 3189–98.
14. Banks, R.E., A.J. Stanley, D.A. Cairns, J.H. Barret, P.C. Clarke, D. Thompson, and P.J. Selby, (2005) Influences of Blood Sample Processing on Low-Molecular-Weight Proteome Identified by Surface-Enhanced Laser Desorption/Ionization Mass Spectrometry. *Clinical Chemistry*, **51**(9): 1637–49.
15. West-Nielsen, M., E.V. Hogdall, E. Marchiori, C.K. Hogdall, C. Schou, and N.H. Heegaard, (2005) Sample handling for mass spectrometric proteomic investigations of human sera. *Anal Chem*, 77(16): 5114–23.
16. Findeisen, P., D. Sismanidis, M. Riedl, V. Costina, and M. Neumaier, (2005) Preanalytical impact of sample handling on proteome profiling experiments with matrix-assisted laser desorption/ionization time-of-flight mass spectrometry. *Clin Chem*, **51**(12): 2409-11.
17. Karsan, A., B.J. Eigl, S. Flibotte, K. Gelmon, P. Switzer, P. Hassell, et al., (2005) Analytical and preanalytical biases in serum proteomic pattern analysis for breast cancer diagnosis. *Clin Chem*, **51**(8): 1525–8.

Chapter 21

Biobank Informatics: Connecting Genotypes and Phenotypes

Jan-Eric Litton

Abstract

The sequencing of the human genome, completed at the dawn of the twenty-first century, allows researchers to integrate new data on genetic risk factors with demographic and lifestyle data collected via modern communication technologies. The technical prerequisites now exist for merging these cascades of molecular genetic information, not only to national health registers, but also to epidemiology and clinical data.

Long-term storage of biological materials and data is a critical component of any epidemiological or clinical study. In designing Biobanks, informatics plays a vital role for the handling of samples and data in a timely fashion. Biobank Informatics contains important elements concerning definition, structure, and standardization of information that has been gathered from a multitude of sources from population-based registries, biobanks, patient records, and from large-scale molecular measurements.

Keywords: Biobank, Biological samples, Data integration, Biobank Information Management System, Informatics

1. Introduction

The move toward a universal information infrastructure for population-based biobanking is directly connected to the issues of semantic interoperability through standardized message formats and controlled terminologies. Incorporating genetic, medical, and lifestyle information will position biobanking as a powerful resource to help researchers unravel the origins of important diseases. Data sources with different formats have been set up to support various aspects of genomics, proteomics, epigenetic, lifestyle information, etc. The database infrastructure has become a critical component in life-sciences research. The explosion of genotype data requires that data are properly

Joakim Dillner (ed.), *Methods in Biobanking*, Methods in Molecular Biology, vol. 675,
DOI 10.1007/978-1-59745-423-0_21, © Springer Science+Business Media, LLC 2011

loaded, accessed, managed, queried, analyzed, and shared. Longitudinal research over a long period of time, for generations of researchers, demands completely new methods and systems for gathering and storing genotype and phenotype information. The population-based biobanks bring to the fore the problems concerning the need for standardized research data and a long-term storage strategy.

To collect biological samples like blood, tissues, and whole cells is not enough for competitive biobanking; we need to organize the information about the samples, including registration of disease-related information in order to investigate the properties of the organism as a whole and of the molecular and cellular makeup of the tissues. The combined use of lifestyle surveys, associated biological samples, and relevant registers will help us identifying possible links between genetic disposition and disease. The successful and systematic collection of demographic and lifestyle data is central to the process of any epidemiological or clinical study. However, traditional surveys have difficulties with capturing the changes in our lifestyle. Tools that are better suited for dynamic populations are needed. The Internet presents a powerful alternative for collection of data, but several intrinsic features remain unexplored.

The lack of standardization is a general problem, which restricts the utilization of many biological samples. In spite of several large-scale projects and global achievements in standardization, there are still niche areas of informatics that are largely isolated. The awareness of the problems being tackled, the progress being made, and the possible solutions that each niche could offer to one another in support of common goals is limited. Specific key issues include proteomic profiles, sampling procedures, and storage conditions of samples, etc. Standard protocols for different types of samples, so that these can be utilized also in the future, as well as making the protocols public are essential. The same is true for lifestyle factors where standard protocols will make comparisons of results from different cohorts simpler.

An important purpose of biobank informatics is identifying the complete scope of information structures needed and analyzing how available nomenclature and coding systems can be used for storing and retrieving biobank information. Several controlled terminologies and coding systems may be used for organizing the information about biobanks. Some of them were originally created for other purposes and encompass only parts of the information needed for comprehensive coverage. If these data-management requirements can be met, each biobank will be a supreme resource for epidemiological and clinical studies.

2. Biobank Information Management Challenges

The nature and format of data associated with biobank present significant scientific and informatics challenges in the development of an integrated coherent dataset. Moreover, the size of data collected in large-scale genetic epidemiology studies amounts to hundreds of billions of genotypes, together with extensive epidemiological outcome data. To deal with the complexity of biobanking, biobank informatics can be divided into six different parts; Sample management, Data integration, Data collection, Data query, Security and Administration management, and Data analysis (Fig. 1).

The challenge facing the biobank community is well described in a special study published by IDC (1).

2.1. Sample Management Systems

In a biobank, thousands or millions of samples are stored in different kinds of physical stores, e.g., nitrogen tanks, low-temperature freezers, refrigerators, and room-temperature stores. To be able to handle the information, a LIMS (Laboratory Information Management System) is used to keep track of samples and link them to the study donors (2). The LIMS can also be used to handle analysis results. Some LIMS download data directly from the analysis equipment and link the results to the samples in the database. It can also be used to control the robots handling the samples.

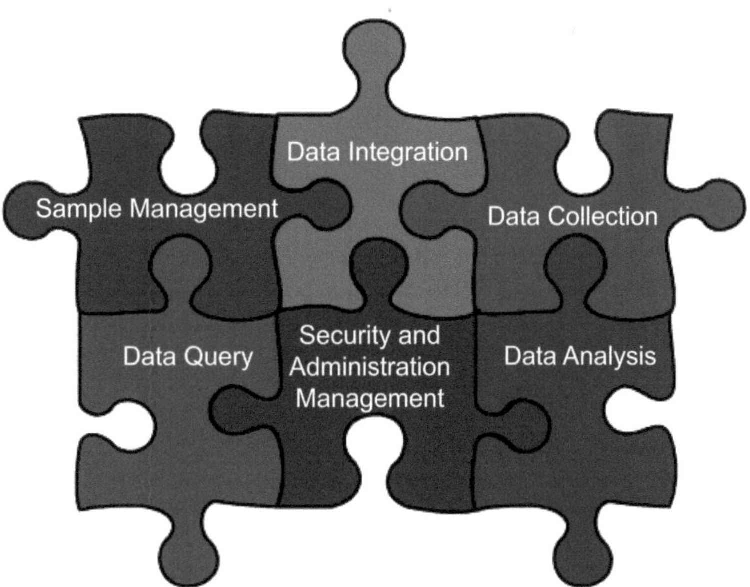

Fig. 1. Biobank informatics is defined by the six different parts: Sample management, Data integration, Data collection, Data Query, Security and Administration management, and Data analysis.

In the LIMS, each entity (e.g., sample, sample donor) must have a unique identity: these should be given automatically. The use of barcodes minimizes the risk of introducing errors during sample handling. Barcodes also enable automatic handling of the samples by robots (aliquoting, storing, etc.). To be able to keep track of the origin and keep sample identities unique, it could be important to introduce an identifier for each participating country/organization, so that all identifiers are unique in a world-wide perspective. The recommended approach to identifier roots is to use an internationalized root, such as an ISO country code, followed by a unique organization or biobank identifier.

In large-scale sample collection studies, standardized operating procedures (SOPs) are necessary for ensuring the quality of the samples and the data produced. A sample management system, like LIMS, can be a standalone system in a biobank, or be clustered in a network of many LIMS sharing a common controlled terminology, SOPs, and coding systems. Such a network can connect freezers in an institute or hospital, locally or nationally, but can also connect information between countries. LIMS with web functionality independent of a common Intranet facilitates such a system. This is described below in Subheading 3.

2.2. Data Integration

Data integration is one of the major challenges in all studies, where an extensive amount of genotype and phenotype data are combined from different data sources located in different places and computers. When integrating data from many data sources, a range of issues must be handled; some of these are listed with a short description below. We will focus on two different data integration aspects since these play a vital role in the information management of medical data.

A major concern when integrating data from different data sources is that the information is structured in completely diverse ways. Data models usually range from fully normalized relational-data models to very simple spreadsheet-like models or plain text with no relations. Mechanisms must be built to provide secure, harmonized, and fair data exchange and compute infrastructure, so that the database system is kept flexible and can be adapted easily to various requirements.

There are two main routes in integrating data for sharing:

- The first is integrating data from different data sources that should be *stored for a long time*, for the next generation of research – a biobank repository like the Karolinska Institutet Biobank Information Management System (BIMS) (3)
- The second is integrating a lot of data from different data sources *for a limited time* – for a project, like the GenomEUtwin project, which uses TwinNET (4)

In this article we will delve into these two approaches. However, both systems use *data federation* (5).

Traditionally, data have been collected in studies, like UK Biobank (6) or the Mayo Clinic (7), into one centralized repository, a data warehouse, using a strict data-submission protocol. In the BIMS and TwinNET, a complementary approach was chosen where data could be accessed on demand using direct database connections. This strategy gives more flexibility in how data can be distributed and how data management and harmonization work can be shared. In the federated database system, remote database tables, or data sources in general, are mapped into corresponding local database objects, Fig. 2. This is advantageous since remote data sources can be accessed from the same-relation database management system using standard SQL (Structured Query Language) and application programming interfaces. This simplifies system management and application development significantly.

In database federation, all features, which are integral parts of modern database systems like the DB2, Oracle or SQL Server, are available without extra investments. Sophisticated query optimizers can be used to find optimal query paths by weighting connection speeds with other statistics collected from local and remote systems. The database federation complements data warehousing. Data are collected into one database management system and often transformed in a way that data can be queried rapidly. Database federation offers users the opportunity to combine data from multiple sources in a single query. By "wrapping" the actual sources, extensibility and encapsulation are provided as well. Data sources could be both structured (relation database, Excel, XML (8), etc.) and/or unstructured data (medical records, etc.).

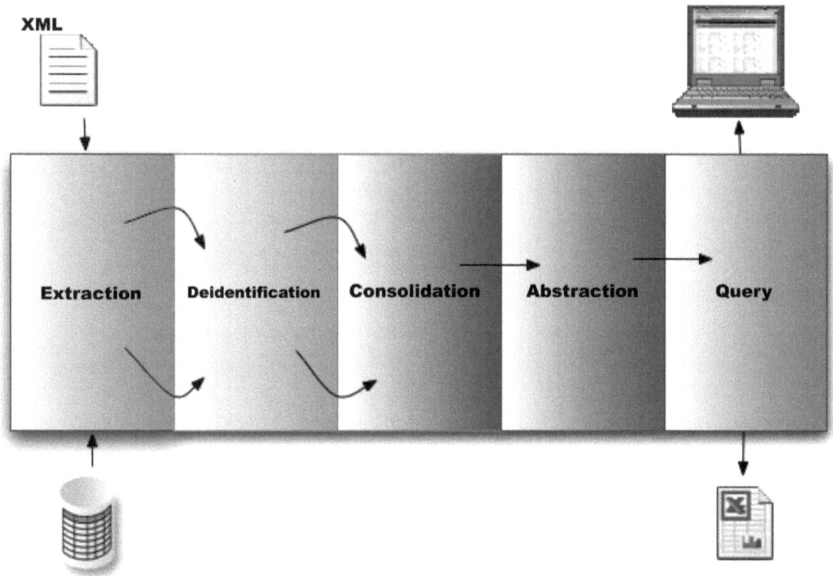

Fig. 2. *Left* The flow of data through BIMS. Starting on the, information is processed through the extraction, deidentification, consolidation, abstraction, and query steps.

2.2.1. BIMS

The BIMS (3) offers services to support the collecting, managing, searching, and querying of research data for long-term storage. A key service of BIMS is to allow researchers to post queries via a graphical user-friendly web interface toward several of BIMS data sources simultaneously. The BIMS solution supports a number of sequential processes that together constitute the BIMS framework (Fig. 3):

- Extraction
- Deidentification
- Consolidation
- Abstraction and query

Extraction: Data input comes from a number of original data sources not under direct control of the biobank administration. The data sources can be in different formats, for example, relational databases, Excel spreadsheets, SAS, R and common text files, and be physically spread over several locations. The information itself

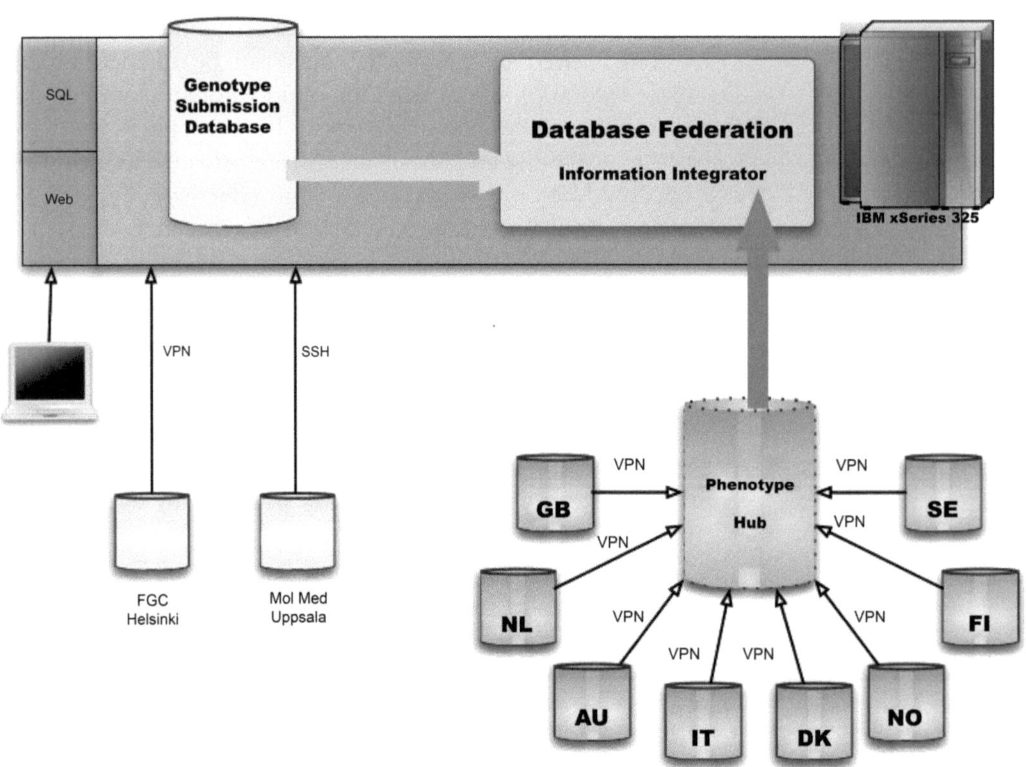

Fig. 3. General topology of TwinNET. Twin registries are the data provides and are connected to a Hub using direct database connections. The Hub provides a single database access point for the data using the DB2 and Discovery Link (also called Information Integrator) bundle, together called the WebSphere Federation Server. Federated data forms remote databases that can be shared through the DB2 database management system as any other data stored locally into a relational database.

can be organized in vastly different data models. In addition, there are huge differences in the expertise of people responsible for the data, in the data sources and in how often the data are updated with new information.

New data are added to the system using either of the two methods:

- Federated database connections
- XML (Extensible Markup Language) documents

Data from a federated data source are copied directly and stored inside the BIMS repository using IBM WebSphere Information Integrator (9). The federated approach is appropriate if the source data are already present in a relational database where the information will be continually updated and maintained and where the data owners are willing to allow this kind of access.

However, the more common scenario is that research data are not contained in sophisticated databases but instead reside in a variety of text-file formats. In order to allow these data into the system, a set of specific XML schemas, representing different data types, is provided. Researchers can then use XML documents conforming to these schemas to send their data into the system. The major advantage of this data input interface is that it moves the responsibility of the data cleaning work from the biobank administration to the persons that know the data best, the researchers themselves. An added benefit is that the schemas allow for certain types of validation before the data enter the system.

There are two ways that the XML documents can enter the system. Either the user manually uploads them or they are sent automatically from other systems connected to the BIMS web service interface. Acknowledging the fact that not all researchers are used to XML, the BIMS also provides a set of simple tools that can convert their text data into XML documents. The XML documents are uploaded with a secure connection and are protected by encryption.

Deidentification: There are many definitions of deidentification (10), but generally it involves the process of removing or substituting information that can identify a physical person. This process can be either reversible or irreversible.

The irreversible removal of personal identifiers from data or samples, such that no specific individual can be identified, is usually called anonymization.

Using a third party for deidentification, in which an external organization is responsible for the deidentification process, is not implemented in most system. However, the architecture should accommodate such a solution, if needed from a legal or ethical viewpoint.

Consolidation: Consolidation means bringing different data sets together in a source-specific model. It involves two different data manipulation steps:

- Harmonization
- Synchronization

To generalize, the shared model is heavily normalized and uses many model parts for the sake of clarity and flexibility. Such a model gives a structure that facilitates the possibility to combine research data from different studies. Regardless of the purpose, data need to be synchronized, a process designed to connect data from the various data sources to the physical persons to which they belong.

Abstraction and queries: The process closest to the end-user in the BIMS is the query interface. It consists of a web interface accessible from the Internet that allows the user to construct questions and pose them to the database repository. Even without knowing anything about SQL, a researcher can do complex queries on data originally present in multiple data sources. It does so by applying the rules specified in the abstraction layer and essentially finds the connection between the abstract entities the user is interested in and the physical database tables. Query results can then be exported for further scientific analysis.

Through the mechanism of abstraction using the Data Discovery and Query Builder solution (DDQB) (11), the data can be made easier to understand, for example by renaming underlying database variables to descriptive names. In addition, by using a number of different abstraction layers mapping on top of the same data, several customized views of the data can be given.

A number of measures have been taken to guarantee the integrity of the data. Foremost is that BIMS has to reside in a network zone with its own IP addresses and firewall. User access to the BIMS demilitarized zone (DMZ) can only be granted to the user by presenting a valid digital certificate that acts as login broker to the web portal. Once logged in, the user is restricted to what services can be accessed in the portal. In the case of the Query web interface, the type of queries that the user is allowed to perform is regulated on several levels. More on security see Subheading 2.4.

2.2.2. Twin-NET

GenomEUtwin (12), http://www.genomeutwin.org, is a European Commission-funded collaboration between Twin Registries in the Netherlands, Denmark, Norway, Sweden, Finland, Italy, UK, and Australia. By pooling potentially over 300,000 twin pairs, the collaboration aims to identify genetic variants associated with common diseases. Many of these twin-cohorts include longitudinal, life-long data. Furthermore, most participating twin-cohorts have permission to link the study samples to health outcomes registries such as the Inpatient Registry, the Cancer

Registry and the Cause of Death Registry, which makes GenomEUtwin an epidemiologic goldmine for different projects.

There are three steps in the data integration:

- Data are extracted and harmonized into a common format
- Data are transferred into a data mart, a Twin mart
- Data are loaded into a federated database

The network architecture of TwinNET are Hub and Spoke, where the Hub is the integration node and Spokes are data providing centers, the twin registries, as shown in Fig. 4. Genotype data, generated in older studies or produced by genotyping centers without appropriate database backing, are maintained and collected by the Finnish Genome Center at the University of Helsinki. Phenotype data are accessed directly from the twin centers. To maximize security, all unneeded connections and network protocols are eliminated. A safe connection using virtual private network (VPN) (13) is initiated from the data-providing centers or Hub side, and an opened VPN tunnel is used to carry the appropriate database protocol.

The server is located in the TwinNET demilitarized zone (DMZ), as show in Fig. 5, and it can be disconnected from the local area network (LAN) to simplify security management. The local database, the Twin mart, is updated by pushing data from operational databases using transient secure connections. It is also possible to access live data directly from the LAN given that it is permitted by the local security rules. This can be done by, for example, rerouting database traffic using the local database federation without compromising the security of the LAN. Currently users can access the Hub using a web interface and Windows terminal services.

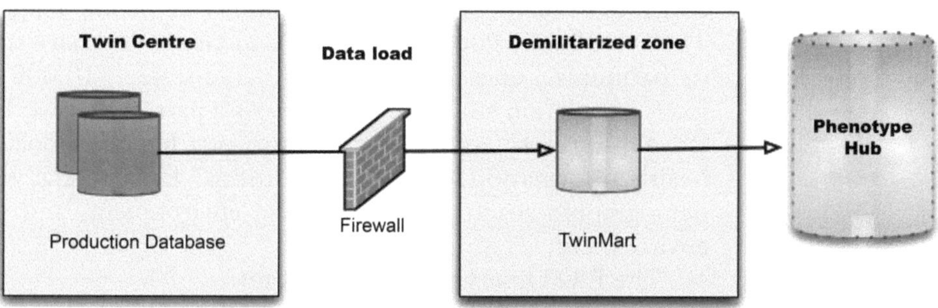

Fig. 4. Data is harmonized and pushed from the Twin Center's production databases into a database (TwinMart) located in a demilitarized zone of the TwinNET. The TwinMart data sets are implemented on study bases and optimized for data integrity and query purposes.

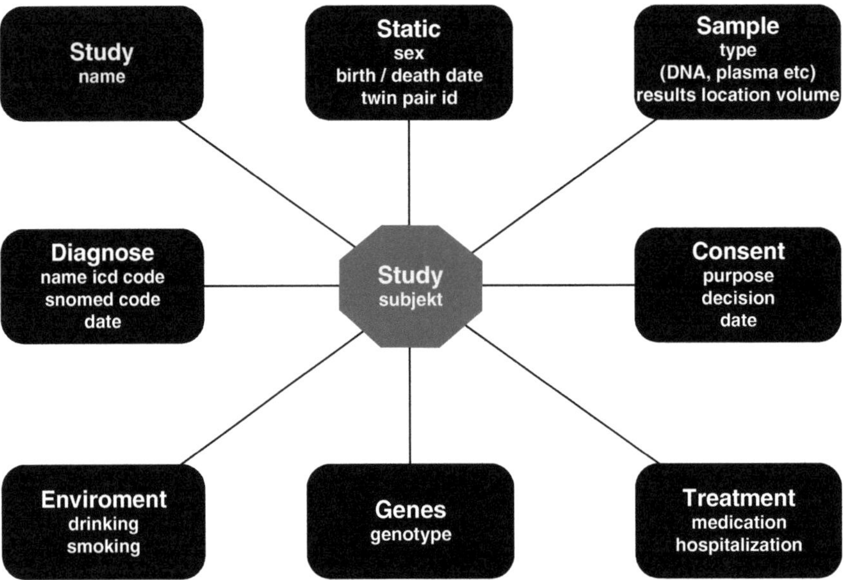

Fig. 5. The data is organized into Data Categories. These categories are there to help the user to identify where the data is located. Instead of organizing the data in terms of the underlying data sources, it is structured into a common data model for all data belonging to the same study.

2.3. Data Collection

The amount of information that medical researchers encounter has increased exponentially. Particularly in the research field of epidemiology, there is a need to manage the masses of information garnered from the study subjects' questionnaires, medical records, registries, genotypes, etc. Genetic variables in large population-based studies will be difficult to hold in a standard table in a relation database. Due to the very nature of the information, the resulting informatics problems usually manifest clearly although similar problems exist in all information-intensive enterprises.

One of the major challenges of integrating data is the varying quality of data. Very few scientific data sources are regulated by a quality system, resulting in difficulties when comparing data. The P^{3}G (Public Population Projects in Genomics) have started to harmonize questionnaires from large-scale (>10,000 subjects) population studies worldwide. As a part of that work, the P^{3}G Observatory, www.p3gobservatory.org, has been looking at health information, physical conditions, family data, socio-demographic characteristics, physical environment, and social environment.

The P^{3}G Observatory mission is to:

- Provide tools that support researchers in the harmonization, development, and implementation of epidemiological and genomics studies

- Disseminate scientific and technical information developed or suggested by P^{3}G collaborators
- Make feasible both comparison and sharing of information and data between studies

The P^{3}G's work program has started with prospective harmonization. It will also provide a template for the assessment of retrospective harmonization. Finally, it will support and enhance the design of emerging biobanks, thereby increasing the scientific utility of individual studies. The generic dataset should not be seen as a prescriptive list of all the variables that should be collected by a biobank. It will include a core set of approximately 100 data items and about 100 optional items. It is not anticipated that the core dataset will ever represent *all* of the data items that may be collected by a given biobank.

2.3.1. Terminologies and Coding Systems

In order to postulate a general model for collected data, a large number of terms and codes will be needed. These terminologies and coding systems will preferably be assembled from existing sources to avoid unnecessary work and reinvention. Extensive terminologies and coding systems have been developed during the last 25 years on both sides of the Atlantic for the use in health care in general and in laboratory medicine in particular (14).

Biobank information will certainly include information regarding the samples stored, procedures performed on those samples, and also properties of the samples, although the full information model will be a result of the project. Information regarding samples will need terms for the anatomical parts from which the sample was taken and also terms for any procedures performed on the samples such as preparation, conservation, etc. Properties of the samples will need terms representing types of property examined and terms for nominal property values.

Apart from general purpose terminologies and coding systems such as SNOMED CT (15), ICD (16) or the meta taxonomy UMLS (17, 18), several other domain-specific systems exist and are likely to be useful, e.g., the Foundational Model of Anatomy for representing anatomy (FMA) (19, 20), C-NPU (21) and LOINC kind-of-property codes (22) for representing properties, MGED (23) for describing some experimental processes, Gene Ontology (GO) (24) for genes, etc. However, it is already known that important properties of patients, experimental processes, or the repositories themselves cannot be expressed using even a combination of several available terminologies. There is often incomplete standardization of clinical information, and free text is subject to misinterpretation. Promising technologies such as text mining may provide a solution, but these technologies are still at an exploratory stage and only used in very advanced clinical environments. Furthermore, the concurrent use of more than one terminology requires careful management of any semantic overlap between terminologies, and between all the terminologies and the information model. Methodologies for describing and managing such overlaps are only just beginning to be described.

2.3.2. e-epidemiology

As the traditional methods of communication such as paper and telephone are growing more and more inefficient, technologies such as the Internet, mobile phones/text messaging, digital TV are increasingly used for communication across borders. However, the scientific society, although having integrated the new technology, has yet to make use of the possibilities inherent in these technologies to the full extent. The time has come for the Internet to enter the scene of data collection tools. The dream of collecting phenotypes on demand seems plausible only with means such as the web and mobile phones, which allow for rapid and cost-efficient assembly of data on determinants for lifestyle and health, thus representing an important tool for future epidemiological surveys. These and information on conditions and morbidities that are not recorded and retrievable from hospital records can be collected in a longitudinal fashion. We define this area, known as e-epidemiology, as "the science underlying the acquisition, maintenance and application of epidemiological knowledge and information using digital media such as the Internet, mobile phones, digital paper, and digital TV. E-epidemiology also refers to the large-scale epidemiological studies that are increasingly conducted through distributed global collaborations enabled by the Internet" (25).

Among the new communication techniques, mobile phones are the most widely used technique in the general population. SMS could complement web-questionnaires and potentially increase participation, above all from individuals with less access to the Internet, such as pensioners and people on sick-leave. With more mobile phones using JAVA, web-like applications in the mobile phones are made possible. E-epidemiology tools can also be used for the establishment of a population-based network of sentinel individuals for the monitoring of infectious disease occurrences.

Web-questionnaires are becoming more and more common as a means for collecting health data in epidemiological studies (26, 27). During the last decade, studies have been performed to establish the feasibility of this mode. The Internet also allows for interactive data capture with rapid checking of responses, adjustments of unforeseen problems, and the removal or adding of questions (28). For example, as new issues arise based on preliminary results, follow-up questions can be added. Instant feedback in the form of analysis of, for example, nutritional or summary statistics of individual responses given, is another advantage that could potentially increase response rates in web-based surveys (29, 30).

An interesting feature of the web tool is real-time randomization of survey questions and/or answers and rotation of items, which offers new complex experimental designs for methodological research (31, 32). A major concern often cited by the adversaries of the web mode is non-response bias. However, this is a constant concern irrespective of method and might not be larger in web-studies. This was shown by Ekman, et al., in a population-based cohort study of 50,000 women (33). Although there were some suggestions of a non-response bias, there was no difference between the web and paper mode used, which has also been confirmed in other studies (34–36).

2.4. Security and Administration Management

The security of any data comprised within a biobank is of primary importance to the trust of biobanks. A fundamental prerequisite for all biobank research is access to data. Because of the sensitive nature of the data, the unauthorized access to data is out of the question; hence a biobank system must meet tough security requirements. When integrating data from two data sources, the combined data set may contain more information than the original sources. Thus, it is important to be able to restrict authorization on a very detailed level. As an example, a user could have access to data source X and Y respectively, but should not have access to X and Y *at the same time.* When integrating data from data sources with different owners, it is important to be able to control and precisely define the ownership and control of the new dataset (3).

One of the most popular methods for authenticating users in web applications is a password, pass phrase or PIN-code. The greatest advantage of such a method is its simplicity. Unfortunately, passwords are vulnerable to a number of attacks. They can be guessed, sniffed over the wire, brute-forced, stolen, or coaxed out through social engineering. This is why standard password authentication is considered to be weaker than using one-time passwords, hardware tokens or others forms of authentication. There is a stronger method of authenticating users: client certificates or personal certificates for each user. With this method, we can authenticate web users by issuing a certificate and a private key connected to the client certificate.

Administration Management handles identifiers and studies. The purpose of an identifier is to describe a real world object in a globally unique and timeless manner. Some identifiers may be regarded as personal or particularly sensitive. The use of different identifier structures involves a balance between the usability and the security of the systems used.

Study Management supports the management of administration in connection with research projects. The functionality of such a system includes contact history, consent information, and other administrative information. Essentially the needs are identical to a traditional Customer Relationship Management (CRM) system used in many companies and organizations.

Study Management comprises the following:

- Log consent status of study subject
- Store contact information for study subject
- Log contact and events with/for study subject
- Log study subject status, i.e., response pending, response received, subject dropped-out
- Report on stored data such as number of samples collected, etc.
- Offer custom meta-data fields specific to certain research projects

Many large projects in genetic epidemiology differ from general biobank projects in the need to include a family identifier in the ID-number – or alternatively in the description of the phenotype. This meets the need to uniquely identify family as well as mother or father status. Provisions for a family identifier should be made early in the process of creating identifiers. In certain designs, such as case-control studies, this identifier will remain blank. In designs where the family identification is meaningful, its inclusion is crucial.

2.5. Data Query and Analysis

Data Query and Analysis is one the main services of biobank informatics. The main purpose of this component is to support the process of querying and analyzing collected scientific data across different research projects, i.e., build and store data queries, run adhoc or predefined data queries, store query results and export data queries for further analysis. Furthermore, implementation of state-of-the-art algorithms in genetic epidemiology and genetic statistics, used to analyse genetic data sets, and quantitative phenotype information generated by large scale population-based biobank projects, is important. Some of these algorithms could be included in the biobank solution some could be exported using standard protocols.

Vast amounts of data reside today in specialized data sources, with specialized query processing capabilities. Data from one source often must be combined with data from other sources to give users the information they desire. There are database systems that extract data from multiple sources in response to a single query. BIMS is one such system, providing the users with a virtual database to which they can pose arbitrarily complex queries. The actual data needed to answer the query may originate from several different sources, and none of these sources may be capable of answering the query by itself. Researchers can use a single SQL (Structured Query Language) statement through a web interface, for example, "*Cervix cancer, age 35–60,*" and not only retrieve information about who may have tissue and blood samples in a biobank, or branch of another biobank, but also information about contact persons and (with the right permissions) information about other phenotype/genotype databases connected to the samples.

A prerequisite for the Data Abstraction Model in Subheading 2.2.1 is integration and federation of data sources. Essentially all research data can be connected to a study subject and different schema can be used for categorization. The following (Fig. 6) categorization is an example of how the data can be categorized.

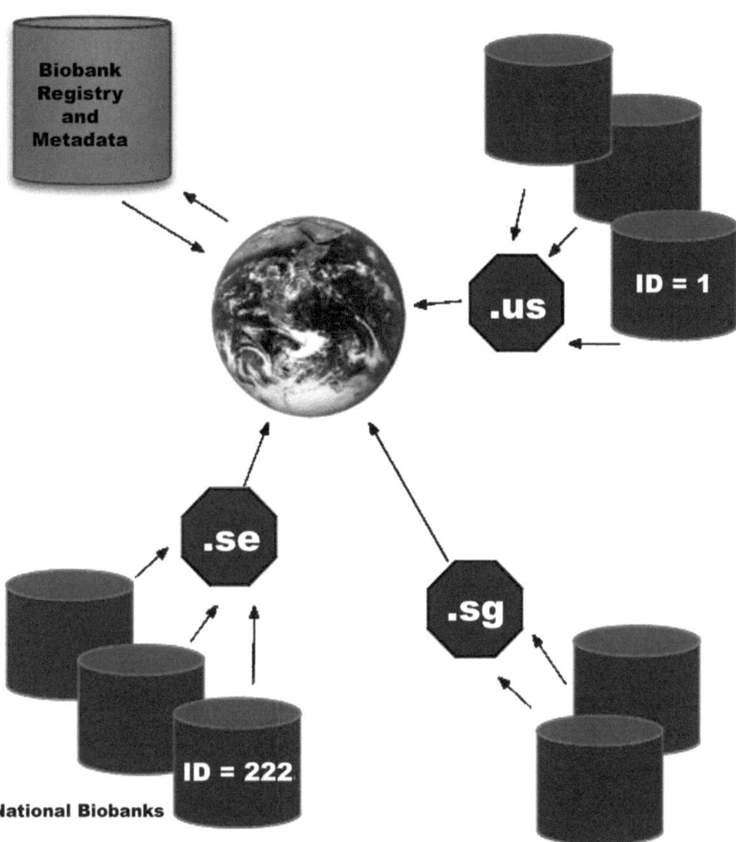

Fig. 6. In order to achieve the communication strategy described above, we are proposing the use of Web Services. This is a highly standardized platform based on the SOAP protocol, which in turn is based on the widely-spread XML standard. Web services provide an optimal platform for communication between biobanks since it is entirely built on open standards and not connected to any particular vendor or programming language. Thus, it will enable biobanks with completely different IT infrastructures to communicate with a common protocol.

3. Biobank Networking

By itself, a biobank can be very useful for many types of studies. However, the power of biobank research will increase enormously if multiple biobanks are connected to enable sharing of information and samples. A cost-effective and harmonized network of population-based studies and longitudinal cohort studies across the globe would support studies on a wide range of disease-environment-lifestyle-genetic relationships. It would enable studies of genetic and non-genetic determinants of the onset and natural history of established diseases. This infrastructure, and new information generated from it, would create invaluable opportunities for research into prevention, diagnosis, and management of a wide range of important complex diseases with major implications for public health throughout the world.

The first requirement that has to be fulfilled to enable biobank communication is a unique identity for each biobank; second, a common nomenclature is needed in order to communicate between biobanks. The final requirement is a well-defined protocol for sending information between biobanks as well as finding other biobanks. Each biobank joining the communication network will use well-defined methods to register metadata about the biobank contents in terms of samples and information. In a later stage, assuming a biobank wants to perform a specific search, it sends a request to the central biobank registry. Based on this request, the registry will send a reply with information describing the biobanks containing such information. Finally, the requesting biobank sends the actual query to these biobanks in order to get results.

The potential advantages of harmonizing biobanks and cohort studies to strengthen the foundation of public health will be to:

- Enhance the effective sharing and synthesis of information, thereby addressing the need for very large sample sizes and helping to promote collaborative international genetic epidemiological and clinical research
- Avoid the expensive mistakes and inefficiencies that can arise when individual initiatives repeatedly "re-invent the wheel," thereby saving funders and researchers a lot of time and money
- Promote communication within and between major biobanking initiatives thereby helping to overcome existing fragmentation of population genomic research

4. Conclusion

Many countries are setting up nationally coordinated biobank structures, which seek to gather clinical phenotypic data together with various biological blood, tissue, and genetic data over extended periods of time.

The lack of general information management systems for biobanks is a bottleneck for processing the increasingly complex data structures incurred in modern clinical and epidemiological research. By defining biobank informatics, phenotype and genotype can be merged, using a BIMS. Setting up a common, safe storage infrastructure and computer strategy will have implications for future database handling and processing for population-based biobanks and for global epidemiologic collaborative research. The BIMS is the key to increasing research information and data integration. Vast amounts of data reside today in specialized data sources, with specialized query processing capabilities. Data from one source often must be combined with data from other sources to give users the information they desire. There are database systems that extract data from multiple sources in response to a single query.

The long-term aim of biobank collaboration is to aid research in the search for the causes of common, complex diseases by arriving at a common strategy that will provide an efficient network of databases for population biobanks and cohorts. The challenge is to easily combine information from biobanks and cohorts so that very large sample sizes are generated, providing relevant, high quality and correctly documented sets of data to researchers.

It is our conviction that, effectively used, a system like BIMS could pave a new way in population-based biobank research.

References

1. Betsou, F., Ferguson, M., Jallal, B. and Litton, J.-E. (2004) Biobanks: accelerating molecular medicine, challenges facing the global biobanking community, in *Life science insights* (Z. Zimmerman, M. Swenson and B. Reeve, eds.), IDC, Framingham, MA. pp. 1–31.
2. Mahan, S., Ardlie, K. G., Krenitsky, K. F., Walsh, G. and Clough, G. (2004) Collaborative design for automated DNA storage that allows for rapid, accurate, large-scale studies. *Assay Drug Dev Technol* **2,** 683–9.
3. Ölund, G., Lilndqvist, P. and Litton, J.-E. (2007) BIMS: An information management system for biobanking in the 21st century. *IBM Systems Journal* **46,** 171–82.
4. Litton, J. E., Muilu, J., Bjorklund, A., Leinonen, A. and Pedersen, N. L. (2003)

Data modeling and data communication in GenomEUtwin. *Twin Res* **6,** 383–90.
5. Haas, L. M., Lin, E. T. and Roth, M. A. (2002) Data integration through database federation. *IBM Systems Journal* **41,** 578–96.
6. Ollier, W., Sprosen, T. and Peakman, T. (2005) UK Biobank: from concept to reality. *Pharmacogenomics* **6,** 639–46.
7. Rhodes, R. (2002) A Healthy Approach to Data. *IBM Systems Magazine.*
8. Achard, F., Vaysseix, G. and Barillot, E. (2001) XML, bioinformatics and data integration. *Bioinformatics* **17,** 115–25.
9. IBM Corporation. WebSphere information integrator, http://www-01.ibm.com/software/data/integration 2010. Accessed on 30th Sep. 2010
10. Knoppers, B. M. and Saginur, M. (2005) The Babel of genetic data terminology. *Nat Biotechnol* **23,** 925–7.
11. IBM Corporation. Data discovery and query builder, DB2 for i5/OS, http://www-03.ibm.com/systems/isoftware/db2 2010. Accessed on 30th Sep. 2010
12. Peltonen, L. (2003) GenomEUtwin: a strategy to identify genetic influences on health and disease. *Twin Res* **6,** 354–60.
13. Herscovitz, E. (1999) Secure virtual private networks: the future of data communications. *Int J Network Mgmt* **9,** 213–20.
14. Cimino, J. J. and Zhu, X. (2006) The practical impact of ontologies on biomedical informatics, in *IMIA yearbook of medical informatics* (R. Haux and C. Kulikowski, eds.), Schattauer Verlagsgesellschaft mbH, Stuttgart. pp. 124–35.
15. Wang, A. Y., Barrett, J. W., Bentley, T., Markwell, D., Price, C., Spackman, K. A. and Stearns, M. Q. (2001) Mapping between SNOMED RT and Clinical terms version 3: a key component of the SNOMED CT development process. *Proc AMIA Symp,* 741–5.
16. http://www.who.int/classifications/icd/en/
17. http://www.nlm.nih.gov/research/umls/about_umls.htm
18. Selden, C. R. l. and Humphreys, B. L. (1997) Unified medical language system. *Current Bibliographies in Medicine* **8,** 96
19. Rosse, C., Shapiro, L. G. and Brinkley, J. F. (1998) The digital anatomist foundational model: principles for defining and structuring its concept domain. *Proc AMIA Symp,* 820–4.
20. Rosse, C., Mejino, J. L., Modayur, B. R., Jakobovits, R., Hinshaw, K. P. and Brinkley, J. F. (1998) Motivation and organizational principles for anatomical knowledge representation: the digital anatomist symbolic knowledge base. *J Am Med Inform Assoc* **5,** 17–40.
21. http://www.ifcc.org/index.php?option=com_content&task=blogcategory&id=91&Itemid=150
22. Forrey, A. W., McDonald, C. J., DeMoor, G., Huff, S. M., Leavelle, D., Leland, D., Fiers, T., Charles, L., Griffin, B., Stalling, F., Tullis, A., Hutchins, K. and Baenziger, J. (1996) Logical observation identifier names and codes (LOINC) database: a public use set of codes and names for electronic reporting of clinical laboratory test results. *Clin Chem* **42,** 81–90.
23. Stoeckert, C. J., Jr. and Parkinson, H. (2003) The MGED ontology: a framework for describing functional genomics experiments. *Comp Funct Genomics* **4,** 127–32.
24. http://www.geneontology.org/
25. http://en.wikipedia.org/wiki/Main_Page
26. Manfreda, K. L. and Vehovar, V. (2002) Survey design features influencing response rates in web surveys. in *The International Conference on Improving Surveys,* University of Copenhagen, Denmark.
27. Cook, C., Heath, F. and Thompson, R. L. (2000) A meta-analysis of response rates in web- or internet-based surveys. *Educ Psychol Meas* **60,** 821–36.
28. Wyatt, J. C. (2000) When to use web-based surveys. *J Am Med Inform Assoc* **7,** 426–9.
29. Sax, L. J., Gilmartin, S. K. and Bryant, A. N. (2003) Assessing response rates and nonresponse bias in web and paper surveys, in *Research in higher education, vol. 44,* Springer, Netherlands. pp. 409–32.
30. Miller, E. T., Neal, D. J., Roberts, L. J., Baer, J. S., Cressler, S. O., Metrik, J. and Marlatt, G. A. (2002) Test-retest reliability of alcohol measures: is there a difference between internet-based assessment and traditional methods? *Psychol Addict Behav* **16,** 56–63.
31. Jackob, N. and Zerback, T. (2006) Improving quality by lowering non-response: a guideline for online surveys. in *Quality criteria in survey research VI,* Cadenabbia, Italy.
32. Fricker, R. D. and Schonlau, L. (2002) Advantages and disadvantages of internet research surveys: evidence from the literature. *Field Methods* **14,** 347–67.
33. Ekman, A., Dickman, P. W., Klint, A., Weiderpass, E. and Litton, J. E. (2006) Feasibility of using web-based questionnaires in large population-based epidemiological studies. *Eur J Epidemiol* **21,** 103–11.
34. Tolonen, H., Dobson, A. and Kulathinal, S. (2005) Effect on trend estimates of the difference between survey respondents and non-

respondents: results from 27 populations in the WHO MONICA Project. *Eur J Epidemiol* **20,** 887–98.

35. Fejer, R., Hartvigsen, J., Kyvik, K. O., Jordan, A., Christensen, H. W. and Hoilund-Carlsen, P. F. (2006) The Funen Neck and Chest Pain study: analysing non-response bias by using national vital statistic data. *Eur J Epidemiol* **21,** 171–80.

36. Jousilahti, P., Salomaa, V., Kuulasmaa, K., Niemela, M. and Vartiainen, E. (2005) Total and cause specific mortality among participants and non-participants of population based health surveys: a comprehensive follow up of 54 372 Finnish men and women. *J Epidemiol Community Health* **59,** 310–5.

Chapter 22

A Practical Guide to Constructing and Using Tissue Microarrays

Ian Chandler, Richard Houlston, and Göran Landberg

Abstract

Tissue microarray (TMA) technology is a robust "high throughput" method of tissue analysis, whereby a large number of patient samples can be examined in a short time using a minimum number of slides. In a TMA, cylinders of tissue are cored out of formalin-fixed, paraffin-embedded tissue blocks and slotted in a regular grid pattern into a blank recipient paraffin wax block. The TMA block is then cut using a standard laboratory microtome. Sections generated are suitable for all in situ techniques, such as immunohistochemistry (IHC) and in situ hybridisation, using essentially the same protocols as are used in conventional sections. The principle advantages of TMAs are that they save valuable biological material and ensure more reproducible reaction conditions while at the same time reducing re-agent costs and laboratory processing. Immunohistochemical studies designed to examine the prognostic utility of TMAs compared with large sections have generally found that they are comparable.

Key words: Tissue microarray, Tissue core, Image analysis, Antigen heterogeneity, Immunohistochemistry

1. Introduction

While the technology used in the histological assessment of surgically resected tissue is over 100 years old, it still shows no signs of being replaced by more "state-of-the-art" methods. Formalin-fixed, paraffin-embedded tissue has reproducible good morphological preservation for diagnostic purposes. There is also good antigenic and genomic preservation for ancillary techniques, such as immunohistochemistry (IHC), fluorescent in situ hybridisation (FISH), and chromogenic in situ hybridisation (CISH).

In the "bench to bedside" philosophy of translational cancer research, molecular techniques inevitably have to prove themselves in the testing ground of patient samples and clinical trials

Joakim Dillner (ed.), *Methods in Biobanking*, Methods in Molecular Biology, vol. 675,
DOI 10.1007/978-1-59745-423-0_22, © Springer Science+Business Media, LLC 2011

before they can be said to be relevant and of clinical promise. To use oncology as an example, the rationale behind the large amount of translational research currently underway is that current techniques are not able to completely predict the behaviour and outcome of a tumour in an individual patient (prognostication) or its response to a specific therapy (prediction). Validation of novel prognostic markers requires large patient cohorts with long-term clinical follow-up.

A large amount of archived histopathological material is present in hospital pathology departments. Tissue microarrays (TMAs) can be constructed from this, assuming that such follow-up data is obtainable, and the expression of existing or new biomarkers analysed. One of the main advantages of TMA technology is that it preserves valuable biological material. Tissue samples from patients are precious and cannot readily be replaced once the primary source is exhausted. Cutting conventional sections can quickly use up tissue blocks, particularly when the lesion under study is small. Removing small diameter cores of tissue means that the tumour in the main block is preserved and can be returned to, should this be necessary. Using TMAs ensures more reproducible reaction conditions and reduces inter-batch variation in the quality and intensity of staining produced, while at the same time reducing reagent costs and laboratory processing. Small TMAs can also be constructed with either one or a combination of types of tissue for immunohistochemical control or quality assurance purposes within a diagnostic or research setting (1).

2. Materials

2.1. Tissue

Standard pathological blocks of formalin-fixed, paraffin-embedded tissue are used as the starting material for constructing TMAs. Any tissue can in theory be cored and placed in a TMA. Most studies, to date, have been conducted using cancer samples, but there is no reason why any other tissue or disease types cannot be used, including muscle and brain. Tissues fixed in acid-based fixatives, such as Bouin's, are not really suitable as these cause excessive molecular degradation within cells. Removing small diameter cores of tissue means that the tissue in the main block is preserved. Taking such cores does not damage the structure of the tumour in the donor block in any significant way, and it can be returned to for further conventional sectioning, should this be necessary.

TMAs can also be constructed using material from cell lines. The protocol for this is described in the Methods section.

2.2. Coring Machine

There are a number of machines commercially available for constructing tissue arrays. Two commonly used systems are made by Beecher Instruments and Alphelys. Both manually operated and automatic devices are available. How these machines are operated is discussed in the Methods section.

2.3. Software

TMAs allow for the immunohistochemical analysis of hundreds of tissue samples with multiple antibodies on a relatively small number of slides. Typically, three to four tissue cores are generated per patient sample, and therefore there are literally thousands of points of information to be handled. The staining intensity or quality may vary from one core to another within an individual case, and so either a manual or computerised decision has to be made regarding the overall score for the specimen. Added to this, there will be information on the adequacy of the tissue, quality of staining, and possibly scores by more than one reader, so it is clear that specialist software tools are required. Many software systems are available and can be downloaded from the Internet. They are generally similar in basic design, consisting of multiple worksheets upon which the above information can be entered and then exported to Treeview, SPSS or similar software. Some systems have statistical analysis software incorporated.

One of the uses of TMAs, besides finding novel prognostic markers, is in finding novel ways of classifying tumours by the proteins they express. Hierarchical cluster analysis groups tumours together based on the relatedness of their RNA or protein expression profiles. It was first used for cDNA microarray data but has successfully made the transition to TMAs. Hierarchical clustering based on immunostaining is able to subgroup tumours in biologically or prognostically important ways other than by morphology alone (2).

2.4. Digitising Images

In addition to collecting scoring results from the staining, it is important to capture digital images. This can be potentially memory demanding (over a terabyte) and requires some forethought, and perhaps liaison with your local IT department. Saving images is important because the interpretation of staining on small tissue spots is subjective and may require review. It also allows the verification of a result by comparing spots from the same tumour. When TMAs are stained by more than one laboratory or scored by more than one pathologist, comparisons can be made with the quality of the results. In the future, to allow comparison between studies and to improve the reliability of results, it may become a commonplace to make available on the Internet more images of TMA staining than can be included in a printed article (3). Several systems for image capture are commercially available. These can be divided into those that are simply slide scanners, and those that incorporate image analysis software for the quantification of staining.

2.5. Visual Assessment vs. Image Analysis

Visual assessment of immunohistochemistry, even by a trained histopathologist, has its limitations. It is difficult to resolve subtle differences in staining intensity, and so a few ordinal categories, such as "weak", "moderate" and "strong" are used. Estimates of percentages or other continuous values tend to cluster around round figures. The high-throughput nature of microarray assessment means that scoring literally thousands of tissue spots is time consuming, tedious and fatiguing, and therefore image analysis software has certain theoretical advantages. In the past, image analysis was cumbersome as it involved manually drawing around cells of interest, and this was time consuming. Modern methods are faster and provide high-quality data. One such system has been described and termed Automated Quantitative Analysis (AQUA) (4). Briefly, this involves co-localising the cell of interest with a specific antibody labelled with a fluorescent tag (for example, carcinoma cells with a cytokeratin antibody) and at the same time applying the fluorescently labelled antibody of interest. Using a TMA derived from 340 node-positive breast carcinoma patients, a high level of correlation was found between a pathologist's and the automated system's assessment of ER status. On multivariate analysis, automated ER scoring was prognostically significant and similar to the pathologist. Applied Imaging® makes the Ariol microscope system, which has image analysis capabilities.

Given the relatively standardised methodological conditions associated with immunohistochemical staining of TMAs, automated analysis has not only theoretical advantages over visual scoring, but has also been shown in some publications to have practical ones too. It has to be stressed though that automated image analysis is in its infancy, and more work is required on what role it will have in TMA-based research.

3. Methods

3.1. Optimum Tissue Selection

Histopathological input at all stages of construction of a TMA is vital if it is going to contain the optimal amount of representative tissue. First, a decision has to be made on what tissue and which samples are going to be arrayed out. The collected blocks should be checked against their respective pathology reports and haematoxylin and eosin (H&E) stained slides to ensure that they are suitable, i.e. they contain a sufficient quantity of the tumour of interest that is well fixed. As well as being an issue of simply the quantity of the tissue in a block, the issue of exactly the tumour type that is desired and present needs to be thought through at this stage. For example, when arraying prostatic carcinoma, a decision has to be made on which Gleason grade is required. Coring areas of just one Gleason score may be preferable, or areas of different score can be cored and recorded separately. Renal cell carcinoma or germ cell tumours of the gonads are examples of where morphologically different areas are frequent, and these will probably need to be identified separately. In all tumour types, areas rich in inflammatory cells or necrosis need to be identified and excluded. At this stage, the pathologist circles with a pen on the associated H&E slide areas he deems suitable for coring out.

A method has been described for making TMAs with material from core biopsies by cutting small lengths of the biopsy out of the donor block and re-embedding them vertically in a new block (5) although this is rarely done.

3.2. Limitations on What can be Cored

There are certain practical limitations to what can be cored. Pre-malignant lesions, such as carcinoma in situ (CIS) of the prostate or breast, may be difficult to target if lesions are small. Carcinomas with a large stromal component, such as pancreatic ductal adenocarcinoma, or large mucinous component, such as colorectal adenocarcinoma, which contains by surface area a small proportion of viable cells, can be problematic. With arraying machines, cores of up to 3 mm can be taken, and this is useful if having a representative amount of the tumour of interest in a smaller core is a problem. Warming the donor and recipient blocks for 20 min in a 37°C incubator aids the transfer of larger cores (6).

3.3. Potential Problems with Core Contents

In the authors' experience, when assessing an individual tumour core where architectural clues are lost, differentiating between poorly differentiated carcinoma and activated lymphocytes or macrophages can be problematic. With certain antibodies, where the loss of protein expression is being looked for, such as those used to detect the expression of the DNA mismatch repair (MMR) proteins MLH1 and MSH2, normal stromal fibroblasts or lymphocytes are used as an internal positively staining control. A 0.6 mm core of cancer may contain insufficient stroma to act as such, which makes staining interpretation difficult.

3.4. Preparing Cell Lines for TMA

For this protocol, it is assumed that the cells are being grown in a standard T175 cell culture flask (or similar). If cells are confluent when harvested, this will provide a final volume of around 5 ml in the bottom of an eppendorf tube.

1. Wash with versene.
2. Add non-enzymatic cell dissociation solution (Sigma, St Louis, MO) and leave at 37°C for several minutes until the cells are free.
3. Spin at 1,000 rpm for 5 min.
4. Wash in PBSA.
5. Re-spin and remove PBSA.
6. Re-suspend pellet in 1 ml 10% formalin and allow fixing overnight at room temperature.
7. Embed in paraffin wax in the usual manner.

4. Array Construction

4.1. TMA Size and Orientation

The size of the TMA made can vary from 10×10 up to 30×20, with 1 mm spaces between cores. Excessively large TMAs have the disadvantage that antibody coverage may not reach the edges adequately. To aid orientation when assessing the slides, the grid of cores is laid out in an asymmetrical fashion, often in a "Swedish flag" pattern.

4.2. Control Tissue in TMAs

TMAs do not have to contain control tissue – this can be stained on a separate slide – but it can be helpful. Control tissue can be the benign counterpart of the tumour tissue of interest, or a different kind of tissue altogether (normal liver or kidney, for example). The advantage of using control tissue that is the same as the tissue of interest is that this provides controls for the in situ methodology used, but a different type of tissue aids in orientation, assuming that they are placed asymmetrically. Exactly what controls to use has to be made on a case-by-case basis.

4.3. Performing the Coring

Tissue cylinders are cored out by comparing the donor block with its marked H&E slide and coring out the desired area. In this sense, the coring is done in a "semi-blind" manner. Being able to recognise the subtle topography of the tumour within a block is therefore vital, and the job of coring can really only be done by either an experienced biomedical scientist or pathologist.

Typically, three cores are taken from each donor block, and this is discussed in more detail in the section on tumour heterogeneity. Cores are generally 0.6 mm diameter, and are punched out cleanly from the donor block. All coring machines have a similar basic construction. They consist of a metal base, on to which fits a block holder, which snugly holds the tissue block in position. Over this is an arm that moves up and down and this arm bears two punches with stylets. The most commonly used punch size is 0.6 mm, but sizes up to 3.0 mm are available. The position of the arm can be finely adjusted in the X and Y co-ordinates. To make a TMA manually, a 0.6 mm diameter core of wax is first punched out of a recipient blank wax block by lowering the moveable arm. A tissue cylinder is then cored out by comparing the donor block with its marked H&E slide and coring out the desired area. This tissue core is then slotted in to the empty hole in the donor block. The position of the arm is adjusted by 1 mm at a time and the process repeated. Using automatic devices, the area of the donor block to be cored is pre-selected before the coring begins.

It takes approximately 2 days to construct an average-sized TMA. Some authors have described making two microarrays from the same tumour and using the results obtained from the combined staining scores of them both (7).

4.4. Finishing the Array

After the array has been constructed, the ends of the cores tend to slightly protrude from the surface of the block. If the block is gently warmed in a 40°C incubator for around 30 min (until a corner can be gently indented with the pad of a finger), then these cores can be gently pushed into the block using the flat side of a glass slide.

4.5. Cutting TMA Sections

Cores are usually around 3 mm deep, and this means that multiple serial sections of the completed TMA can be cut for analysis of many different markers. If cut carefully by an experienced technician, at least 200 sections are potentially obtainable. Prior use of the donor block may reduce this number, as the cores obtained will be shorter. If carefully performed, the cores stay intact and the tissue in the completed TMA has remarkably good architectural and morphological preservation. When sectioning TMA blocks for analysis, always orientating them the same way on the slide is enormously helpful.

The 0.6 mm diameter spots produced can be fragile and prone to "floating" during antigen retrieval, and so Tape Transfer methods can be used to reduce this (Instrumedics Inc, NJ).

4.6. Potential Problems with Core Contents

Targeting the desired area of tumour can be problematic. An area of the block rich in tumour will almost inevitably contain some normal structures, stroma, necrosis or blood vessels. This problem is increased when using tumours known to be particularly stroma-rich, such as pancreatic carcinoma, or targeting small lesions, such as carcinoma-in-situ of the breast. Some cores may contain normal structures, for example, glandular epithelium of the breast or prostate, which needs to be distinguished from tumour for accurate scoring. For this reason, combined with the loss of a few cores that tends to occur during immunohistochemical antigen retrieval, it can be expected that 15–20% of cores in a completed array will contain insufficient tumour to be assessable. It also follows that scoring of IHC or in situ techniques should only be done by somebody totally familiar with the histopathological appearances.

Tumour within the donor block is a three-dimensional structure, and so is the core. This means that the composition of any individual core may be seen to change as sections are cut. A core may contain tumour cells at one end but only stroma at the other. If arrays are not well constructed and some cores are shorter than others, then some cores, and hence some patient samples, will "cut out" before others.

5. Tissue Heterogeneity

5.1. Underlying Principles of Tissue Heterogeneity

Several studies have addressed the question of whether TMAs are representative of tumours as a whole, and if they are an adequate replacement for a large section. From first principles, the volume of a spherical tumour of diameter 30 mm is 1.4×10^4 mm^3; the volume of a 25×20 mm tissue section is 2 mm^3 and the volume of a 0.6 mm diameter microarray spot is 1.0×10^{-3} mm^3. Therefore, histopathologists are normally assessing in their routine work only a tiny fraction of a tumour. Whether a TMA is an adequate replacement for a large section may be dependent on which antigen is the subject of the study. Some antigens will be uniformly expressed across a tumour and others may be under a selective pressure of some kind. Some antigens are obligatorily expressed in order for a cell to function, or are almost always phenotypically characteristic of a particular tumour type, even down to even quite poor levels of dedifferentiation. Some mutations may have arisen early in the carcinogenetic process and be present in all subsequent daughter cells. Others may have been acquired at a later stage and only be present in a subpopulation of cells. Innate physiological stresses that a tumour may undergo, such as central ischaemia, may cause intratumoural variation in mRNA or protein expression. Other phenotypic changes may only be apparent at the advancing edge of a tumour, such as neoangiogenesis or stromalysis. Alternatively, an apparently single tumour mass may have arisen through the collision of two originally separate tumours, which are genetically different. This means that when designing and conducting a TMA study, a conscious decision has to be made regarding where in a tumour cores are taken from, and whether the antigen of interest will be expressed in that part of the tumour in a uniform or non-uniform manner.

5.2. Evidence for intratumoural Variation in Expression

Many studies have provided evidence that there is intratumoural variation at both the genotypic and phenotypic level in breast and colorectal cancer. In breast carcinoma, comparative genomic hybridisation (CGH) has demonstrated cytogenetic variation within individual tumours (8), and regions with different staining characteristics for pRb and p53 have been found (9). In colorectal carcinoma, intratumoural variation in ploidy status measured using flow cytometry (10), and mutation status variation in the p53 and k-ras genes (11), have been described. Microsatellite instability status has been found to be uniform within CRC however (12).

5.3. Tissue Heterogeneity and TMAs

Immunohistochemical studies designed to examine the prognostic utility of TMAs compared with large sections have generally found that they are comparable.

Torhorst et al. (13) investigated ER, PR and p53 expression in 553 breast carcinomas using four TMAs constructed from central and peripheral parts of the tumours. They found that the frequency of ER positivity ranged from 79 to 81% in the four TMAs and this closely agreed with the rate on large sections (80%). The frequency of PR positivity ranged from 41 to 53% in the TMAs but was 60% in the large sections. Interestingly, the proportion of tumours showing heterogeneity of expression was 9% for ER but 29% for PR. Loss of either ER or PR expression was strongly associated with poor prognosis both for an individual TMA, the combination of the four of them or the large section. The frequency of p53 positivity ranged from 15 to 21% among the four TMAs but was 43% in the large sections. When survival curves were calculated, the TMA results correlated better with poor prognosis than did the large sections. The explanation given by the authors for this was that tumours in which weak patchy staining was called positive in the large sections were negative on the TMA, and this type of staining did not correlate with poor patient outcome.

Aaltonen et al. (14) studied cyclin A expression in representative areas of 200 breast cancers that had been arrayed into two duplicate TMAs and compared them with large sections. For the two assessing pathologists, the kappa values for positive staining between the TMAs and large sections was good, at 0.62 and 0.75. High cyclin A expression was not significantly associated with overall survival on either TMA or large sections, but it was for metastasis-free survival on TMAs but not large sections.

Several groups have examined the correlation between FISH and either CISH or IHC in the assessment of HER2 amplification on TMAs (15–20). Agreement of up to 99% has been reported, and in the future, this technique may be used for laboratory quality control purposes, although more work needs to be done before it can be concluded that the correlation is high enough between large sections and TMAs for it to be used in a diagnostic setting.

5.4. Conclusion on Tissue Heterogeneity

It is difficult to extrapolate from the studies of hormone receptor status and cell cycle-associated proteins to all proteins that a tumour cell expresses that might conceivably be a prognostic or predictive marker. When considering novel marker studies, it is important to take potential heterogeneity of expression into account in the experimental design and the interpretation of results. Ultimately, TMAs are useful as a screening tool but have their limitations and probably cannot be used in all situations.

6. Helpful Hints

1. When designing a study, consider what clinical or pathological biases may have unintentionally crept in to the system.
2. Try to keep pathological reports and outcome data in a physically separate place to where the laboratory staff perform the analyses, to minimise bias.
3. Histopathological input at all stages of construction of a TMA is vital if it is going to contain the optimal amount of representative tissue. Consider whether the tissue type to be cored will be well represented in a 0.6 mm diameter core, or whether larger cores should be used.
4. Arranging cores asymmetrically into smaller squares within the grid, for example 5 × 5, makes orientation and reading easier.
5. Tissue spots tend to be fragile and can float off during antigen retrieval. Using tape transfer methods reduces this.
6. Some tissue tends to be lost during "trimming in" when cutting sections. Batching section cutting reduces this, although these need to be used relatively quickly afterwards (within weeks) to avoid oxidation of sections.
7. When designing studies, consideration needs to be given to whether the protein of interest is uniformly or non-uniformly expressed in the tissue. Non-uniform expression may be physiological, pathological or artefactual.

References

1. Gulmann, C., Loring, P., O'Grady, A., and Kay, E. (2004) Miniature tissue microarrays for HercepTest standardisation and analysis. *J Clin Path* **57**, 1229–1231.
2. Nielsen, T.O., Hsu, F.D., Jensen, K., Cheang, M., Karaca, G., Hu, Z., et al. (2004) Immunohistochemical and clinical characterisation of the basal-like subtype of invasive breast carcinoma. *Clin Ca Res* **10**, 5367–5374.
3. Deutsch, E.W., Ball, C.A., Bova, G.S., Brazma, A., Bumgarner R.E., Campbell, D., et al. (2006) Development of the minimum information specification for *in situ* hybridization and immunohistochemistry experiments (MISFISHIE). *OMICS* **10**, 205–208.
4. Camp, R.L., Chung, G.G., and Rimm, D.L. (2002) Automated subcellular localization and quantification of protein expression in tissue microarrays. *Nature Medicine* **8**, 1323–1328.
5. Jhavar, S., Corbishley, C.M., Dearnaley, D., Fisher, C., Falconer, A., Parker, C., et al. (2004) Construction of tissue microarrays from prostate needle biopsy specimens. *Br J Cancer* **93**, 478–482.
6. Yang, X.R., Charette, L.A., Garcia-Closas, M, Lissowska, J., Paal, E., Sidawy, M., et al. (2006) Construction and validation of tissue microarrays of ductal carcinoma in situ and terminal duct lobular units associated with invasive breast carcinoma. *Diagn Mol Pathol* **15,** 157–162.
7. DiVito, K.A., Berger, A.J., Camp, R.L., Dolled-Filhart, M., Rimm, D.L., and Kluger, H.M. (2004) Automated quantitative analysis of tissue microarrays reveals an association between high Bcl-2 expression and improved outcome in melanoma. *Cancer Res* **64**, 8773–8777.
8. Aubele, M., Mattis, A., Zitzelsberger, H., Walck, A., Kremer, M., Hutzler, P., et al. (1999) Intratumoral heterogeneity in breast carcinoma revealed by laser-microdissection and comparative genomic hybridisation. *Cancer Genet Cytogenet* **110**, 94–102.
9. Trudel, M., Mulligan, L., Cavenee, W., Margolese, R., Cote, J., and Gariepy, G. (1992)

Retinoblastoma and p53 gene product expression in breast carcinoma. *Hum Pathol* **23**, 1388–1394.

10. Quirke, P., Dyson, J.E.D., Dixon, M.F., Bird, C.C., and Joslin, C.A.F. (1985) Heterogeneity of colorectal adenocarcinomas evaluated by flow cytometry and histopathology. *Br J Cancer* **51**, 99–106.
11. Losi, L., Baisse, B., Bouzourene, H., and Benhattar, J. (2005) Evolution of intratumoral genetic heterogeneity during colorectal cancer progression. *Carcinogenesis* **26**, 916–922.
12. Samowitz, W.S., and Slattery, M.L. (1999) Regional reproducibility of microsatellite instability in sporadic colorectal cancer. *Genes Chromosomes Cancer* **26**, 106–114.
13. Torhorst, J., Bucher, C., Kononen, J., Haas, P., Zuber, M., Kochli, O.R., et al. (2001) Tissue microarrays for rapid linking of molecular changes to clinical endpoints. *Am J Pathol* **159**, 2249–2256.
14. Aaltonen, K., Ahlin, C., Amini, R.M., Salonen, L., Fjallskog, M.L., Heikkila, P., et al. (2006) Reliability of cyclin A assessment on tissue microarrays in breast cancer compared to conventional histological slides. *Br J Cancer* **94**, 1697–1702.
15. Bhargava, R., Gerald, W.L., Li, A.R., Pan, Q., Lal, P., Ladanyi, M., et al. (2005) EGFR gene amplification in breast cancer: correlation with epidermal growth factor receptor mRNA and protein expression and HER-2 status and absence of EGFR-activating mutations. *Mod Pathol* **18**, 1027–1033.
16. Loring, P., Cummins, R., O'Grady, A., Kay, E.W. (2005) HER2 positivity in breast carcinoma: a comparison of chromogenic in situ hybridisation with fluorescent in situ hybridisation in tissue microarrays, with targeted evaluation of intratumoural heterogeneity by in situ hybridisation. *Appl Immunohistochem Mol Morphol* **13**, 194–200.
17. Kay, E., O'Grady, A., Morgan, J.M., Wozniak, S., and Jasani, B. (2004) Use of tissue microarray for interlaboratory validation of HER2 immunocytochemical and FISH testing. *J Clin Pathol* **57**, 1140–1144.
18. Press, M.F., Slamon, D.J., Flom, K.J., Park, J., Zhou, J.Y., and Bernstein, L. (2002) Evaluation of HER-2/neu gene amplification and overexpression: comparison of frequently used assay methods in a molecularly characterized cohort of breast cancer specimens. *J Clin Onco* **20**, 3095–3105.
19. Ricardo, S., Milanezi, F., Carvalho, S., Leitao, D., and Schmitt, F. (2006) HER2 evaluation through the novel rabbit monoclonal antibody SP3 and CISH in tissue microarrays of invasive breast carcinomas. *J Clin Pathol* **60**(9), 1001–1005.
20. Sapino, A., Marchio, C., Senetta, R., Castellano, I., Macri, L., Cassoni, P., et al. (2006) Routine assessment of prognostic factors in breast cancer using a multicore tissue microarray procedure. *Virchows Arch* **449**, 288–296.

Chapter 23

Breast Cancer Genomics Based on Biobanks

Asta Försti and Kari Hemminki

Abstract

Attempts to find genes contribution to complex diseases, such as cancer, require new study designs which incorporate an efficient use of population resources and modern genotyping technologies. We describe here two approaches, used by us for the study of breast cancer, both of which take the use of biobanks. One uses a cancer registry as a source of case information, which is then linked to a biobank on blood DNA. The biobank provides also samples from matched controls. After genotyping, clinical data are retrieved from hospital records, and the results can be presented for genotype-specific cancer risks, or similarly for genotype-specific clinical and survival parameters. The second approach uses registered data on cancer in families or among twins. On defined groups of patients, paraffin tissue is collected by contacting the pathology departments of the hospitals where the patients were diagnosed. Tumor and healthy tissue is prepared and used for mutation, the loss of heterozygosity, or copy number analysis. We believe that in the era of whole-genome genotyping technologies, the importance of well-characterized sample sets cannot be overemphasized. Samples rather than technologies limit the rate of gene discovery in complex diseases.

Key words: Cancer registries, Linkage, Genotyping, Twin research, Familial breast cancer, Whole-genome genotyping

1. Introduction

Cancer, like many other common diseases, is characterized by a small Mendelian component, a somewhat larger familial (non-Mendelian) component and a large sporadic component. The genetic bases of many Mendelian cancers have been resolved using pedigree-based linkage studies. The Mendelian cancers constitute the hereditary cancer syndromes, which account for a small proportion of all cancers. Among the most prevalent hereditary cancer syndromes, BRCA1 and BRCA2 combined account for 2% of breast cancers (1); for ovarian cancer the combined

Joakim Dillner (ed.), *Methods in Biobanking*, Methods in Molecular Biology, vol. 675,
DOI 10.1007/978-1-59745-423-0_23,

attributable fraction of BRCA1 and BRCA2 may be over 10% (2); hereditary nonpolyposis colorectal cancer (HNPCC) accounts for 1–3% of colorectal cancers (3). However, the attributable fractions of hereditary syndromes depend on the frequency of the disease variants in the population, which may be highly variable. The above figures refer to some Western European and North American white populations. Twin data suggest that heritable causes of cancer extend far beyond the currently known genes (4); for example, 27% of the variation of breast cancer among twins could be accounted for by heritable factors, which is ten times more than the combined effect of the known high-risk genes. Several low-risk genes have been associated with breast cancer, but their role cannot be estimated at this point of time because of variable levels of validation.

The changing paradigms from monogenic to polygenic diseases have led to changes in concepts about disease causation and in study designs. The risks of common alleles are likely to be small, but because they affect a large proportion of the population, the resulting attributable risk may be high (5). This is one of the reasons why association studies between DNA variants and diseases have become widely used. Another reason is the lack of feasible alternative approaches and the experienced difficulties with linkage studies. The third reason is the readily available data on millions of single nucleotide polymorphisms (SNPs) in the human genome, including confirmed SNPs produced by the HapMap project (http://www.hapmap.org). However, there is a continuing scientific debate on the optimal strategy, in which the classical experience on rare Mendelian diseases (many rare disease alleles) is contrasted with the common diseases – common variant hypothesis (5–7). The underlying assumptions of these hypotheses are beyond the scope of the present article. Suffice to mention that the practical implications have a strong bearing on the selection of genetic techniques for study. The common diseases – common variant hypothesis lends to genome-wide association studies, utilizing the linkage between the disease allele and the marker allele (linkage disequilibrium) as a mapping tool, whereas the more classical paradigm continues to emphasize family-based approaches. An important neglected aspect of this scientific debate is the availability of human samples, the theme of the present volume. We describe two approaches using biobanked material for genetic studies, the first one from an unselected patient series and the second one on family-based sampling.

2. Genetic Association Studies Using Samples from a Biobank

A general outline on the use of samples collected by a biobank for genetic association studies is shown in Fig. 1. By record linkage to the cancer registry cancer, cases can be identified from the biobank.

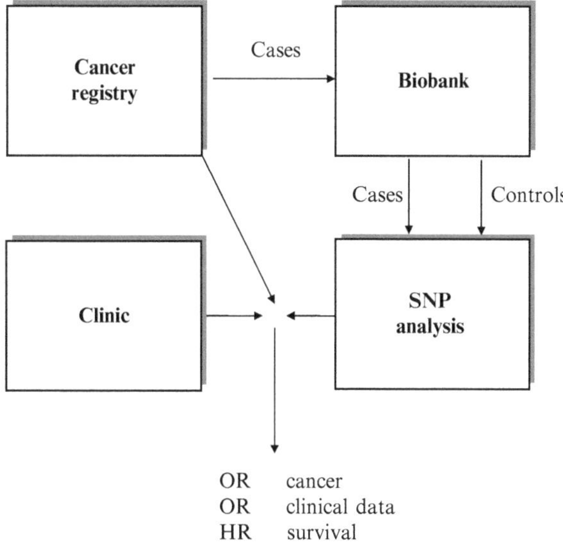

Fig. 1. Genetic studies using case data from a cancer registry and samples from a biobank.

Controls, matched for age, gender, ethnicity, and geographical background, are then selected from the same biobank. Clinical data for the cancer cases are retrieved from a clinical registry, and the date and cause of death from the cancer registry or the population registry. DNA is isolated from the blood samples, samples are divided randomly on the plates and the SNP analyses are carried out. Genotyping results are linked to the case-control status, clinical and survival data. Odds ratios (ORs) and 95% confidence intervals (95% CIs) for associations between genotypes and cancer risk or tumor characteristics are calculated by logistic regression analysis or similar methods. The Kaplan–Meier method can be used to estimate the survival probabilities with the log-rank test to test differences between subgroups. The hazard ratios (HRs) are calculated using the Cox regression analysis.

We have used a large, well-characterized breast cancer series of 959 cases and 952 matched controls from Umeå, Sweden to examine whether genetic variation in the genes related to cancer progression, regulating angiogenesis and extracellular matrix degradation, is associated with susceptibility to and prognosis of breast cancer (8–10). The cases with the age and gender matched controls were drawn from two large, well-characterized, population-based cohorts: the Västerbotten intervention cohort and the mammary screening cohort, which contain blood samples collected between January 1990 and January 2001 from an ethnically homogenous population living in a geographically defined region in North Sweden (11). The cases were identified from the cohorts by record linkage to the regional cancer registry. In 261 cases, the breast cancer was diagnosed after sampling (incident cases), and in 402 cases before sampling (prevalent cases). During the same time period, 257 samples were collected consecutively from untreated patients referred to the Department of Oncology (Norrlands University Hospital, Umeå, Sweden) for newly diagnosed breast cancer. Their controls, also matched for age and gender, were selected from the Västerbotten intervention cohort. Clinical data for the breast cancer cases were retrieved from the registry managed by the Northern Sweden Breast Cancer Group. Information about the date of death was collected from the Swedish population register with a follow-up until September 2005.

Angiogenesis is a crucial step for the development of fatal cancer, and it is necessary for primary tumor growth, invasiveness, and metastasis (12). Vascular endothelial growth factor (VEGF) is important for the initiation of angiogenesis, and it is the major regulator of breast cancer angiogenesis (13). Although the biological effects of VEGF are mediated by two receptors (VEGF receptor-1 or Flt-1 and VEGF receptor-2 or Flk-1/KDR), the interaction between VEGF and the kinase domain receptor (KDR) is believed to be the most important one for angiogenesis during tumor development (12). In human breast cancer, functional KDR has been found to be overexpressed and correlated with proliferation, and it is co-expressed with VEGF, providing thus evidence that the VEGF/KDR system plays a functional role in the progression of breast cancer (14–16). Recently, overexpression of the periostin (POSTN) gene by human breast cancers has been shown to lead to an enhanced tumor progression and angiogenesis (17). Elevated serum levels have been observed in patients with bone metastases from breast cancer (18). The induction of angiogenesis by POSTN has been shown to derive partly from the upregulation of the KDR gene (17).

Functional polymorphisms, which affect the regulation of gene expression or the function of the coded protein, can contribute to the differences between individuals in susceptibility to and severity of a disease. The effect may be caused by the polymorphism alone, or in combination with other polymorphisms, or it may be a marker of an ancient haplotype with a distinct functional domain. We investigated the relationship between genetic polymorphisms in the VEGF, KDR, and POSTN genes and breast cancer (8, 9). In the VEGF gene, none of the studied SNPs were associated with the risk of breast cancer, but the –634 CC genotype and the –2578/–634 CC haplotype were significantly associated with high tumor aggressiveness (large tumor size and high histological grade). Both the –634 CC and the –2578 CC genotypes have been associated with higher VEGF production than the other genotypes (19, 20). In the POSTN gene, the A allele of the T-953A SNP both alone and in a haplotype was associated with an increased risk of breast cancer. However, no correlation with the tumor characteristics was observed. On the other hand, the G allele of another POSTN SNP, C-33G, correlated with tumor characteristics associated with worse prognosis (high histological grade and estrogen receptor negative tumors). The number of homozygotes was, however, small ($n = 7$) and no significant effect on survival could be noted.

Extracellular matrix degradation mediated by matrix metalloproteinases (MMPs) is a critical step in tumor invasion and metastasis (21). The MMPs are synthesized as inactive zymogens to be activated by a proteinase cleavage; their activities are regulated by endogenous tissue inhibitors of metalloproteinases (TIMPs) (22); and a reversion-inducing cysteine-rich protein with a kazal motif (RECK) (23, 24). The expression of MMPs is increased in almost all human cancers, most likely by transcriptional regulation rather than by genetic alterations (25). Recent studies have revealed a broader role of MMPs in tumor progression than originally assumed. They can regulate cell proliferation, apoptosis, angiogenesis, invasion, and metastasis, as well as the immune response to cancer ((25) and the references therein). In addition, the expression of MMPs and TIMPs has been correlated with the progression of breast cancer (25). Also the expression of RECK has been reported as a promising prognostic factor in several cancers, including breast cancer (26). We examined the correlation of a set of promoter SNPs in MMPs and their inhibitors with susceptibility to and progression of breast cancer (10). We observed marginal associations between the MMP9 and TIMP3 genotypes and breast cancer susceptibility, but the results are difficult to interpret: even though the rare MMP9 TT genotype carriers and the C allele carriers of the TIMP3 SNP seemed to have a slightly increased risk, the carriers of both these risk genotypes were not at risk. On the other hand, the carriers of the more common opposite genotype combination, MMP9 CC and TIMP3 TT, seemed to be slightly protected against breast cancer. The other SNPs involved in this study showed no association with breast cancer risk. We observed also a gene–gene interaction between the SNPs in MMP2 and MMP9, and in RECK and TIMP3 with the progesterone and the estrogen receptor status, but the scarcity of the patients carrying the two risk genotypes limits the prognostic significance of the finding. Similarly, the observed better survival among the RECK –402 CT genotype carriers should be taken with caution because no association between this genotype and prognostic tumor characteristics was detected.

Since our study had a power of more than 80% to detect an OR of 1.5, when 10% of the controls carry the risk genotype, the results suggest that most of the SNPs in the genes related to angiogenesis and extracellular matrix degradation are not associated with breast cancer susceptibility. However, genetic variation, especially in the genes involved in angiogenesis, may have an effect on breast cancer progression.

3. Use of Tumor Tissue Samples in Genetic Studies

A Family-Cancer Database, containing information on family relationship and cancers, can be used to select families with several affected family members, and paraffin-embedded tumor material can be retrieved from the pathology departments of the hospitals of diagnosis as outlined in Fig. 2. After inspection by a pathologist, tumor tissue can be dissected from the paraffin blocks; DNA will be isolated and can be used for mutation analysis. Additionally, histologically normal tissue can be dissected from the same paraffin blocks as the tumor tissue, and the paired tumor and normal tissue can be used for allelic imbalance analyses, including the loss of heterozosity (LOH) or copy number change analyses.

Use of familial cancer cases increases the power to detect low-penetrance susceptibility alleles in association studies (27, 28). We used the Swedish Family-Cancer Database to identify breast cancer cases from families with at least two first-degree relatives diagnosed with breast cancer; at least one of them was diagnosed with bilateral breast cancer (29). Women with a family history of bilateral breast cancer have about three-fold increased the risk to develop breast cancer (29). Paraffin-embedded tissue samples were retrieved from 86 cases from the hospitals of diagnosis. Microdissected, histologically normal tissue DNA was used in the following association studies. The high-risk familial cases together

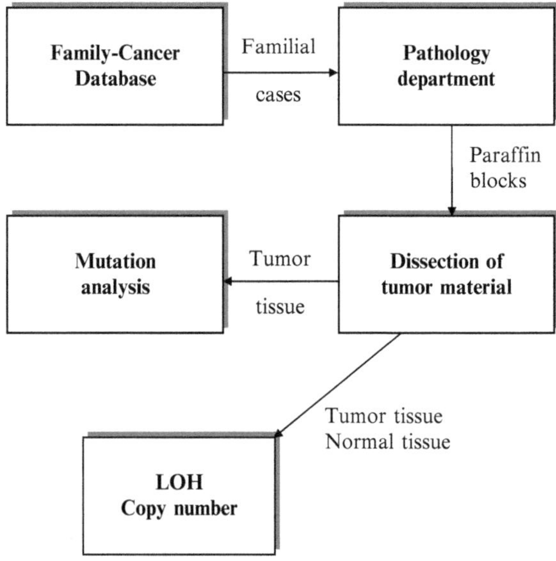

Fig. 2. Use of family history data to collect tumor material from several effected family members.

with other familial cases collected in Gliwice, Poland by the Chemotherapy Clinics, the Genetic Councelling Service and the Surgery Clinics and in German by the German Consortium for Hereditary Breast and Ovarian Cancer were used to evaluate the importance of published potentially functional polymorphisms on the risk of breast cancer (30–33) Only the insulin-like growth factor binding protein 3 A-202 polymorphism, which is known to affect circulating IGFBP3 levels (34), was associated with an increased risk (33). With a power of 90% to detect a 1.5-fold increased risk, our studies excluded a major effect of the suggested susceptibility alleles in the sex hormone-binding globulin gene, the integrin β_3 gene and the transforming growth factor β1 and its receptor genes on the risk of breast cancer (30–32).

Our results from the Swedish Family-Cancer Database have shown a strong clustering of breast cancers and melanomas at the population level (35). We collected paraffin-embedded tissue samples from several members of 46 affected families (36). DNA was extracted from the tumor and the surrounding benign tissues and used for mutation screening of the B-RAF gene. In 48% of the melanoma samples, a somatic mutation was detected in codon 600, which is located in the kinase domain of the protein and known to affect the basal kinase activity. No mutations in the breast cancer samples were observed. These results suggest that despite common genetic susceptibility factors that predispose individuals to melanoma and breast cancer within familial clusters, the two tumor types do not share somatic mutations in the B-RAF gene, which remain specific to melanoma.

We have also linked the Swedish Twin Registry to the Swedish Cancer Registry to identify monozygotic twins concordant for breast cancer (37). Paraffin-embedded tissue samples were collected from the hospitals of diagnosis. One randomly chosen sample of tumor DNA from 12 concordant twin pairs was analyzed for BRCA1 and BRCA2 mutations (38). Tumor tissue samples and at least one normal tissue sample per twin pair were available from nine monozygotic twin pairs. These pairs were used for the LOH analysis in an approach to map putative tumor suppressor genes (38, 39). If LOH occurs in a genomic region containing a tumor suppressor gene that confers a strong predisposition to breast cancer, concordant LOH with a loss of the same allele would be expected in tumor DNA of both twins because monozygotic twins share identical haplotypes. No BRCA1 mutations were found, although frequent LOH was observed at 17q. In BRCA2, two polymorphisms and two missense changes were detected. The overall LOH frequency on chromosomes 13 and 17 was not genetically determined, which could be tested in the twin model. Our LOH analyses suggested that another tumor suppressor gene is located between the BRCA2 and the Rb-1 regions at chromosome 13q. On chromosome 16, the telomere of the long arm showed concordant LOH more often than expected. On chromosome 17, the region telomeric to the p53 locus seemed to be more important than p53 in the development of breast cancer. Although BRCA1 seemed not to be involved in the development of breast cancer in the twins studied, another tumor suppressor gene near BRCA1 may be involved.

Monozygotic twins concordant for breast cancer, which were identified by linking the Swedish Twin Registry and the Swedish Cancer Registry, were also used to identify copy number changes in the tumor tissues compared to the normal tissues (40). Comparative genomic hybridization was used to screen the entire tumor genome for gains and losses in 12 samples from 11 patients, including three twin pairs. Six samples exhibited DNA copy number changes. Gains (76%) were more common than losses (24%). Gains or high-level amplifications at 8q were present in all but one of the cases with DNA copy number changes. The most frequent loss, detected in three cases, was at 1p32-pter. One twin pair showed similar changes in four chromosomal locations involving the loss of 1p32-pter and gains at 1q25-qter, 5 and 8q.

4. Conclusions

These examples show the usefulness of the cancer databases in the identification of cancer cases among the individuals, whose biological samples are collected in the biobanks. Matched controls can then be selected from the biobanks. After DNA isolation, genetic studies, including the association studies and mutation analyses, can be carried out, in order to explain the etiology of cancer development and progression as well as to predict response to treatment or cancer survival. In the era of whole-genome approaches, the importance of well-characterized sample sets becomes even more evident, in order to use the resources in the most efficient way.

A now well-recognized problem of many earlier candidate gene studies has been the small sample size (41). Large biobanks will help to solve some of the sample size issues, exemplified by national efforts, such as the UK Biobank (42) or internationional efforts, such as the Nordic biobanks collaboration described in this volume or the European Prospective Investigation into Nutrition and Cancer (EPIC). Large case collections are expected in such studies allowing studies on binary interactions (gene–gene and gene–environment). These are one of the tenets of the multifactorial disease concept and they need to be addressed in studies providing a sufficient statistical power. Some of the large cohorts may allow the selection of familial cases as a particular target for association studies. As familial aggregation, however, is rare (less than 10% in breast cancer (43)), specific case collections from clinically counseled or mutation-tested individuals may afford a more cost-effective way of reaching a suitable sample size. Moreover, the reliability of a family history is much better for data collected in a clinical setting compared to nonfocused interviews. A basic, largely unrecognized dilemma in common disease genomics in the western countries is that heritable causes will be difficult to find because the largely undefined environmental factors have increased the background incidence over ten times the level found in the developing countries (5). Genomics of diseases that are common in western countries (e.g., most cancers, including breast cancers) could probably be more effectively addressed in populations where these diseases are rare, yet noting that gene–environment interactions may be quite different in low- and high-risk populations.

Acknowledgments

Supported by Deutsche Krebshilfe, the Swedish Cancer Society and EU LSHC-CT-2004-503465.

References

1. Syrjakoski, K., Vahteristo, P., Eerola, H., Tamminen, A., Kivinummi, K., Sarantaus, L., Holli, K., Blomqvist, C., Kallioniemi, O.P., Kainu, T. and Nevanlinna, H. (2000) Population-based study of BRCA1 and BRCA2 mutations in 1035 unselected Finnish breast cancer patients. *J Natl Cancer Inst*, **92**, 1529–1531.
2. Risch, H.A., McLaughlin, J.R., Cole, D.E., Rosen, B., Bradley, L., Kwan, E., Jack, E., Vesprini, D.J., Kuperstein, G., Abrahamson, J.L., Fan, I., Wong, B. and Narod, S.A. (2001) Prevalence and penetrance of germline BRCA1 and BRCA2 mutations in a population series of 649 women with ovarian cancer. *Am J Hum Genet*, **68**, 700–710.
3. Lynch, H.T. and de la Chapelle, A. (2003) Hereditary colorectal cancer. *N Engl J Med*, **348**, 919–932.
4. Lichtenstein, P., Holm, N.V., Verkasalo, P.K., Iliadou, A., Kaprio, J., Koskenvuo, M., Pukkala, E., Skytthe, A. and Hemminki, K. (2000) Environmental and heritable factors in the causation of cancer – analyses of cohorts of twins from Sweden, Denmark, and Finland [see comments]. *N Engl J Med*, **343**, 78–85.
5. Hemminki, K., Lorenzo Bermejo, J. and Forsti, A. (2006) The balance between heritable and environmental aetiology of human disease. *Nat Rev Genet*, 7, 958–965.
6. Buchanan, A.V., Weiss, K.M. and Fullerton, S.M. (2006) Dissecting complex disease: the quest for the Philosopher's Stone? *Int J Epidemiol*, **35**, 562–571.
7. Wang, W.Y., Barratt, B.J., Clayton, D.G. and Todd, J.A. (2005) Genome-wide association studies: theoretical and practical concerns. *Nat Rev Genet*, **6**, 109–118.
8. Forsti, A., Jin, Q., Altieri, A., Johansson, R., Wagner, K., Enquist, K., Grzybowska, E., Pamula, J., Pekala, W., Hallmans, G., Lenner, P. and Hemminki, K. (2007) Polymorphisms in the KDR and POSTN genes: association with breast cancer susceptibility and prognosis. *Breast Cancer Res Treat*, **101**, 83–93.
9. Jin, Q., Hemminki, K., Enquist, K., Lenner, P., Grzybowska, E., Klaes, R., Henriksson, R., Chen, B., Pamula, J., Pekala, W., Zientek, H., Rogozinska-Szczepka, J., Utracka-Hutka, B., Hallmans, G. and Forsti, A. (2005) Vascular endothelial growth factor polymorphisms in relation to breast cancer development and prognosis. *Clin Cancer Res*, **11**, 3647–3653.

10. Lei, H., Hemminki, K., Altieri, A., Johansson, R., Enquist, K., Hallmans, G., Lenner, P. and Forsti, A. (2006) Promoter polymorphisms in matrix metalloproteinases and their inhibitors: few associations with breast cancer susceptibility and progression. *Breast Cancer Res Treat*, **103**(1), 61–69.
11. Kaaks, R., Lundin, E., Rinaldi, S., Manjer, J., Biessy, C., Soderberg, S., Lenner, P., Janzon, L., Riboli, E., Berglund, G. and Hallmans, G. (2002) Prospective study of IGF-I, IGF-binding proteins, and breast cancer risk, in northern and southern Sweden. *Cancer Causes Control*, **13**, 307–316.
12. Ferrara, N., Gerber, H.P. and LeCouter, J. (2003) The biology of VEGF and its receptors. *Nat Med*, **9**, 669–676.
13. Morabito, A., Sarmiento, R., Bonginelli, P. and Gasparini, G. (2004) Antiangiogenic strategies, compounds, and early clinical results in breast cancer. *Crit Rev Oncol Hematol*, **49**, 91–107.
14. Kranz, A., Mattfeldt, T. and Waltenberger, J. (1999) Molecular mediators of tumor angiogenesis: enhanced expression and activation of vascular endothelial growth factor receptor KDR in primary breast cancer. *Int J Cancer*, **84**, 293–298.
15. Nakopoulou, L., Stefanaki, K., Panayotopoulou, E., Giannopoulou, I., Athanassiadou, P., Gakiopoulou-Givalou, H. and Louvrou, A. (2002) Expression of the vascular endothelial growth factor receptor-2/Flk-1 in breast carcinomas: correlation with proliferation. *Hum Pathol*, **33**, 863–870.
16. Ryden, L., Linderholm, B., Nielsen, N.H., Emdin, S., Jonsson, P.E. and Landberg, G. (2003) Tumor specific VEGF-A and VEGFR2/KDR protein are co-expressed in breast cancer. *Breast Cancer Res Treat*, **82**, 147–154.
17. Shao, R., Bao, S., Bai, X., Blanchette, C., Anderson, R.M., Dang, T., Gishizky, M.L., Marks, J.R. and Wang, X.F. (2004) Acquired expression of periostin by human breast cancers promotes tumor angiogenesis through up-regulation of vascular endothelial growth factor receptor 2 expression. *Mol Cell Biol*, **24**, 3992–4003.
18. Sasaki, H., Yu, C.Y., Dai, M., Tam, C., Loda, M., Auclair, D., Chen, L.B. and Elias, A. (2003) Elevated serum periostin levels in patients with bone metastases from breast but not lung cancer. *Breast Cancer Res Treat*, 77, 245–252.
19. Awata, T., Inoue, K., Kurihara, S., Ohkubo, T., Watanabe, M., Inukai, K., Inoue, I. and Katayama, S. (2002) A common polymorphism in the 5′-untranslated region of the VEGF gene is associated with diabetic retinopathy in type 2 diabetes. *Diabetes*, **51**, 1635–1639.
20. Shahbazi, M., Fryer, A.A., Pravica, V., Brogan, I.J., Ramsay, H.M., Hutchinson, I.V. and Harden, P.N. (2002) Vascular endothelial growth factor gene polymorphisms are associated with acute renal allograft rejection. *J Am Soc Nephrol*, **13**, 260–264.
21. Crawford, H.C. and Matrisian, L.M. (1994) Tumor and stromal expression of matrix metalloproteinases and their role in tumor progression. *Invasion Metastasis*, **14**, 234–245.
22. Lambert, E., Dasse, E., Haye, B. and Petitfrere, E. (2004) TIMPs as multifacial proteins. *Crit Rev Oncol Hematol*, **49**, 187–198.
23. Noda, M., Oh, J., Takahashi, R., Kondo, S., Kitayama, H. and Takahashi, C. (2003) RECK: a novel suppressor of malignancy linking oncogenic signaling to extracellular matrix remodeling. *Cancer Metastasis Rev*, **22**, 167–175.
24. Rhee, J.S. and Coussens, L.M. (2002) RECKing MMP function: implications for cancer development. *Trends Cell Biol*, **12**, 209–211.
25. Egeblad, M. and Werb, Z. (2002) New functions for the matrix metalloproteinases in cancer progression. *Nat Rev Cancer*, **2**, 161–174.
26. Span, P.N., Sweep, C.G., Manders, P., Beex, L.V., Leppert, D. and Lindberg, R.L. (2003) Matrix metalloproteinase inhibitor reversion-inducing cysteine-rich protein with Kazal motifs: a prognostic marker for good clinical outcome in human breast carcinoma. *Cancer*, **97**, 2710–2715.
27. Antoniou, A.C. and Easton, D.F. (2003) Polygenic inheritance of breast cancer: implications for design of association studies. *Genet Epidemiol*, **25**, 190–202.
28. Houlston, R.S. and Peto, J. (2003) The future of association studies of common cancers. *Hum Genet*, **112**, 434–435.
29. Hemminki, K., Vaittinen, P. and Easton, D. (2000) Familial cancer risks to offspring from mothers with 2 primary breast cancers: leads to cancer syndromes. *Int J Cancer*, **88**, 87–91.
30. Försti, A., Jin, Q., Grzybowska, E., Soderberg, M., Zientek, H., Sieminska, M., Rogozinska-Szczepka, J., Chmielik, E., Utracka-Hutka, B. and Hemminki, K. (2002) Sex hormone-binding globulin polymorphisms in familial and sporadic breast cancer. *Carcinogenesis*, **23**, 1315–1320.

31. Jin, Q., Hemminki, K., Grzybowska, E., Klaes, R., Soderberg, M. and Forsti, A. (2004) Re: Integrin beta3 Leu33Pro homozygosity and risk of cancer. *J Natl Cancer Inst*, **96**, 234–235; author reply 235.
32. Jin, Q., Hemminki, K., Grzybowska, E., Klaes, R., Soderberg, M., Zientek, H., Rogozinska-Szczepka, J., Utracka-Hutka, B., Pamula, J., Pekala, W. and Forsti, A. (2004) Polymorphisms and haplotype structures in genes for transforming growth factor beta1 and its receptors in familial and unselected breast cancers. *Int J Cancer*, **112**, 94–99.
33. Wagner, K., Hemminki, K., Israelsson, E., Grzybowska, E., Soderberg, M., Pamula, J., Pekala, W., Zientek, H., Mielzynska, D., Siwinska, E. and Forsti, A. (2005) Polymorphisms in the IGF-1 and IGFBP 3 promoter and the risk of breast cancer. *Breast Cancer Res Treat*, **92**, 133–140.
34. Deal, C., Ma, J., Wilkin, F., Paquette, J., Rozen, F., Ge, B., Hudson, T., Stampfer, M. and Pollak, M. (2001) Novel promoter polymorphism in insulin-like growth factor-binding protein-3: correlation with serum levels and interaction with known regulators. *J Clin Endocrinol Metab*, **86**, 1274–1280.
35. Plna, K. and Hemminki, K. (2001) Re: High frequency of multiple melanomas and breast and pancreas carcinomas in CDKN2A mutation-positive melanoma families. *J Natl Cancer Inst*, **93**, 323–325.
36. Gast, A., Forsti, A., Soderberg, M., Hemminki, K. and Kumar, R. (2005) B-RAF mutations in tumors from melanoma-breast cancer families. *Int J Cancer*, **113**, 336–337.
37. Ahlbom, A., Lichtenstein, P., Malmström, H., Feychting, M., Hemminki, K. and Pedersen, N.L. (1997) Cancer in twins: genetic and nongenetic familial risk factors. *J Natl Cancer Inst*, **89**, 287–293.
38. Försti, A., Luo, L., Vorechovsky, I., Söderberg, M., Lichtenstein, P. and Hemminki, K. (2001) Allelic imbalance on chromosomes 13 and 17 and mutation analysis of BRCA1 and BRCA2 genes in monozygotic twins concordant for breast cancer. *Carcinogenesis*, **22**, 27–33.
39. Försti, A., Jin, Q., Sundqvist, L., Soderberg, M. and Hemminki, K. (2001) Use of monozygotic twins in search for breast cancer susceptibility loci. *Twin Res*, **4**, 251–259.
40. el-Rifai, W., Tarmo, L., Hemmer, S., Forsti, A., Pedersen, N., Lichtenstein, P., Ahlbom, A., Soderberg, M., Knuutila, S. and Hemminki, K. (1999) DNA copy number losses at 1p32-pter in monozygotic twins concordant for breast cancer. *Cancer Genet Cytogenet*, **112**, 169–172.
41. Wacholder, S., Chanock, S., Garcia-Closas, M., El Ghormli, L. and Rothman, N. (2004) Assessing the probability that a positive report is false: an approach for molecular epidemiology studies. *J Natl Cancer Inst*, **96**, 434–442.
42. Davey Smith, G., Ebrahim, S., Lewis, S., Hansell, A.L., Palmer, L.J. and Burton, P.R. (2005) Genetic epidemiology and public health: hope, hype, and future prospects. *Lancet*, **366**, 1484–1498.
43. Hemminki, K. and Czene, K. (2002) Attributable risks of familial cancer from the Family-Cancer Database. *Cancer Epidemiol Biomarkers Prev*, **11**, 1638–1644.

Chapter 24

Monitoring, Alarm, and Data Visualization Service on Sample Preparing and Sample Storing Devices in Biobanks

Halla Hauksdóttir, Kristín Jónsdóttir, and Andres Thorarinsson

Abstract

An important feature in "Good Biobanking Practices" is to monitor and log conditions of sample storing devices. The Institute of Laboratory Medicine at Landspitali University Hospital in Reykjavik, Iceland, has installed a temperature monitoring and alarm system for freezers, incubators, and refrigerators as a part of its quality program. This paper describes the key features of the system, how it works, and what has been learned.

Key words: Electrical monitoring of storing devices, Good biobanking practices, Quality program

1. Introduction

It is well known that there is a quality issue regarding the operation of incubators, coolers, and freezers in laboratories. The main problem has been out-of-limit temperatures caused by cabin doors left slightly open, unstable control system, or power failure. Therefore, the Institute of Laboratory Medicine at Landspitali University Hospital has been seeking solutions to monitor and log the temperatures and other parameters of the operation and to issue secure alarms when values are out of limit. A logging and alarm system was installed 2 years ago in one department, and this year the software part of the system was upgraded and many more devices in two other departments were connected.

Joakim Dillner (ed.), *Methods in Biobanking*, Methods in Molecular Biology, vol. 675,
DOI 10.1007/978-1-59745-423-0_24,

2. Technical Aspect

2.1. Location of Equipment

The laboratories are located in several buildings of the Landspitali University Hospital in Reykjavik, see Fig. 1.

All the buildings of the Landspitali University Hospital are interconnected with a high-speed local area network (LAN). The Vista Data Vision (VDV[1, 2, 3]) applications are installed on a PC-server in the computer center and pull data from data loggers at each location.

2.2. Data Logging System

At each of the new locations configured in 2007, a data logger of type Campbell Scientific CR1000 was installed (www.campbellsci.com). The datalogger is able to connect to temperature sensors type PT100 and thermistor in incubators, coolers, and freezers, and to 4–20 mA current signals from CO_2 sensors. The CR1000 has 16 inputs for sensors, and use a multiplexer to increase the number of inputs up to 48. The CR1000 is programmed to make

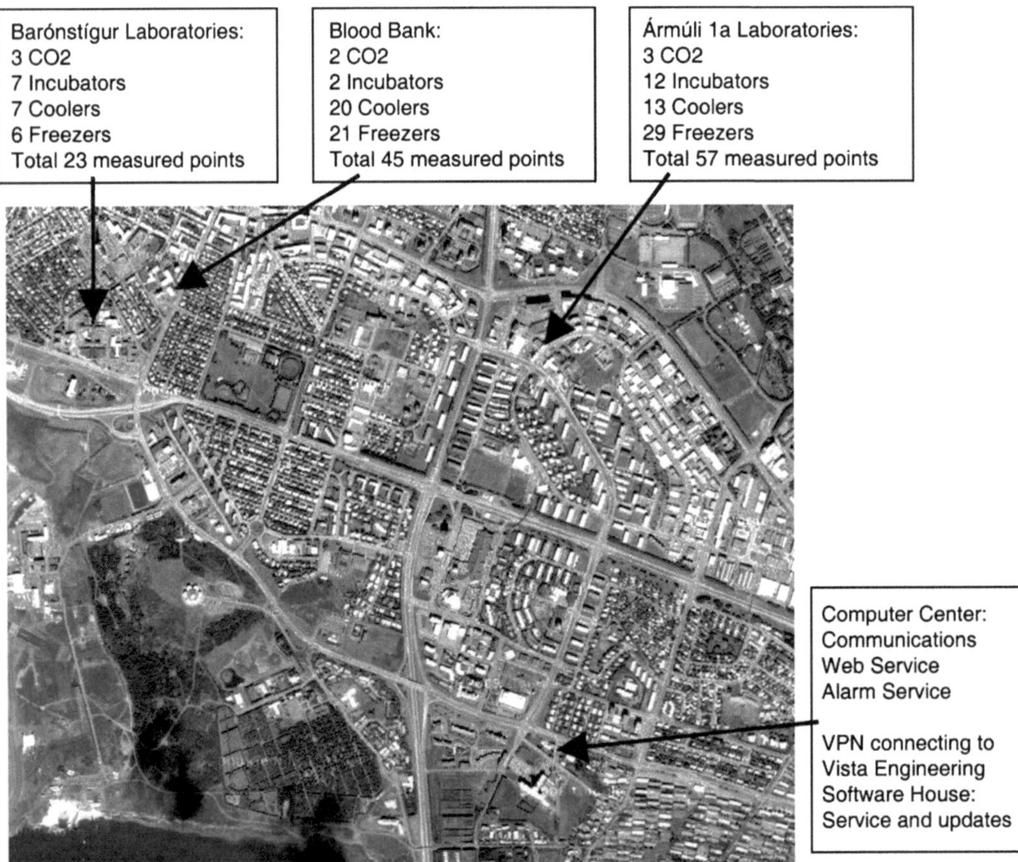

Fig. 1. Overview of the locations of the laboratories and the computer center.

a measurement every minute and to store the average value of each sensor in its memory every 5 min. Each CR1000 has a backup power.

2.3. Data Retrieval, Storage, Alarm, and Visualization System

As for data retrieval, LoggerNet software (www.campbellsci.com) is used to read data from each data logger every 5 min. For all other data handling, VDV is used (www.vistadatavision.com). VDV has three main applications:

- db.robot.c for reading new data from LoggerNet, check values for alarm limits, and store data in relational database;
- db.data.browser for configuration of graphs and pages;
- db.web.browser for set up of interactive Web service and running a Web service.

2.4. Sensors and Installation in Incubators, Coolers, and Freezers

The Campbell Scientific type 107 thermistor is used in all incubators, coolers, and freezers down to −35°C, the range is from +50°C to −35°C.

For freezers colder than −35°C, the PT100 type temperature sensor manufactured by Pentronics (www.pentronics.se) is used. These sensors were selected as its housing and cabling is able to withstand −80°C and even colder temperatures.

The Vaisala type GMT221 (www.vaisala.com/instruments) is used for CO_2 sensing with a 0–20% CO_2 range and incorporate a numeric display.

The temperature sensors are fitted inside the incubators, coolers, and freezers, and mounted at the side of the interior. The cabling is routed through the same grommets as other cables into the interior. Only in few instances, a cable hole had to be drilled.

3. Web-Based Service

3.1. Main Display

All staff members have access to sensor trend lines using an access-controlled Web site. Figure 2 shows the main display, where graphs are organized in sites and pages. Each page can display up to 6 graphs, each with up to 6 sensors or a total of 36 sensors on display simultaneously. There are three main sites (operation in three buildings) and the site on display is organized using four sub-pages.

During a normal working day, measurements show how values change when work start in the morning and end in the afternoon. A good example is the CO_2 measurements in the upper left graph. CO_2 value is steady during the night but drops when work start in the morning and the cabin door is opened and closed. Once working day comes to an end, the CO_2 values rise again.

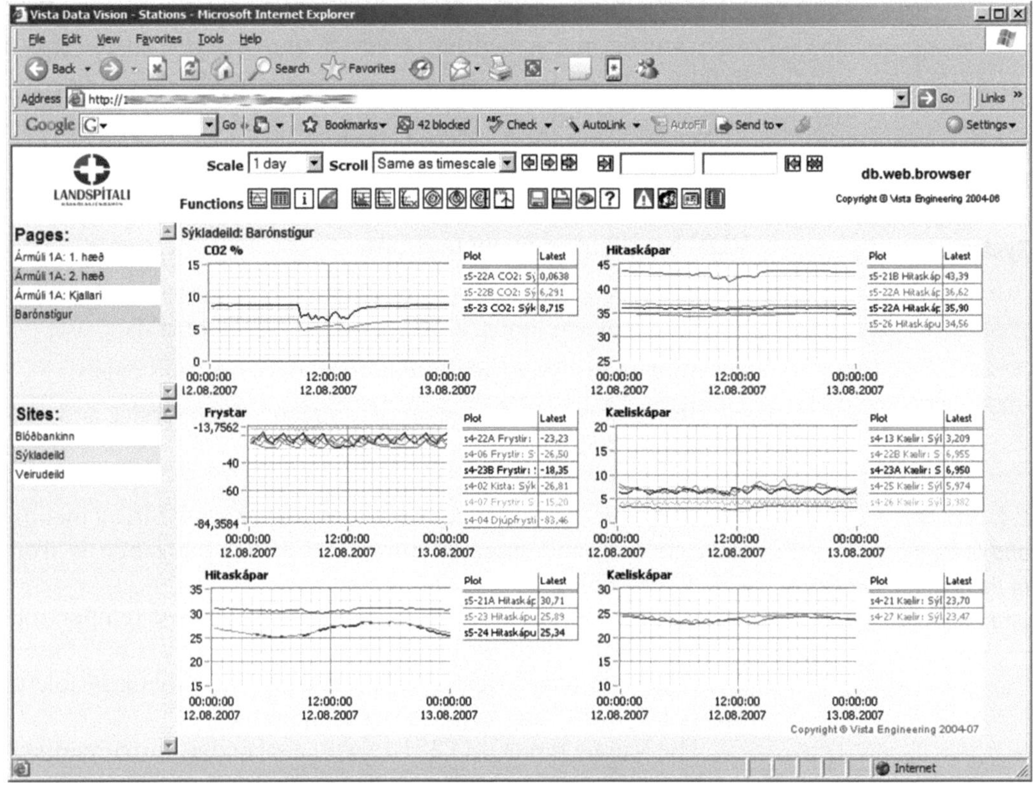

Fig. 2. Main display of the interactive Web site for measurements. On display is a single day data from one of the laboratories.

3.2. Single Graph Display

Clicking on any graph allows the user to envision full size of the graph (Fig. 3). Not only are the graphs more easy to read, but mini report is also displayed to the right, showing maximum, average, and minimum value for each sensor on display and for the time period selected. The check boxes to the right of the graph (five check boxes in vertical row) are used to uncheck trend lines to make them become invisible and, therefore, to unclutter the graph when needed.

3.3. Web Maps

A Web map is used to display a large number of measured values alongside information about the alarm status. Figure 4 shows a background floor plan of a laboratory, markings for each incubator, cooler, and/or freezer, and the latest measured value. Background color for each value indicates the present alarm status, where *green* is ok, *yellow* is outside the first limits, and *red* is outside the second limits. In Fig. 4, there are five items with a green background, one item with a yellow background, and one item with a red background. The supervisor uses the Web map to get a clear overview of the current status and the location of the problems, and it is used as an opening window for Biobank

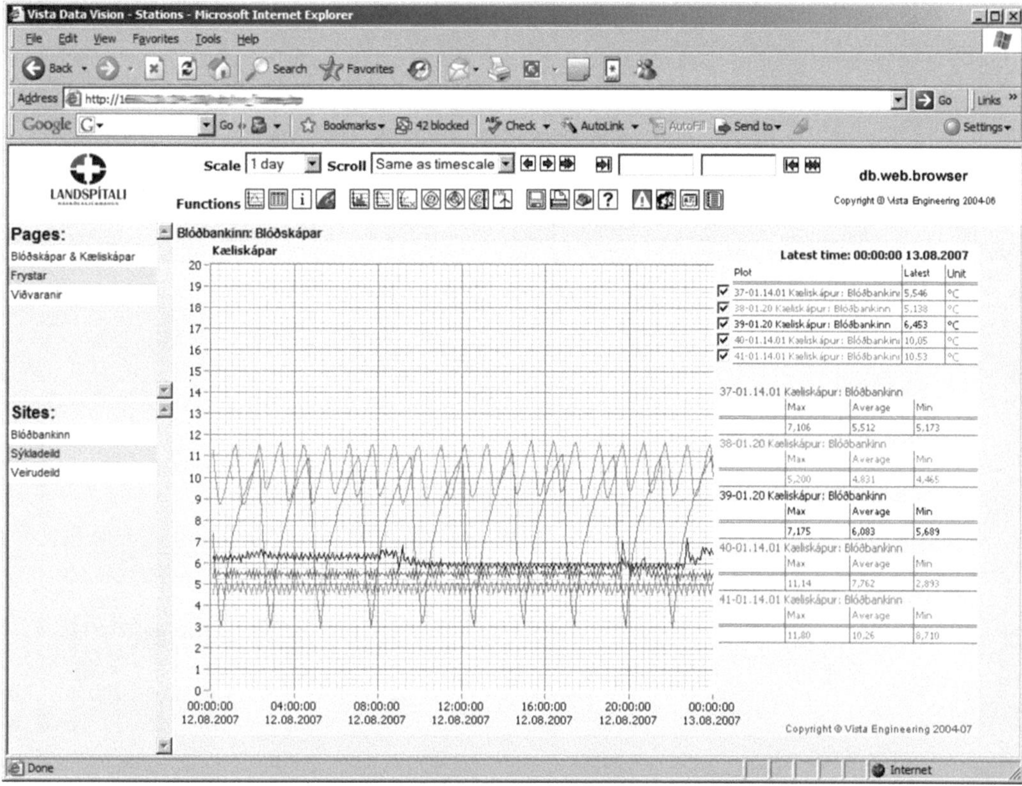

Fig. 3. Single graph display showing single day trend lines for five sensors.

personnel to make it easier to find out where the apparatus with alarm signal is stationed.

3.4. Alarms

An important part of the monitoring of temperature values is the Alarm Service, see Figs. 5 and 6. For each measured value, an alarm threshold may be configured, with L and H marks for inner thresholds and LL and HH for outer thresholds. If a measured value drifts outside inner or outer thresholds, an alarm is sent as an email and/or SMS text message to selected staff members. Alarms can be acknowledged on a Web page by clicking on the confirm button. There is a provision to write explanation text when an alarm is acknowledged (confirmed), which is stored in the database alongside data from that sensor.

All alarm thresholds for all sensors can be activated and adjusted on an access-controlled Web page, see Fig. 6. Alarms are sent according to the settings on "Send Low & High Warning" and "Send LowLow & HighHigh Warning." The user has the option to set up groups of staff members and have the VDV applications send alarms according to a simple or complex schema.

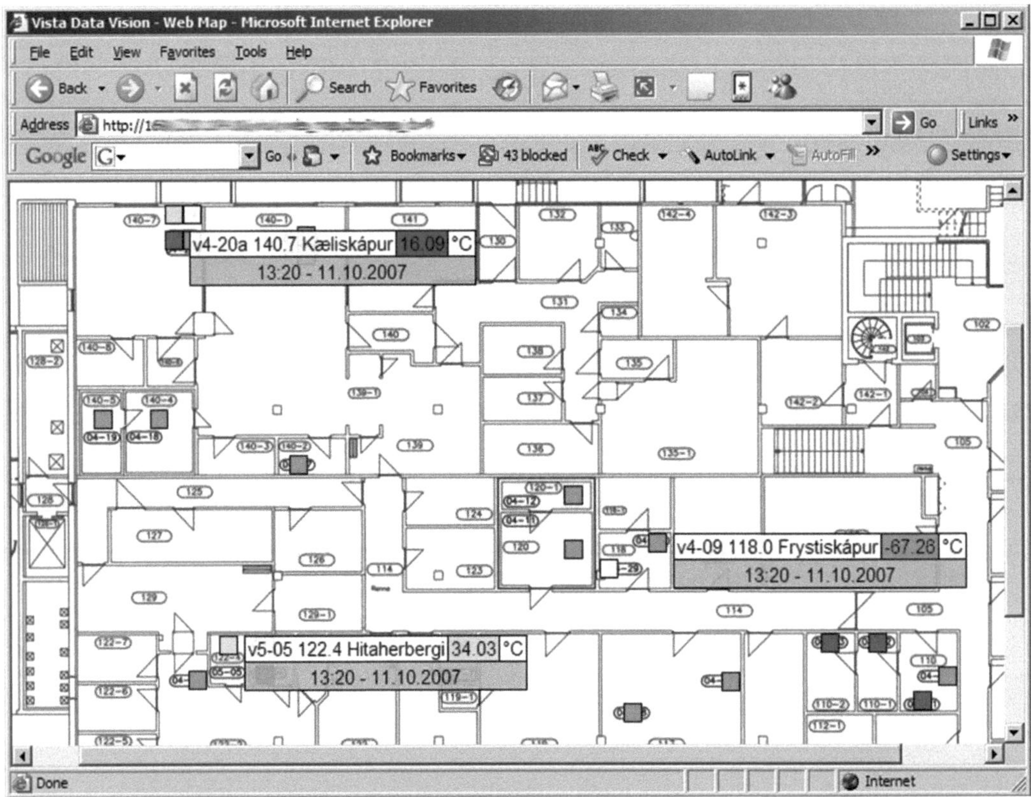

Fig. 4. A Web map gives an excellent overview of the current status of a large number of incubators, coolers, and freezers. This is a partial image of the Web map display.

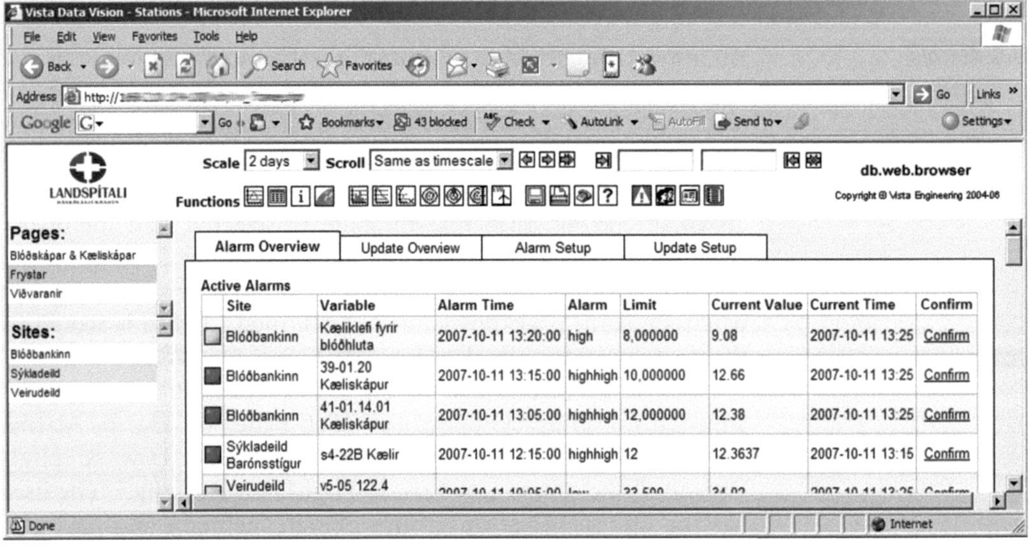

Fig. 5. A Web access to alarm overview and acknowledgment of alarm situation.

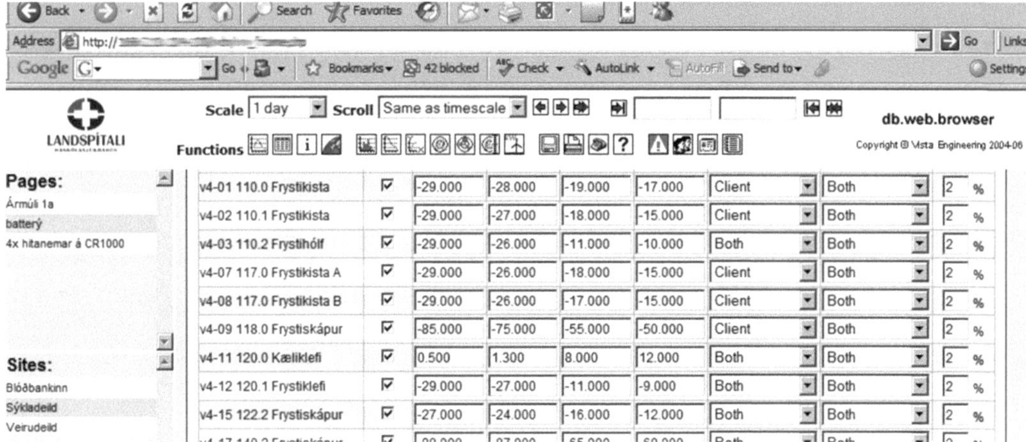

Fig. 6. Configuration of alarm thresholds is done on an interactive Web page.

4. What has been Learned

4.1. Normal Operation

As expected, daily routine in the laboratories affects temperatures in incubators, coolers, and freezers. During the nighttime, temperatures and CO_2 values become stable but when work starts in the morning, there is a noted change in temperature. Some examples are shown in Figs. 7–10.

4.2. Operational Faults

Several operational faults have been logged during the time the monitoring system has been running. Figure 11 shows a trend line for a freezer that went out of order. This was before the alarm system for this freezer had been activated. The temperature rises from –20°C up to +9°C. Luckily, samples inside did not completely thaw, but it is important to have this instance logged in case these samples will be used in the future.

When freezers are opened, some types warm up quickly. The graph in Fig. 12 with five trend lines shows a steady condition in most of the devices, with the exception of one sensor for –80°C freezer, which has crossed the –60°C limit several times during daily operation of inserting and extracting samples. All the other sensors show normal situation. De-icing can be seen on the lines as peaks in temperature every 6 and 12 h.

A graduation of temperature measurements and due rectification at each spot must be done before the alarm system is activated, as there can be a slight difference between measured values and actual temperature.

4.3. The Temperature Controls in Stock Incubators, Coolers, and Freezers

It has been interesting to learn that the temperature controls in incubators, coolers, and freezers are most different from one another. Some temperature controls keep a steady temperature, but most often the temperature fluctuates around the set point.

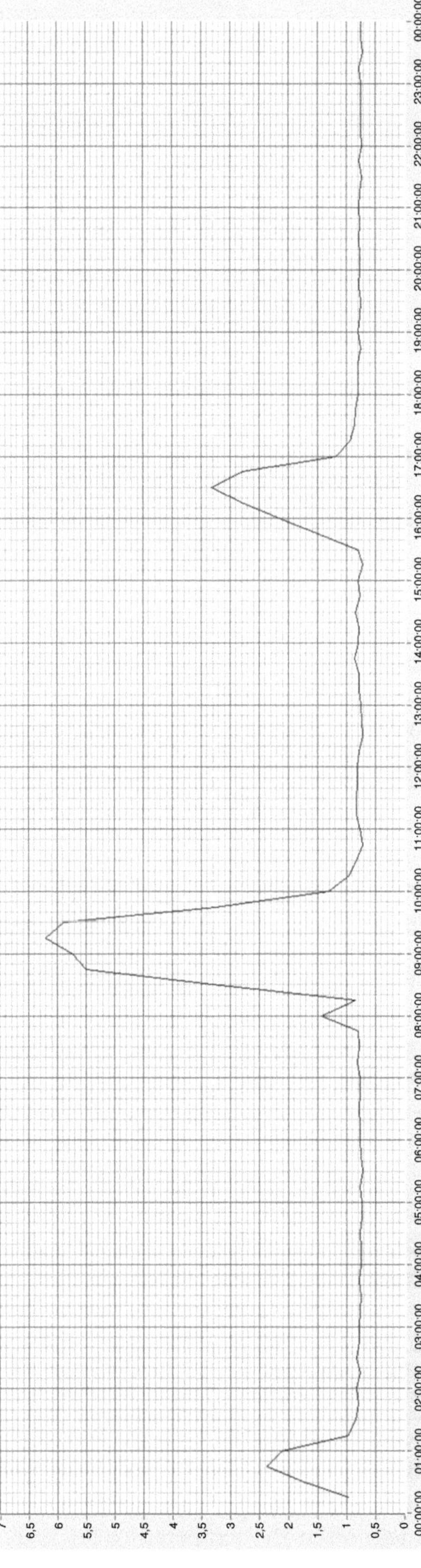

Fig. 7. A temperature trend line in a cooler showing temperature during daytime laboratory work and the de-freezing operation. The graph shows 24 h, starting at midnight.

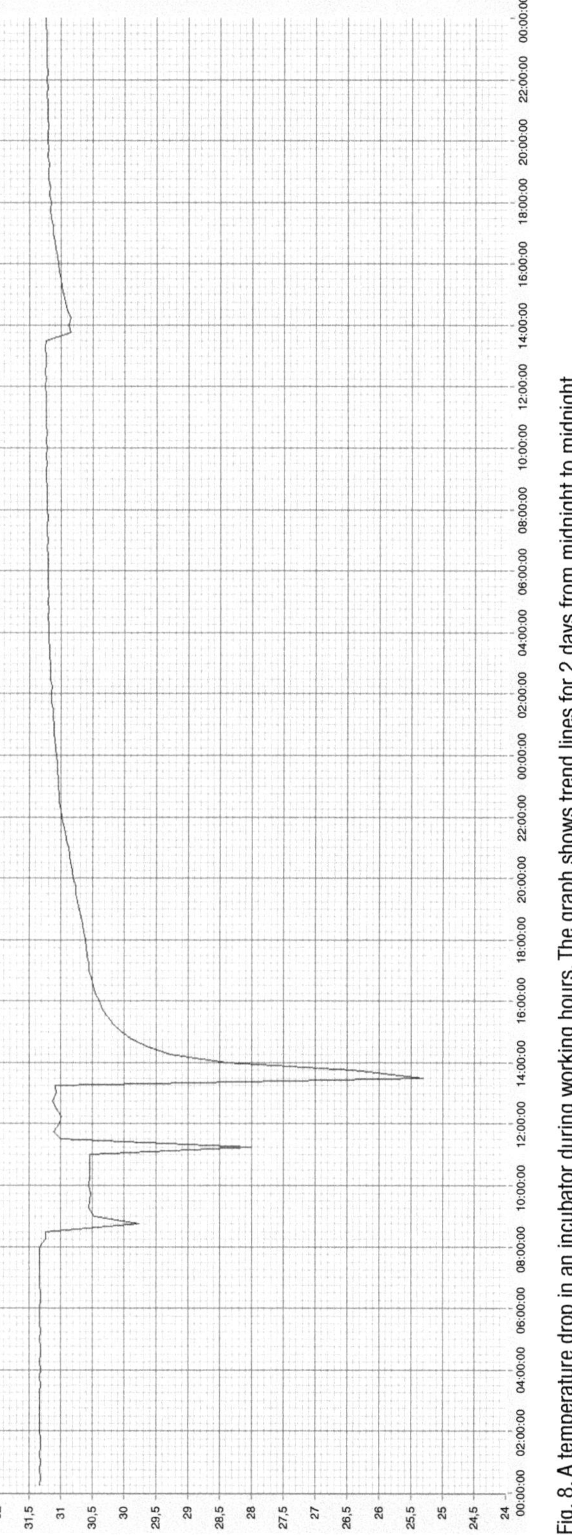

Fig. 8. A temperature drop in an incubator during working hours. The graph shows trend lines for 2 days from midnight to midnight.

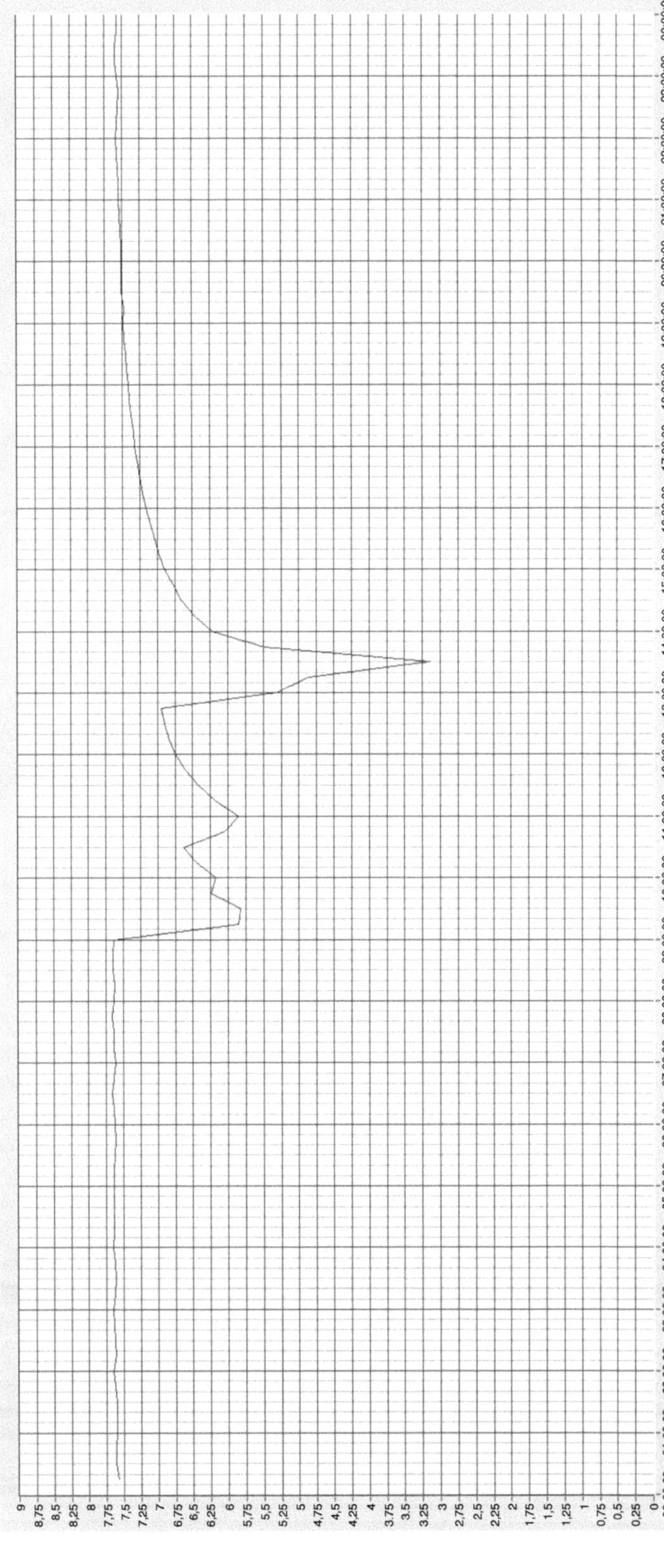

Fig. 9. The CO_2% value is steady during nighttime but drops as soon as work begins in the morning. The graph shows 24 h, starting at midnight.

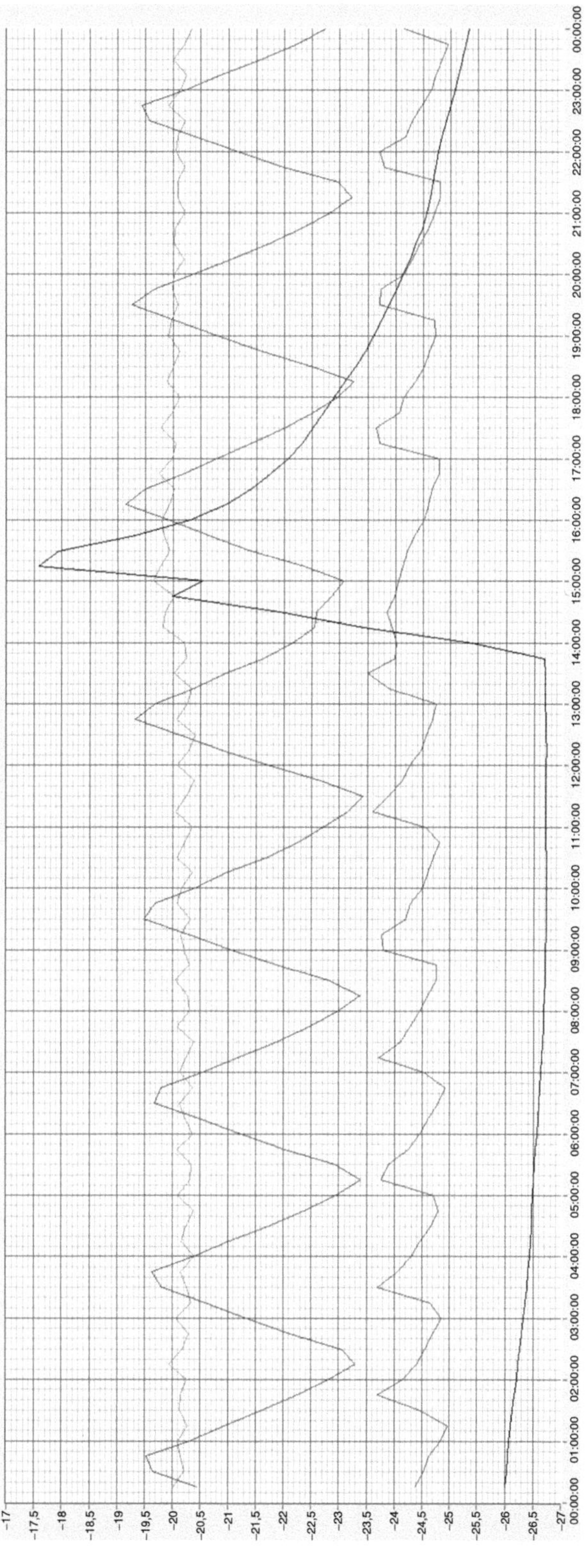

Fig. 10. The trend lines show temperatures in four freezers. The graph shows 24 h, starting at midnight.

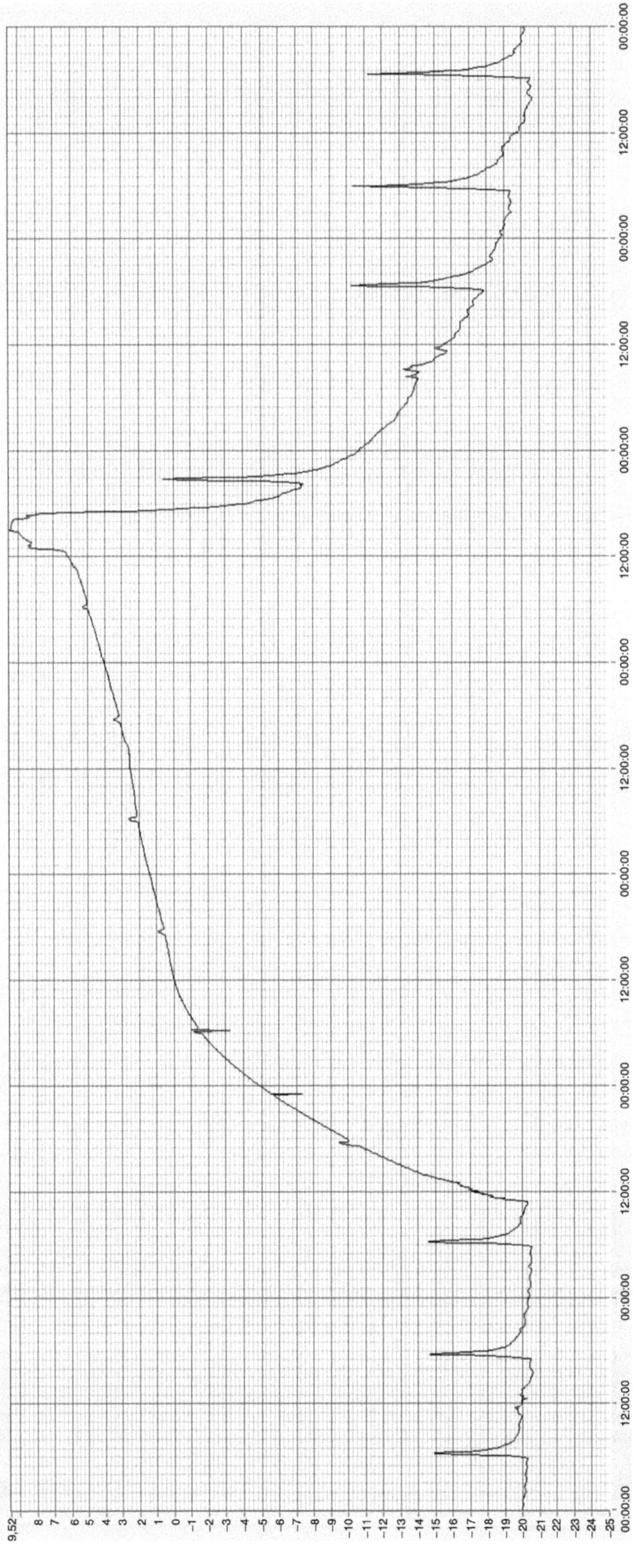

Fig. 11. Temperature rises in a freezer due to compressor fault. The graph shows 7 days.

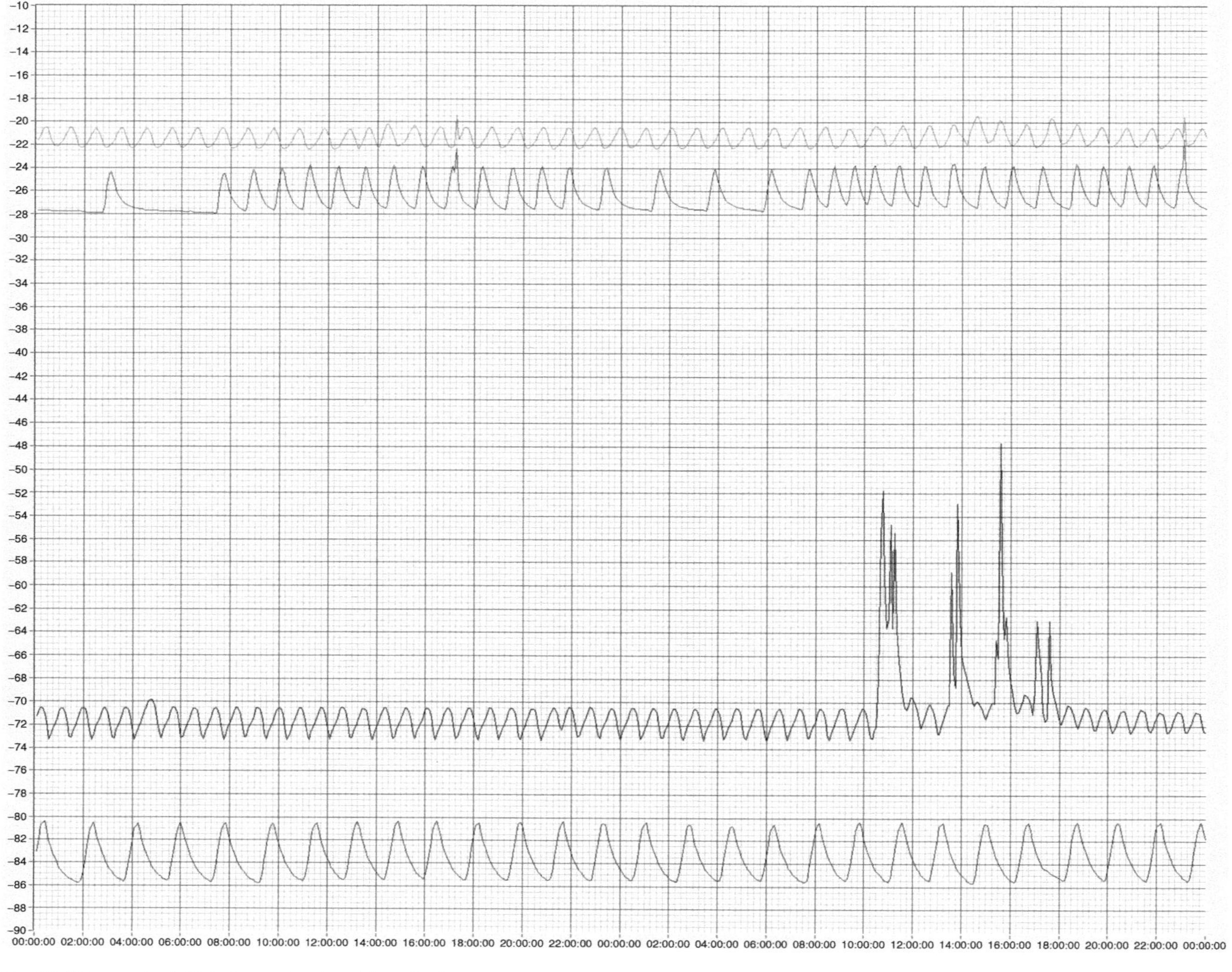

Fig. 12. This graph of 48 h shows four trend lines for freezers.

There are also temperature controls that seem to have no fixed set point but move around seemingly only controlled by how often the cabin door is opened. In Figs. 13 and 14, two examples are given alongside with an explanatory text.

5. Notes

1. Before activating an alarm system like this, a thorough observation of real temperature variations of each apparatus must be made. There is a big difference in chosen temperature variations and actual variations due to type and age of equipment. If real deviations are not discovered and taken into account before the alarm system is activated, hundreds of false alarms will be sent out 4.3; Fig. 6.

6. Conclusion

The benefits of a full-fledged monitoring and alarm system to be used to monitor the temperatures of incubators, coolers, and freezers in the laboratory landscape at a Landspitali University Hospital seem clear: it makes the operation safer, lowers the risk of losing valued samples by mistakes or equipment failure, and greatly reduces the time used to check and log temperatures. It also helps in evaluating good or bad brands in equipment making. An interesting result of the measurements is that the variance in quality of incubators, coolers, and freezers is easily distinguished. Some stock equipment are doing well while others have extended temperature fluctuations that are not acceptable. By using the temperature and CO_2 long-term logged measurements, quality equipment can be identified and therefore, during equipment renewal, lower quality can be avoided. In short, this monitoring system saves money and time for the Biobanks and the departments as a whole.

The system eliminates all paper documentation of temperature monitoring and has saved much time for the laboratory staff. The original system installed in 2005 has in fact paid for itself. Time otherwise spent on checking and recording temperature values can now be spent on other important laboratory work.

Automatic data logging into a database and the alarm service are most important for quality management of a Biobank. Also, easy Web access to all trend lines and data and the alarm handling are of great value. It becomes the backbone of a streamlined quality data handling system and ensures highest quality management of sample storage at the lowest operational cost for years to come.

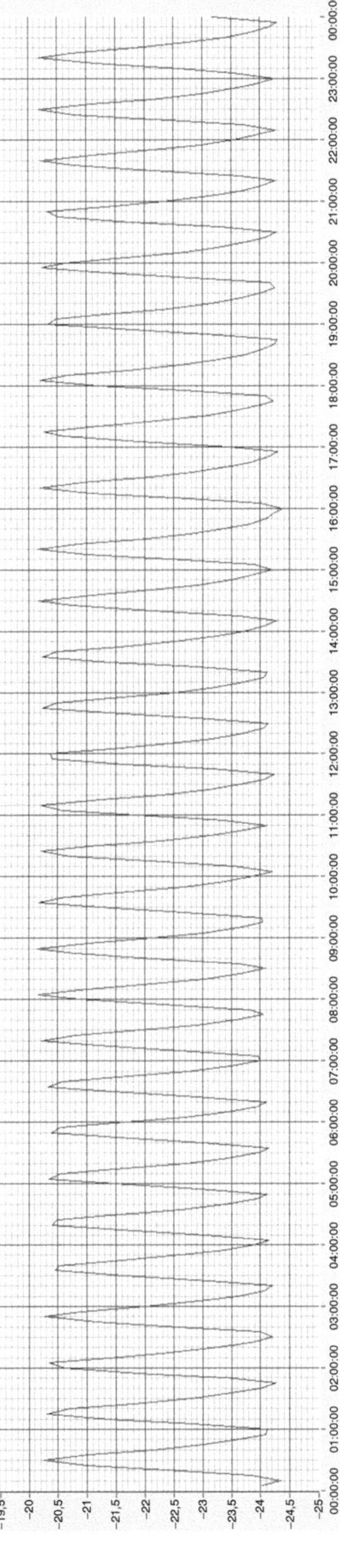

Fig. 13. This freezer has a control period of approximately 45 min. During this period, the temperature drops to −21°C, then the controller starts the freezer that cools the interior down to −25°C. It seems like the isolation of the freezer is so thin that the temperature rises 4°C in 25 min, and then it takes 20 min to cool down again.

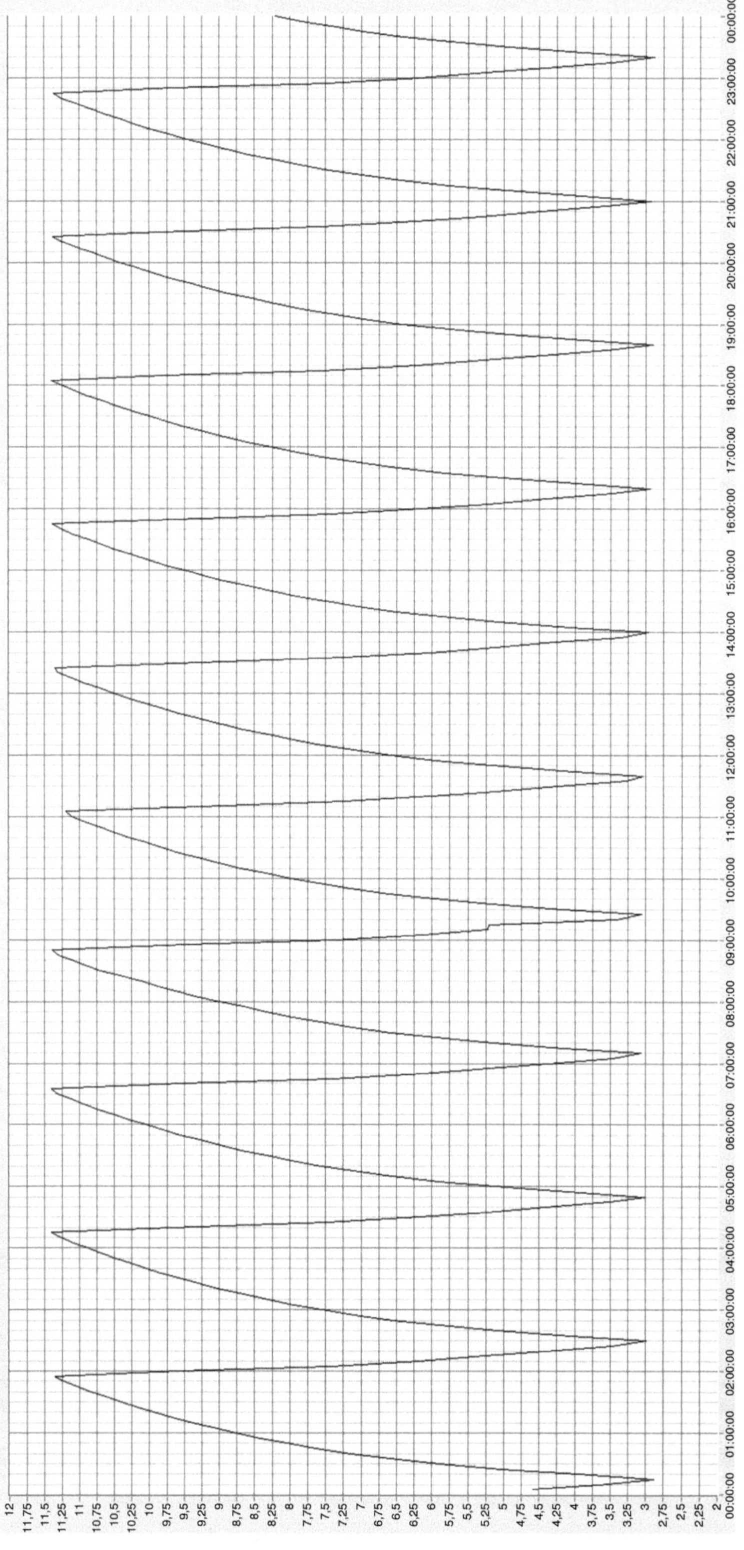

Fig. 14. This cooler has a control period of approximately 2 h. Once the temperature has reached the trigger temperature of 11°C, the cooling engine starts and cools the interior down to 3°C in about 30 min. Then there is a wait time of 90 min with no cooling. Due to the monitoring log, this refrigerator was considered broken and a new one has been installed.

References

1. Thorarinsson, A. (2007). "Methods for automatic storage, visualization and reporting in datalogging applications". Paper accepted for presentation at the 7th International Symposium on Field Measurements in GeoMechanics.
2. Bessason, B., Eiriksson, G., Thórarinsson, O., Thórarinsson, A. and Einarsson S. (2008) Automatic detection of avalanches and debris flows by seismic methods *Journal of Glaciology* **53**(182), 461–472.
3. Elefsen, S. Ó., Haraldsson, H., Gíslason,S. R. and Kristmannsdóttir, H. (2002). "Real-time monitoring of glacial rivers in Iceland." In: *The Extremes of the Extremes: Extraordinary Floods*, Snorrason, Á., Finnsdóttir, H.P. and Moss, M.E. (eds.). (Proc. Reykjavík, Iceland Symp. July 2000), 199–204, IAHS Publ. no. 271.

Chapter 25

Fresh Frozen Tissue: RNA Extraction and Quality Control

Johan Botling and Patrick Micke

Abstract

Since RNA is believed to be the most vulnerable molecular component of unfixed tissue, preserved RNA integrity can be used as a general quality indicator in fresh frozen tissue biobanks. As the size of samples and biopsies often is small, in the range of millimeters or milligrams, it is important to implement quality control procedures adapted to minute the amounts of tissue. To this end, we here describe RNA extraction from one or a few frozen tissue sections and subsequent analysis of structural RNA integrity by microcapillary gel electrophoresis.

Key words: Biobank, Fresh frozen tissue, RNA extraction, RNA quality, Bioanalyzer, RIN

1. Introduction

Storage of tissue with intact morphology, proteins, DNA, and RNA for use in research or diagnostics is the ultimate objective of human tissue biobanks (1). Avoiding RNA degradation represents a major challenge in this process. Many of us share the mystic experience that RNA is able to vanish from a working sample in an unpredictable manner. In correspondence to these subjective laboratory perceptions, there are numerous mechanisms for RNA decay at work along the chain of biobanking logistics – in vivo during surgery, ex vivo during specimen transport to the biobank, and during thawing and manipulation after snap-freezing. For most organ tissues, no published hard data is available regarding susceptibility to these RNA degradation activities.

Physical trauma and warm ischemia during surgery result in significant gene expression changes (2). This regulation represents a stress response in metabolically active cells in part mediated by targeting certain mRNA transcripts for degradation in the exosome. These physiological mechanisms are important to

Joakim Dillner (ed.), *Methods in Biobanking*, Methods in Molecular Biology, vol. 675,
DOI 10.1007/978-1-59745-423-0_25, © Springer Science+Business Media, LLC 2011

consider when evaluating gene expression signatures from clinical biobank samples. From a biobank perspective, these peri-operative variables cannot be controlled, but prompt cooling of surgical specimens on wet ice after devitalization is recommended to inhibit these processes (see Chapters 16 and 19). To describe the next phase in biobanking routines – transfer of a cooled specimen at ±0 to +4°C to the biobank facility – the term "cold ischemia" has been borrowed from the field of organ transplantation. Notably, the impact of cold ischemia time, in the context of transplantation research, is usually studied in tissues after reperfusion and reheating to +37°C. Therefore, the described effects on RNA-quality and gene expression patterns are probably not relevant for biobanking. Instead, we and others have shown that RNA and gene expression patterns are surprisingly stable in fresh unfixed specimens during transport before snap-freezing (3–6). Other groups report a significant influence of lag time between surgery and freezing on RNA-quality and gene expression patterns (7–9). However, these effects were seen in tissues left at room temperature postoperatively and may not be informative for transported cooled specimens. After sample storage at –80°C, thawing presents an important, but controllable, threat to the RNA (10). As cell membranes are disrupted by a freeze-thaw cycle the intrinsic "RNase activity," present to a variable extent in all tissue extracts, will degrade RNA transcripts in an autolytic fashion independent of cellular metabolism. Therefore, thawing should be avoided in all routine biobank and research procedures. If thawing cannot be avoided, for example, during laser microdissection and in-situ hybridization, RNA-stabilization should be considered. RNAlater, ethanol- and ZnCl-based fixation have been used successfully in different protocols (11). A general use of RNAlater in biobanking manuals is debated. Mutter et al. showed that RNAlater did not significantly influence RNA expression profiles on gene expression arrays (12). Nevertheless, we have seen deterioration of tissue morphology and an influence on gene expression analysis by quantitative PCR (3).

In any event, results based on the analysis of partially degraded RNA are most often not reliable and must be interpreted with caution. For use in real-time quantitative PCR, expression microarrays, ribonuclease protection assays, in vitro translation and cDNA library construction, the integrity of the input RNA will be pivotal (10, 13). Therefore, validated assays for RNA quality need to be integrated in all biobanking practices aimed at RNA-based research. An RNA-quality test should optimally be simple, cheap, and reliable. Importantly, it should only dissipate a small fraction of a precious clinical sample and thawing must be avoided. Here, two reliable RNA extraction procedures are described together with an accurate technique to evaluate structural RNA integrity by microchip gel electrophoresis.

2. Materials

2.1. RNA Isolation from Frozen Tissue: Qiagen RNeasy Mini Kit

1. RNeasy Mini Kit (Qiagen, Hilden, Germany)
2. RNasin (20 U/μl, Promega, Madison, WI, USA)
3. Cryogel (O.C.T., Tissue-Tek, Sakura Finetek, Zoeterwoude, Netherlands)
4. RNaseZAP (Ambion, Austin, TX, USA)
5. Cryostat; β-Mercaptoethanol; Ethanol (99.5%); Forceps or syringes; Eppendorf tubes (1.5 ml)

2.2. RNA Isolation from Frozen Tissue: Trizol Method

1. Trizol reagent (Invitrogen, Carlsbad, CA, USA)
2. PelletPaint (Novagen, San Diego, CA, USA)
3. RNase-free water (Ambion)
4. RNasin (20 U/μl, Promega)
5. RNaseZap (Ambion)
6. Cryogel (O.C.T., Tissue-Tek, Sakura)
7. Cryostat; Ethanol (99.5%); Forceps or syringes; Eppendorf tubes (1.5 ml); Chloroform; Isoamylalcohol; Isopropanol; Forceps or syringes; Vortex

2.3. RNA Quality: Agilent 2100 Bioanalyzer

1. Agilent 2100 Bioanalyzer (Agilent Technologies, Santa Clara, CA, USA)
2. Chip Vortex mixer (IKA, Agilent Technologies)
3. Chip Priming Station (Agilent Technologies)
4. RNA 6000 ladder (Ambion). Should be aliquoted and stored at −80°C until use.
5. RNA 6000 Nano Assay Kit (Agilent Technologies). The gel has to be filtered before applying the dye (see manufacturer's protocol). After filtering, the gel is aliquoted and can be stored at 4°C for 4–8 weeks. Protect the dye concentrate from light. Light exposure may cause decomposition of the dye
6. RNase-free water (Ambion)

3. Methods

3.1. RNA Isolation from Frozen Tissue: Qiagen RNeasy Mini Kit

For tissues with a section area of approximately 5 × 5 mm RNA yields are typically in the microgram range. Carefully read the Qiagen RNeasy Mini handbook before using this protocol.

In advance:

- Clean the cryostat chamber and cool to −20°C (see Note 1).

- Check that ethanol has been added to Buffer RPE. If not, add four volumes of ethanol (99.5%) to the Buffer RPE bottle.
- Label Eppendorf tubes (1.5 ml) and add 600 μl Buffer RLT. Put the tubes on ice.

1. Retrieve the vials or blocks containing frozen tissue from the –80°C freezer and keep them on dry ice until cryosectioning.
2. Transfer a vial or block to the cryostat chamber. Let the sample equilibrate for a few minutes at –20°C. Fasten tissue or block to cryosection holder by the use of cryogel.
3. Trim the tissue and cut 10–20 μm sections. Using a sterile syringe or a clean forceps to transfer the frozen sections to the Eppendorf tube containing 600 μl Buffer RLT. The frozen section should not thaw until immersed in Buffer RLT. The standard procedure is to prepare RNA from 3 to 10 sections.
4. Vortex the tube thoroughly and place it on wet ice (±0°C). Clean the cryostat knife and instruments and proceed with the next sample (step 2). When all samples are dissolved in Buffer RLT, continue with step 5 (or store at –80°C for proceeding with the isolation procedure at a later time point).
5. Add 6 μl β-mercaptoethanol. From now on, all steps are performed at room temperature.
6. Add one volume (600 μl) of 70% ethanol and mix by pipetting.
7. Immediately add 600 μl of the sample to an RNeasy column placed in a 2 ml collection tube (supplied in kit). Centrifuge for 15 s at >8,000 g. Discard the flow-through. Add the remaining 600 μl to the column and centrifuge for 15 s at >8,000 x g. Discard the flow-through.
8. Add 700 μl Buffer RW1 to the column. Centrifuge for 15 s at >8,000 g. Discard the flow-through. If performing an on-column DNase digestion (see Appendix E in the handbook), instead add 350 μl Buffer RW1 to the column. Centrifuge for 15 s at >8,000 g and discard the flow-through. Add 10 μl DNase I stock solution to 70 μl Buffer RDD. Mix gently. Add 80 μl of the DNase I incubation mix to the column and incubate for 15 min at room temperature. Add another 350 μl Buffer RW1 to the column. Centrifuge for 15 s at >8,000 g and discard the flow-through. Continue with step 9.
9. Transfer the column to a new collection tube (supplied in kit). Add 500 μl of Buffer RPE to the column. Centrifuge 15 s at >8,000 g. Discard the flow-through.
10. Add another 500 μl of Buffer RPE to the column. Centrifuge for 2 min at >8,000 g.

11. Place the column in a new collection tube and centrifuge at full speed for 1 min.
12. Transfer the column to a new Eppendorf tube. To elute, add 30 μl RNase free water (supplied in kit) to the column. Centrifuge for 1 min at >8,000 g. *(Optional.* To obtain a higher concentration of RNA, repeat the elution step using the first eluate.).
13. Keep the eluted RNA on ice. Add 1 μl of RNase inhibitor before long time storage at -80°C.

3.2. RNA Isolation from Frozen Tissue: Trizol Method

This section describes the isolation of RNA from frozen tissue using the Trizol method. The RNA yield will be in the microgram range for cellular tissues with a section area of about 5 × 5 mm.

In advance:

- Clean and cool the cryostat chamber to -20°C.
- Label eppendorf tubes (1.5 ml) and fill them with 300 μl Trizol. Put these tubes on ice.
- Label other sets of eppendorf tubes for the later steps in the protocol.
- Mix an appropriate volume of chloroform: isoamylalcohol 24:1.

1. Retrieve the vials or blocks containing frozen tissue from the -80°C freezer and keep them on dry ice until cryosection.
2. Transfer a vial or block to the cryostat chamber. Fasten tissue or block to cryosection holder by cryogel.
3. Trim tissue and make 10–20 μm sections. Transfer each frozen section separately by the use of a sterile syringe or clean forceps, and put it in an eppendorf tube with 300 μl Trizol. The frozen section should not thaw until immersed in Trizol. The standard procedure is to prepare RNA from 3 to 10 sections.
4. Shake and vortex tube and put back on ice. Repeat until tissue is completely dissolved. Clean cryostat knife and instruments and go on to the next sample – **step 2**. When all samples are dissolved in Trizol move on to **step 5**. From now on, all steps are performed at room temperature.
5. Add 60 ml chloroform isoamylalcohol to each tube. Mix and vortex.
6. Separate the phases by centrifugation at approximately 11,500 g (13,000 rpm if the rotor radius is 6 cm) for 10 min.
7. Transfer the upper aqueous phase to a new tube. Add 2 μl coprecipitant (PelletPaint) and 160 μl isopropanol. Mix and incubate for 5 min.

8. Precipitate by centrifugation at <11,500 g for 10 min.
9. Identify the pellet as a red spot at the bottom of the tube. Carefully remove supernatant by pipetting. Wash pellet by carefully pipetting 200 µl 70% ethanol onto the pellet. Incubate for 1 min. Remove ethanol by careful pipetting. Repeat wash.
10. Leave the tubes open and let the pellet dry for 15 min. Cover the tubes by an unfolded kleenex sheet.
11. Resuspend pellet in 30 ml RNAse free water. Put the tube on ice.
12. Add 1 µl RNasin to each tube.
13. Store at –80°C.

3.3. RNA Quality: Agilent 2100 Bioanalyzer

This procedure describes the use of microchip gel electrophoresis in order to analyze RNA concentration and transcript size distribution on the Agilent 2100 Bioanalyzer using the RNA 6000 Nano Assay. One microliter of RNA sample in a concentration range of 5–500 ng/µl is applied on an RNA 6000 Nano chip. In one assay, 12 RNA samples can be analyzed simultaneously in less than 60 min. The measurement is performed strictly according to the manufacturer's protocol.

In advance:

- Allow all reagents to equilibrate to room temperature 30 min before use.
- Let a prepared aliquot of filtered gel (65 µl) equilibrate to room temperature.
- Set a heating block to 70°C.
- Switch on the Bioanalyzer and start the computer software.
- Clean the electrodes of the Bioanalyzer by applying a cleaning chip filled with 350 µl RNaseZAP for 1 min followed by another cleaning chip filled with 350 µl RNase-free water for 10 s. Leave the lid open for another 10 s to permit the water to evaporate from the electrodes.
- Vortex all reagents and spin down briefly.
- Place a new RNA Nano chip on the chip priming station.

1. Preparation of RNA samples and ladder: Heat denature the samples and an aliquot of RNA 6000 ladder for 2 min at 70°C in a heating block. Spin down and place the RNA samples and ladder on ice until final measurement. To avoid evaporation or drying of the applied reagents, the analysis should be performed without any interruptions.
2. Add 1 µl of RNA 6000 Nano dye concentrate (blue cap) to a 65 µl aliquot of filtered gel. Vortex gel-dye mix thoroughly and spin at 13,000 g for 10 min at room temperature.

3. Pipette 9 μl of the gel-dye mix into the well marked G (according to the manufacturer's illustration). Close the chip priming station and pressurize the gel-dye mix for exactly 30 s. Depressurize slowly. Add 9 μl gel-dye mix to each of the wells marked G.
4. Pipette 5 μl of RNA 6000 Nano Marker (green cap) into the 12 sample wells and the ladder well. (If one or more sample wells are not utilized, add 6 μl of RNA 6000 Nano Marker to each unused well. Do not leave any well empty.).
5. Add 1 μl of the heat denatured RNA 6000 ladder to the ladder well.
6. Add 1 μl of sample to the 12 sample wells.
7. Remove the chip from the chip priming station and place it carefully in the vortex mixer. Vortex the chip for 1 min at the vortex set point (2,400 rpm).
8. Immediately transfer the RNA Nano chip to the Agilent 2100 Bioanalyzer and start the measurement (Eukaryote Total RNA Nano Assay).
9. Start assay run. Sample information (sample name, dilution, and preparation) may be added in the chip summary table.
10. Results will be obtained after approximately 30 min. The results are shown as electropherograms (up to 12 sample curves, 1 ladder curve). Save the run and print out the results (see Note 2). Clean the electrodes by applying a cleaning chip filled with 350 μl RNase-free water for 10 s.

4. Notes

1. It is important to avoid RNase contamination. Wear gloves and a clean lab coat at all times. Clean the work area properly. Only use RNase-free tubes and filter pipette tips. Carefully clean the cryostat before and after use. Clean cryostat knife and other instruments with 70% ethanol. Do not use RNaseZAP on the cryostat knife and other metal surfaces (corrosion).
2. Interpretation of Bioanalyzer results. The RNA 6000 ladder curve should show one marker peak at the beginning of the curve and five ladder peaks. Samples with intact RNA display two distinct ribosomal peaks representing 18S and 28S rRNA, respectively (Fig. 1). The ratio between the area under the curve of the 28S and 18S peaks, the so called ribosomal ratio, has been widely used as an objective RNA quality indicator. The size of the 28S peak should be larger than the 18S peak in good

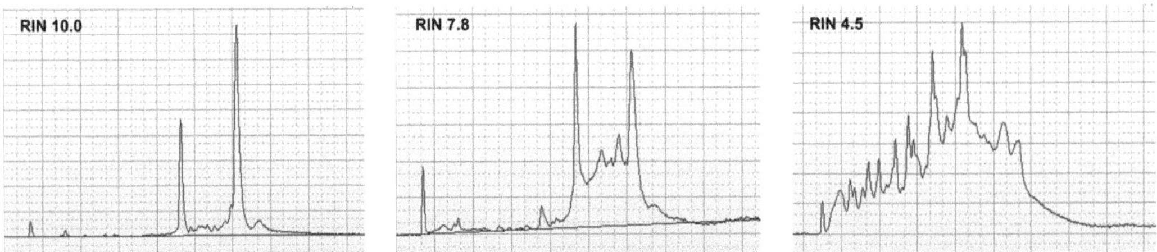

Fig. 1. Examples of different Bioanalyzer curves. A perfect RNA sample is shown in the *left panel*. The 28S ribosomal peak (to the *right*) is taller than the 18S peak – RIN 10. The *middle panel* depicts an RNA extract with slightly degraded RNA. The 28S peak is reduced in comparison to the 18S peak and a baseline elevation between the peaks is evident – RIN 7.8. In the *right panel*, the RNA is significantly degraded. The 28S peak is barely visible, and there is a build up of small RNA fragment peaks in the left side of the curve – RIN 4.5.

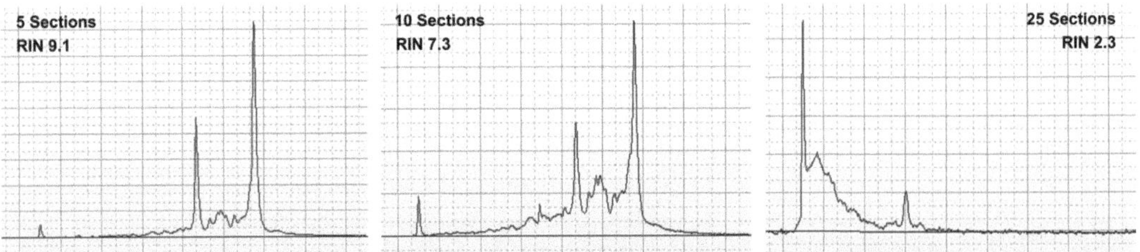

Fig. 2. Overloading the extraction column. RNA was prepared a fresh frozen block of lung cancer tissue using the Qiagen kit. As the tissue input is increased, from 5 to 10 and 25 frozen sections (12 μm), respectively, the RNA quality appear to deteriorate – RIN values decreasing from 9.1 to 7.3 to 2.3. The result in the *right panel* will falsely indicate that this lung cancer sample is degraded.

quality RNA resulting in ratios above 1, up to around 2 in pristine samples. However, ribosomal ratios have been shown to be too crude when it comes to predicting the performance of slightly degraded RNA in downstream applications. To this end, a software algorithm have been designed that in addition to the shape of the 28S and 18S peaks takes other aspects of the electropherogram curve into account, i.e., baseline shape and aberrant peaks composed of small degraded-RNA fragments. A RNA integrity number (RIN) is thus assigned to the sample, ranging from 10 (perfect quality) to 0 (totally degraded) (14).

The RNA 6000 Nano assay has a quantitative range of 25–500 ng/μl and a qualitative range of 5–500 ng/μl. If the Bioanalyzer chip is overloaded poor separation and an artifactual decrease in RIN will ensue. If the RNA concentration exceeds the analytical range of the assay, the sample should be diluted and reanalyzed.

Overloading extraction columns: In our experience, it is essential not to overload the Qiagen kit (Fig. 2). If the tissue input is too large, the quality of the extracted RNA will be poor. In our hands, the amount of tissue that can be extracted properly by this procedure will vary between organ types. Thus, we recommend that this is titrated for each type of tissue before starting a large-scale project on a cohort of patient samples.

Acknowledgments

The skillful assistance of Karolina Edlund is gratefully acknowledged. The Fresh Tissue Biobank project at the Department of Pathology, Uppsala University Hospital was supported by the Swedish National Biobank Platform funded by Wallenberg Consortium North and Swegene.

References

1. Morente, M.M., Mager, R., Alonso, S., Pezzella, F., Spatz, A., Knox, K., Kerr, D., Dinjens, W.N., Oosterhuis, J.W., Lam, K.H., Oomen, M.H., van Damme, B., van de Vijver, M., van Boven, H., Kerjaschki, D., Pammer, J., Lopez-Guerrero, J.A., Llombart Bosch, A., Carbone, A., Gloghini, A., Teodorovic, I., Isabelle, M., Passioukov, A., Lejeune, S., Therasse, P., van Veen, E.B., Ratcliffe, C., and Riegman, P.H. (2006) TuBaFrost 2: Standardising tissue collection and quality control procedures for a European virtual frozen tissue bank network. *Eur J Cancer* **42,** 2684–2691.
2. Lin, D.W., Coleman, I.M., Hawley, S., Huang, C.Y., Dumpit, R., Gifford, D., Kezele, P., Hung, H., Knudsen, B.S., Kristal, A.R., and Nelson, P.S. (2006) Influence of surgical manipulation on prostate gene expression: implications for molecular correlates of treatment effects and disease prognosis. ***J Clin Oncol 24, 3763–3770.***
3. Micke, P., Ohshima, M., Tahmasebpoor, S., Zhi-Ping, R., Östman, A., Ponten, F., and Botling J. (2006) Biobanking of fresh frozen tissue: RNA is stable in nonfixed surgical specimens. *Lab Invest* **86**, 202–211.
4. Breit, S., Nees, M., Schaefer, U., Pfoersich, M., Hagemeier, C., Muckenthaler, M., and Kulozik, A.E. (2004) Impact of pre-analytical handling on bone marrow mRNA gene expression. *Br J Haematol* **126**, 231–243.
5. Ohashi, Y., Creek, K.E., Pirisi, L., Kalus, R., and Young, S.R. (2004) RNA degradation in human breast tissue after surgical removal: a time-course study. *Exp Mol Pathol* 77, 98–103.
6. Jewell, S.D., Srinivasan, M., McCart, L.M., Williams, N., Grizzle, W.H., LiVolsi, V., MacLennan, G., and Sedmak, D.D. (2002) Analysis of the molecular quality of human tissues: an experience from the Cooperative Human Tissue Network. *Am J Clin Pathol* **118**, 733–741.
7. Huang, J., Qi, R., Quackenbush, J., Dauway, E., Lazaridis, E., and Yeatman, T. (2001) Effects of ischemia on gene expression. *J Surg Res* **99**, 222–227.
8. Florell, S.R., Coffin, C.M., Holden, J.A., Zimmermann, J.W., Gerwels, J.W., Summers, B.K., Jones, D.A., and Leachman, S.A. (2001) Preservation of RNA for functional genomic studies: a multidisciplinary tumor bank protocol. *Mod Pathol* **14**, 116–128.
9. Spruessel, A., Steimann, G., Jung, M., Lee, S.A., Carr, T., Fentz, A.K., Spangenberg, J., Zornig, C., Juhl, H.H., and David, K.A. (2004) Tissue ischemia time affects gene and protein expression patterns within minutes following surgical tumor excision. *Biotechniques* **36**, 1030–1037.
10. Botling, J., Edlund, K., Segersten, U., Tahmasebpoor, S., Engström, M., Sundström, M., Malmström P.U., and Micke, P. (2009) Impact of thawing on RNA integrity and gene expression analysis in fresh frozen tissue. *Diagn Mol Pathol* **1**, 44–52.
11. Micke, P., Ostman, A., Lundeberg, J., and Ponten, F. (2005) Laser-assisted cell microdissection using the PALM system. *Methods Mol Biol* **293**, 151–166.
12. Mutter, G.L., Zahrieh, D., Liu, C., Neuberg, D., Finkelstein, D., Baker, H.E., and Warrington, J.A. (2006) Comparison of frozen and RNAlater solid tissue storage methods for use in RNA expression microarrays. *BMC genomis* **5**, 88.
13. Lee, N.H., and Saeed, A.I. (2007) Microarrays: an overview. *Methods Mol Biol* **353**, 265–300.
14. Imbeaud, S., Graudens, E., Boulanger, V., Barlet, X., Zaborski, P., Eveno, E., Mueller, O., Schroeder, A., and Auffray, C. (2005) Towards standardization of RNA quality assessment using user-independent classifiers of microcapillary electrophoresis traces. *Nucleic Acids Res* **33**, e56.

Index

Joakim Dillner (ed.), *Methods in Biobanking*, Methods in Molecular Biology, vol. 675,
DOI 10.1007/978-1-59745-423-0, © Springer Science+Business Media, LLC 2011

D

Q

R

S

MIX
Papier aus verantwortungsvollen Quellen
Paper from responsible sources
FSC® C105338

If you have any concerns about our products,
you can contact us on
ProductSafety@springernature.com

In case Publisher is established outside the EU,
the EU authorized representative is:
Springer Nature Customer Service Center GmbH
Europaplatz 3, 69115 Heidelberg, Germany

Printed by Libri Plureos GmbH
in Hamburg, Germany